DATE DUE

GLOBAL FISSION

GLOBAL FISSION

THE BATTLE
OVER NUCLEAR POWER

JIM FALK

WITH A FOREWORD BY ROBERT JUNGK

University of Charleston Library
Charleston, WV 25304

MELBOURNE
OXFORD UNIVERSITY PRESS
OXFORD   AUCKLAND   NEW YORK

333.79
F186g

OXFORD UNIVERSITY PRESS

London Glasgow New York Toronto
Delhi Bombay Calcutta Madras Karachi
Kuala Lumpur Singapore Hong Kong Tokyo
Nairobi Dar Es Salaam Cape Town
Melbourne Auckland
and associate companies in
Beirut Berlin Ibadan Mexico City

© Jim Falk 1982
First published 1982

This book is copyright. Apart from any fair dealing for
the purposes of private study, research, criticism or
review, as permitted under the Copyright Act, no part
may be reproduced by any process without written
permission. Inquiries should be made to the publishers.

National Library of Australia
Cataloguing-in-Publication data:

Falk, Jim, 1946–.
Global fission.
Includes index.
ISBN 0 19 554315 7.
ISBN 0 19 554316 5 (pbk.).
1. Atomic power—Political aspects. I. Title.
621.48

Computer photocomposed at Griffin Press Limited, Netley, South Australia
Printed by Richard Clay, S.E. Asia
Published by Oxford University Press, 7 Bowen Crescent, Melbourne

CONTENTS

Foreword	1
Acknowledgements	5
Introduction	8

1 From Boom to Crash   11

Beginnings   Triumph, Guilt, and Optimism   The Making of an Ideology   The 'Peaceful Atom'   Heady Expectations   The 1975 Crash   A Stagnated Industry

2 Message from Pennsylvania   29

Fundamentals   In the Control Room   Outside   The Hydrogen Bubble   Evacuation

3 The US Response   43

The Technical Debate   The Radiation Controversy   The Potential for Catastrophe   Design and the Unknown   The Political Fallout Erupting Opposition

4 Political Economics   61

The Problems of General Public Utilities   Operating Costs   The Political Economics of Radiation   The Crucial Capital Costs Consequences   The Brontosaurus   The Plastic Economics of Nuclear Power   Restructuring Economic Reality   The Battle for the Future

5 Origins of the Opposition   90

The Social Context   The Beginning of the Battle

6 The State and the Industry   110

The Slump and the Industry   The Role of the State   Nuclear Power and the Interests of the State   The Terms of Trade   The State and the Opposition   The Nuclear State   Nuclear Power in the Global Economy   Limits to State Power

7 Referred to the People 141

*The Honeymoon Period 1960–73   The Spread of Opposition
1973–75   Divisions within Parliament   The Information
Campaigns 1974–77   The Road to Referenda 1977–79   Final
Manoeuvres   Results   Aftermath*

8 Central Control, Regional Revolt 172

*Central Developments 1950–79   On the Periphery 1970–80   From
Periphery to Centre 1979–80*

9 The International Arena 200

*Spain: Regional Revolt   The Netherlands: A Divided Parliament
West Germany: Conflict on Four Fronts*

10 Counter-currents within the Left 228

*In the USSR   Eastern Europe   China   Re-examination on the
Left   The Unions and Nuclear Power   Bridging the Gap   The
Growth of the Anti-Nuclear Union*

11 The Australian Experience 256

*The Hurdles   Opening Manoeuvres   Citizens and the Labour
Movement   Aboriginals Bar the Way   Preparing for Mining   The
International Issues Come Home*

12 Strategy and Structure 285

*The Role of Example   Improving Communication   Converging
Concern: Nuclear Power and Nuclear Weapons   Convergence in
Japan   Towards a Nuclear-Free Pacific   Support from the Rim
Global Coordination*

13 Beyond Nuclear Power 322

*The Nature of the Battle   From Physical Hazards to Political
Concern   Exploring the Implications   The Point Reached*

Notes 341

Index 396

This book is dedicated to Sue who made it possible, to our baby daughter Anna who made it so personally important, and to those many people around the world who have had the courage to take their future into their own hands and begin to change it.

FOREWORD

The turbulent protests and demonstrations of the 1960s were the beginning – only to peter out pitifully in the 1970s. But only superficially for, since then, the anti-technocrat movement has broadened and deepened throughout the world. Curiously enough, its decentralist character prevents the movement from realizing its own strength. The spectacular demonstration has receded into the background while the movement's varied influences on the spirit of the times has grown. So much that even those who in reality cling to the old maxims of 'more', 'bigger', and 'faster' pretend due respect to the revolution presently taking place and react as though they too believe that we can no longer continue on our present path of destruction in the service of increased rates of growth. 'But have no fear', they placate. '"We" have everything under control.'

And just that is no longer the case. The doubts and misgivings concerning a so-called progress which threatens to endanger survival and lead to new forms of tyranny are now too widespread to ever be repressed. Today reactions are negative to symbols which only yesterday held a positive aura: automation, automobiles, computers, rockets, and atomic power. At first there were only a mere handful of critical thinkers who no longer allowed themselves to be persuaded that pursuit of power and profit and lack of regard for the consequences of developments in industry, science, and technology were desirable goals. They were soon joined by new generations of students and broad sections of the middle classes. Farmers began to understand that rapid technological innovation in agriculture not only endangered the environment but themselves too. And finally – astonishingly late – more and more working people also began to see that the crumbs from the table of increased productivity were poisoned while inflation swallowed up their wage increases which they were now paying for

by more stress, more control, more job insecurity, and more fear of the future.

Thus a new International is evolving against the domination of the concerns and state capitalism organized on large industrial principles in East and West. This movement is broader-based sociologically than all previous movements for change for today the exploited classes comprise the majority of mankind. Clerks and public servants, engineers and those employed in the service sector are just as dependent and often economically worse off than the traditional proletarian workforce in the factories and on the fields. The area of their freedom of decision has become increasingly confined and their economic situation is becoming more and more precarious. In an age of increasing automation and unemployment the job itself has become a blackjack to induce proper behaviour and hypocritical loyalty.

This new proletariat has not yet fully realized the extent to which they are in chains as for a time they still disposed over areas of self-expression and limited power in their own specialized fields. But even these tiny empires disappear as central planning and concomitant rigid discipline increase. Every important decision is taken somewhere above and the centres of power are far away. Those who refuse to knuckle under fall at the wayside. They are usually not even expelled, just left behind. If you don't join the club you won't get the kudos.

This dependency has become unbearable for many. Looking beyond their nine-to-five world they began to understand their whole living environment. Within only a few years the meadows were turned into concrete, trees became street lamps, where we had clean air to breathe we now inhale exhaust gases, and a convivial neighbourhood has been petrified into real estate. Who really made the decisions? In whose interest? Why weren't we consulted? How can we stop it?

Thus began a new struggle against new enemies. Against noise and sickness, ugliness and boredom, against harassment and stress, and finally against threats to survival.

For alongside fear of war – never quite forgotton though people tried – arose a further fear, that of catastrophe. Ill-tidings filled the media, hardly a day went by without the news of oil rig disasters, giant tankers breaking up at sea, chemical plants ablaze, plutonium poisonings, new plagues, thalidomide tragedies, and the like. Earlier science fiction scenarios become reality overnight unbear-

FOREWORD   3

ably threatening not only our own life but that of our children, grandchildren, and a common future.

It is in fact surprising that even more people are not protesting even more energetically against these monstrous developments. For this time, unlike the earlier social movements which fought for improved social and political rights, for higher wages, better information, for increasing participation in the corridors of power, we are now concerned with human existence and the survival of the species.

It proved quite possible to mislead those affected as to how serious the situation really is. Public relations, the media, promises of comfort, and the easy life, the dazzling achievements of science, the offer of more leisure in a colourful, sun-filled paradise – and who wouldn't rather hope than live in fear and anxiety? Well, as Abraham Lincoln knew very well, you can fool some of the people some of the time – but it can't go on forever. People are now beginning to realize the dangers we face from the developments of super-technology: dangers with which we cannot cope, neither socially, ecologically, economically nor, above all, psychologically. Gradually they realize how much of our affluence results from the exploitation of the Third World and hear daily that the billions of disadvantaged in that part of an ever-diminishing planet are staking their claims for a just share of the world's plenitude and that they too are building up a hypertechnical power potential with which to push their demands: in ten years there will be atomic bombs on every continent.

If, as we speed along the most disastrous path in our history, there still should be hope that we can avoid or at least minimize the incomparable dangers of the present crisis caused by the latest developments in weaponry, then such hope is to be found in the growing world-wide resistance of those who urge us to stop and think again. They are already many and their number and influence are increasing. In 1972 the alternative meeting to the Stockholm UN Conference on the Environment made a lasting impression on me. I witnessed there the beginnings of a new ecological, humanistic, solidary, autonomous, and of course anti-capitalist world movement whose decentralized, self-reliant, and varied growth I continue to observe with excitement and hope.

Will this new, informal International grow quickly enough and capture enough power to transform its life-sustaining values and aims into tangible results? It is certainly much easier and more

GLOBAL FISSION

obvious to answer this anxious question with 'no', to resign oneself to the situation, and to leave everything to run its course. This is the attitude of many well-read people who 'won't allow anyone to take me in'. But can we be satisfied with that? Isn't that the beginning of the end?

I have some questions for those who have become pessimistic because of what happened yesterday and is today still happening: Would you have expected fifteen years ago that the 'sceptical generation' would put up such fierce resistance as they began to do only three or four years later? Would you have thought that the 'apolitical consumer' would stand up for his own interests with such tenacity and courage as the citizens rights groups have done? And would you have expected at the beginning of the 1960s that millions of people on Planet Earth would so militantly protest against the development of a new and threatening technology like the so-called 'peaceful use of atomic energy' as they indeed have proved ready to do?

I believe it a grave mistake to underestimate human beings. Once they understand just how much is now at stake then enough of them will attempt to shake off their impotence and find ways to interfere with the new forms of sophisticated technology, their increasing resistance will mitigate against the worst happening, and perhaps even instigate the beginnings of positive reversals of present trends in thought and practice. The breadth of this new movement is not at present usually realized, we forget how many young people belong to it but also that those who still defend present power groupings and believe in their unlimited extension into the future are rapidly becoming less sure of themselves and less and less credible. This must surely lead to a history-making trend reversal. It is also one conclusion to be drawn, and clearly to be seen, from the international perspectives revealed so challengingly in this book.

But even more important for those of us so concerned about the future than the signs of such new beginnings already present is to realize what could and even will happen; yet we should not resign but work for the necessary changes with all the strength at our disposal. Everything depends on each and everyone of us. And what each one thinks, what each one says, and each one does will bear fruit when it is dedicated to the cause of our survival.

Robert Jungk
Salzburg, 1981

ACKNOWLEDGEMENTS

The nuclear conflict is marked by a profusion of comments from many parties on all sides but far less researched analysis of the conflict itself. To deeply research the course of events in each country would be a rewarding but life-long task which on its completion would have been left far behind by events. Rather than undertake such an exercise, I have based this book primarily on newspaper reports and existing analyses, the literature produced from both sides of the controversy, scientific and nuclear industry journals, and interviews and discussions with a variety of people. These have come from both within the nuclear industry and the opposition to nuclear power, from the USA, Japan, the Pacific, and many countries of Western Europe, who I either visited in 1978 or who have met with me in Australia from 1977 to 1981. Although these sources are noted throughout the text, I would like to record my gratitude to a number of people who have generously given time and energy to make this a better book. There is not room to acknowledge everyone who has helped, but some are listed below.

First, my thanks to Peter Jones, who not only sent me a steady stream of materials from North America and several other countries, but also added valuable comments on the text. As well, I am indebted to Sue Rowley, Joe Camilleri, Les Dalton, Nancy Stockdale, Linda Rubenstein, Val Noone, Chris Ryan, and Alan Roberts who read the text (sometimes more than once) and added a multitude of useful comments. While indebted to them, I should stress that the analytical framework and views expressed in this book are my own, and should not reflect on anyone else mentioned here.

6   GLOBAL FISSION

In addition to those who read the text, many others brought or sent me invaluable comments or materials. Included are Leigh Holloway, Andrew Hewett, Anne Wigglesworth, Ed Kapstein, Annie Feith, Carmel Shute, Ken Norling, Boris Frankel, and Henri Jeanjean (Australia); George Wald, Richard Grossman, Gail Daneker, Charlie Komanoff, Anna Gyorgy, and Merrie Watters (USA); Peter Weish (Austria); Wren Green (New Zealand); Heidi Knott, Willi and Nina Gladitz, Rainer Adomat, Petra Kelly, Michael Lucas, Roland Vogt, and Gunter Hopfermuller (West Germany); Pierre Samuel and Annik Paris (France); Ian Wood (Sweden); John Lambert and Frieda Bagare (Belgium); Brian Trench (Ireland); Linda Bradburn and John Leach (UK, and now Australia). My deepest thanks go to all of them, but also to those others who either because one must stop somewhere, or because they would prefer not to have their names mentioned, are not explicitly included above but have answered queries or sent material which has proved useful in filling in the book's detail.

Material from *Atomic Energy in the Coming Era* by David Dietz, which is quoted in chapter 1, is reprinted with the kind permission of Dodd, Mead & Company, Inc. (Copyright 1945 by Dodd, Mead & Company, Inc. Copyright renewed 1973 by David Dietz.)

This book might never have been written if Anne Norton, then of Oxford University Press, had not encouraged me to write it, and it has benefited greatly from Angela Gundert's editing.

Finally, I would like to thank my colleagues in the Department of History and Philosophy of Science at the University of Wollongong who put up with me while I was writing and encouraged me to continue.

Jim Falk

ABOUT THE COVER

The cover design is based on the use of two symbols. In the centre is the international radiation trifoil, representing both the hazards which spark opposition, and the centralized economic and political power inherent in the nuclear industry. The concentric rings of 'smiling sun' symbols represent the international network of dispersed opposition groups which rise to meet it, and the alternatives that they offer.

ACKNOWLEDGEMENTS

The smiling sun symbol was first designed in 1975 by members of Denmark's Organization for Information on Atomic Energy (the OOA), and the author gratefully acknowledges their permission to reproduce it. On their request, some details of its origin are given here. The use of this symbol has spread dramatically since 1975. Only 500 copies of the symbol were first printed, but it became almost instantly popular and another 7 000 were printed in the same year. In 1976, 308 000 were printed, and its use had already spread to West Germany. Since then the symbol has been printed by the OOA in twenty-eight languages and has been used in almost every country touched by the nuclear conflict. It has appeared on badges, stickers, posters, T-shirts, bags, flags, balloons, postcards, stamps, publications, records, films, slides, kites, games, advertisements, and banners.

Badges and stickers bearing this symbol are produced in Denmark and sold with a tiny profit margin to groups around the world. These groups are encouraged to resell these items at a higher price, using the proceeds to help finance their activities. The profits which the OOA receives are used to finance the Smiling Sun Foundation and, through it, the World Information Service on Energy (WISE). Established in 1977, this is a centre for information exchange between groups opposing the nuclear fuel-cycle around the world. WISE's newsletter, published on a subscription basis, may be obtained by writing to WISE, Blasiusstraat 90, 1091 CW Amsterdam, Netherlands. Further details about smiling sun materials may be obtained from the OOA, Ryesgade 19, 2200 Copenhagen N, Denmark.

INTRODUCTION

Why is the nuclear debate so intense, and where is it going? Its range and strength are hard to miss. Since the early 1970s it has resulted in a rapidly growing stream of leaflets, pamphlets, books, and films distributed throughout the Western world, vigorously debating the hazards of nuclear power and the possibilities of alternative energy programmes. But its intensity reflects not only debate but social conflict. Much less has been written on the conflict which underlies the debate over nuclear power.

It has not been the intention of either the nuclear industry or the relevant government agencies to launch a wide-ranging debate over the nuclear issues. But it has sparked anyway as countless citizens in many countries have begun to question the wisdom of the nuclear enterprise – a massive technological, corporate, and government project whose physical and social implications could become a key feature in the future of the industrialized world. The rise of the opposition to this project, and the conflict that has resulted, is itself a phenomenon of great importance. It is the history, prospects, nature, and significance of this conflict which are the subject of this book.

In the course of a decade, the hazards of the nuclear enterprise have been debated by whole populations. Nuclear projects costing billions of dollars have been completed but then abandoned. Governments have fallen under the pressure of opposition to their nuclear programmes, or have been forced to the reluctant expedient of referring them to decision by a national plebiscite. The issues raised have been profound in their consequences and urgent in their time-scale. They have locked communities, political parties, governments, and large corporations into a bitter confrontation

INTRODUCTION

spanning much of the world. At stake is the form the world's future will take, and the safety and style of the lives that we will lead. The outcome will leave few of us untouched.

What then is the nature of this conflict, whose dimensions, urgency, and repercussions carry with them the character more of a battle than a debate? Can we predict its outcome? Will economics finally be the determining factor? If not, what role can the individual play? And, most important of all, where is this battle taking us? These are crucial questions but their answers are far from easy to determine.

Although over the second half of the 1970s and the early 1980s the controversy over nuclear power has surged increasingly strongly, it is difficult to see it in historical perspective. It is even more difficult to distinguish the nature of the alliances and manoeuvres of the contestants. That is true even for the debate occurring in the community or country in which we live. Neither the nuclear enterprise nor the battle is confined to one community or one country. They span the world. News of this broader conflict comes to us more as an echo filtered through snippets thrown up by media which have their own problems about what they should treat as newsworthy. We ignore this broader context at our peril.

Seen in isolation, the fortunes in the battle over nuclear power in any one country or community may fluctuate wildly. But since the concerns of both the opposition and the industry go beyond the effects of any single project it is not these fluctuations which in the end will decide the future. Instead it is the net effect of the world-wide opposition on the entire nuclear enterprise. This can only be judged from a global perspective in which the interacting fortunes of the contestants in each country are traced. That is what I have tried to do in this book.

In doing so I have been forced to mix descriptions of the often dramatic key events which form the fabric of this saga with a more abstract analysis of what it all means. The first four chapters set the scene: the birth of the extraordinarily ambitious nuclear programmes of the post World War II period, and their equally extraordinary failure in the 1970s; the accident at Three Mile Island and its repercussions within the USA, the heartland of the world nuclear industry. In studying these, a central question is raised which is the subject of the next chapter: the relationship between economics and politics in the nuclear debate. Then, the broad framework of the international battle – the nature of the principal

contestants and the power that they wield – is set in the next two chapters. Following this, key features of the conflict are examined by tracing the events in several clusters of countries. Among them is Australia, the country from which this book has been written. It fits naturally between an analysis of the role of the labour movement and the left in chapter 10 and the important developments occurring in the Pacific – one of the key themes of chapter 12. In the final chapter the fundamental threads which have been uncovered are drawn together to give an overview of the significance of both the nuclear enterprise and the opposition in relation to other trends now evident in the global political arena.

Rather than attempt to give each of the parties to the battle 'equal time', the book aims to provide a framework in which the significance of the battle can be revealed. While the actions of the nuclear industry can usually be readily understood as part of more general world-wide political and economic trends, the nature and significance of the opposition is much less well charted territory and requires more time to be adequately explored.

The time-span which the book surveys is the forty years from the beginning of nuclear power to 1981, the entry year to the decade of the 1980s. Running through the book are two themes. One concerns the nature of the power exercised by the contestants in the battle, and in particular the 'legitimation' problems of the nuclear industry and the state. The other concerns the issues which are invoked by the battle beyond those associated with the hazards, or advantages, of nuclear power.

It is only proper that the reader be aware that the author is 'against' nuclear power. That said, whether you are against nuclear power or for it, nothing but good can come from starting to examine, not only the issues currently in debate, but the significance of the conflict which lies behind it. Perhaps, as you move through this book you will be struck by one of the conclusions which, whatever your present position on nuclear power, must be as heartening to you as it is to me. It is remarkably simple, yet pregnant with significance. It is this: the nuclear conflict teaches us a lesson, as it has taught millions of others. It is that the future is not fixed. It lies within our hands to mould and make our own.

1
FROM BOOM TO CRASH

BEGINNINGS

It was 8 a.m. on a crystal clear morning, 6 August 1945. A single plane cut through the blue sky high above, its tiny wings shining brilliantly in the dazzling sunshine. From its belly popped a tiny puff of white, a little cluster of parachutes drifting slowly downward. Seconds later the earth shook, an explosion roared, and the sky filled with a searing white light. A blast of air heated to 7 000 degrees Celsius swept outward at 400 kilometres per hour. An atomic bomb had exploded over the Japanese city of Hiroshima. A new dimension had opened on the future of the world.

For the 400 000 people who were the bomb's target the horror was total.[1]

> Soon I noticed that the air smelled terrible. Thinking that the bomb that hit us might be yellow phosphorus incendiary bombs, I instinctively rubbed my nose and mouth hard with my *tengui* [handkerchief] that was tucked at my belt. I felt something strange with my face. Then I was shocked by the feeling that the skin of my face had come off on the *tengui*. Ah, then, the hands and arms, too. Starting from the elbow to the fingertips, all the skin of my right hand came off and hung down grotesquely. The skin of my left hand, all five fingers, also came off.
>
> I saw on the bridge something like a human figure, running. 'Oh, yes! That must be the Tsurmi Bridge. I must hurry to cross it; if not, there will be no way to escape.' I ran like mad toward the bridge jumping over the piles of debris . . .

What I saw under the bridge was shocking: Hundreds of
people were squirming in the stream. Their faces were swollen
and gray, their hair was standing up. Holding their hands
high, groaning, people were rushing to the river. I felt the
same urge because the pain was all over my body which had
been exposed to a heat ray strong enough to burn my pants to
pieces.

This was the testimony of Futaba Kitayama, a housewife then
thirty-three years old, hit 1.7 kilometres from the point over which
the bomb exploded.[2] Nearly 20 000 people died in the instant of
the explosion. A huge, thick cloud of gas erupted and covered the
whole area. At 1.2 kilometres from the centre, people in the open
were burned to death. At 4 kilometres the heat was twenty times
more intense than that from the sun, and exposed wood and human
skin blackened and peeled. Apart from the blast, the heat, the shock
wave that punched out travelling 4 kilometres in ten seconds, the
'black rain' of oily highly radioactive liquid which fell in torrents,
and the fire storm which raged for six hours burning everything in
its path, there was the silent, invisible, and perilous presence of
ionizing radiation. By the end of the year a further 120 000 people
had died from the effects of the explosion.[3]

The day after the bombing, in London the *News Chronicle*
appeared with the banner headline 'On this Bank Holiday the
course of world history may have been altered. FORCE OF NATURE
HARNESSED: ATOM BOMB ON JAPAN. Power equal to 20 000 tons
of T.N.T. Allies Beat Germans in Battle of Science'.[4] Undoubtedly
the claim was true. The course of history had been altered. But
contrary to many subsequent comments it was only in a sense the
birth of the Atomic Age.

The wartime construction of the atomic bomb had been pre-
ceded by a series of experiments with radioactive materials initiated
by Ernest Rutherford, head of the Cavendish Laboratory, Cambridge, in 1919. Using a fluorescent screen he had observed 'alpha
particles' emitted spontaneously by the radioactive decay of radium
colliding with nitrogen. On the screen he had seen produced sparks
of light corresponding to the presence of new particles with much
longer ranges than the alpha particles. These were protons knocked
out from the nucleii of the original nitrogen atoms. He had
smashed the atom.

A wave of experiments was set off by this discovery. The turning

FROM BOOM TO CRASH

point came in 1939 when Otto Hahn and his collaborators at the Kaiser Wilhelm Institute in Berlin bombarded uranium with another particle, the neutron. They found that uranium atoms could be split into two almost equal fragments emitting comparatively large amounts of energy. This was the process of atomic fission. On learning of this discovery, Enrico Fermi, Nobel Prize winning physicist who had fled Mussolini's Italy and was now working at Columbia University, recognized that since neutrons were also emitted when the uranium atom splits, these could be utilized to cause further fissions. A self-sustaining chain reaction could therefore be started. Towards the end of 1939 he and co-worker Leo Szilard began planning an experiment to demonstrate this.

By July 1941 a graphite cube, about 2.5 metres on edge and containing about 7 tonnes of uranium oxide embedded in it (like raisins in a cake), had been constructed in a squash court at Columbia University. On 2 December 1942 this 'atomic pile' produced a self-sustaining nuclear chain reaction.[5] Although without cooling or electrical generator, it was in essence the world's first man-made nuclear reactor.

Although the pile produced substantial quantities of heat, interest in the potential of the new device as a power source was remote. World War II was raging, and a more relevant property of nuclear fission had already been drawn to the attention of US President Franklin D. Roosevelt in October 1939 by a letter from several scientists including Albert Einstein and Leo Szilard. It was the potential of nuclear fission to act as the driving mechanism of a new, vastly more destructive weapon than previously had been imagined possible. As Fermi and Szilard realized, the flux of atomic particles produced by the fission process within their 'pile' would hit uranium atoms, transforming them into atoms of other elements. Among these would be plutonium, a material which seemed ideal for the construction of nuclear bombs.

Even while Fermi's 'pile' had been under construction, a decision by Franklin D. Roosevelt and Winston Churchill to concentrate relevant nuclear research in the USA and Canada, had been taking effect. As early as August 1942 the US War Department had established the Manhattan Project. Already it had set up the steps necessary to produce a second possible material for the construction of nuclear weapons. This was Uranium-235, the 'fissile' component of natural uranium which is a mixture of both

14   GLOBAL FISSION

Uranium-235, and another isotope, Uranium-238. To accomplish this a large-scale separation plant was under construction at Clinch River near Oak Ridge, Tennessee. Within a month of the first demonstration of a chain reaction by Fermi, a second plant, soon to become a huge plutonium production facility, was under construction in Washington State near the small village of Hanford. A third major complex, to house a large staff of scientists, was also begun at Los Alamos, New Mexico. These three secret complexes would form the hub of the gigantic Manhattan Project, which at its peak would employ 150 000 people.[6]

On 16 July 1945 the world's first nuclear bomb, constructed from plutonium from Hanford, was exploded on the top of a steel tower at Alamogordo Air Base, about 200 kilometres north east of Albuquerque. General Groves, who commanded the Manhattan Project, described what happened:

> First came the burst of light of a brilliance beyond any comparison. We all rolled over and looked with dark glasses at the ball of fire. About 40 seconds later came the shock wave followed by the sound, neither of which seemed startling after our complete astonishment at the extraordinary lighting intensity. A massive cloud formed which surged and billowed upward with tremendous power, reaching the sub-stratosphere in about five minutes . . .
>
> Two supplementary explosions of minor effect other than the lighting occurred in the cloud shortly after the main explosion. The cloud travelled to a great height first in the form of a ball, then mushroomed, then changed into a long trailing chimney shaped column.[7]

Emotions caused by this overwhelming experience differed. One observer, General Farrell, reported:

> The effects could well be called unprecedented, magnificent, stupendous, and terrifying. No man-made phenomenon of such tremendous power had ever occurred before. The whole country was lighted by a searing light with the intensity many times of the midday sun. It was golden, purple, violet, gray and blue. It lighted every peak, crevasse and ridge of the nearby mountain range with a clarity and beauty that cannot be described but must be seen to be imagined. It was that beauty the great poets dream about but describe most poorly and inadequately.[8]

FROM BOOM TO CRASH 15

The reaction of Dr Robert Oppenheimer, director of the scientific research effort, was of a totally different character. As he watched the sinister cloud Roman candle into the air a thought bubbled into his mind competing with his initial euphoria at the success of the explosion. It was a line from the Hindu *Bhagavad-gita*: 'I am become death, the shatterer of Worlds'.[9]

Three weeks later, as a similar cloud boiled over the shattered bodies in the scorched ruins of Hiroshima, the newspaper reports of reactions to the bombing displayed the same ambivalence.

TRIUMPH, GUILT, AND OPTIMISM

In England on the day after the bombing, Churchill's statement combined a strange mixture of triumph, guilt, and optimism.

> The whole burden of execution, including the setting-up of the plants and many technical processes connected therewith in the practical sphere, constitutes one of the greatest triumphs of American – or indeed human – genius of which there is record.
>
> Moreover, the decision to make these enormous expenditures upon a project which, however hopefully established by American and British research, remained nevertheless a heart-shaking risk, stands to the everlasting honour of President Roosevelt and his advisers . . .
>
> This revelation of the secrets of nature, long mercifully withheld from man, should arouse the most solemn reflections in the mind and conscience of every human being capable of comprehension.
>
> We must indeed pray that these awful agencies will be made to conduce to peace among the nations, and that instead of wreaking measureless havoc upon the entire globe they may become a perennial fountain of world prosperity.[10]

One day later, the *News Chronicle* in London ran a feature article by the distinguished physicist Sir Lawrence Bragg. It was titled 'Can Man Control the Power he has Unleashed?'[11]

The bombing of Hiroshima was not a scientific turning point in the search for control over natural forces. All the science deployed at Hiroshima had previously been discovered and demonstrated in the long history of experiments that had cumulated in the nuclear test explosion at Alamogordo. Yet the Hiroshima holocaust was

16   *GLOBAL FISSION*

still a turning point. Up to that time the nuclear developments had
been shrouded in secrecy: first in the secrecy of the cloistered halls
of scientific endeavour, shielded behind the mystifying language
and remote actions of theoretical and experimental physicists, and
then behind the impenetrable barrier of allied military security.
Now the research could no longer be kept hidden from the public
consciousness. Its potent implications were laid bare for all to see.

For the thousands of technicians and scientists who had worked
on the project, and for the global scientific community, heady
triumph was mixed with trepidation. On the one hand they were
credited with 'the greatest achievement of organised science in
history' (President Truman),[12] and 'the greatest step forward ever
made by man in his efforts to control nature' (Sir John Anderson,
Coalition Chancellor of the Exchequer).[13] On the other hand,
underlying many of the reports and editorials after the bombing,
was an undercurrent of fear. Newspaper reporter Robert Waitham
described the reactions in the USA:

> In the newspapers and on the radio you read and hear today
> such passages as this: 'Yesterday we clinched the victory in the
> Pacific, but we sowed the whirlwind . . . With such god-like
> power under man's imperfect control, we face a frightful
> responsibility . . . We have narrowed down the choice to one
> of the end of war or the end of humanity'.[14]

THE MAKING OF AN IDEOLOGY

It was natural that politicians and members of the technical community most intimately connected with the development of the
nuclear bomb would be anxious to stress any positive benefits that
might stem from it and to counter any criticism of their work.
Further, the dramatic results of their work, and the rapidity with
which the single-minded endeavour had moved from theory to
awesome practice, imbued an extraordinary degree of scientific
optimism: a confidence that nature could place no long-term
insurmountable barriers before their indomitable will. Even
directly after the bombing flamboyant optimism underlay many
published comments by scientists.

> The horizon now is limitless, and questions which before were
> relegated to the realm of fiction for the first time become
> practicable and even imminent.

FROM BOOM TO CRASH 17

> At the threshold of boundless energy, man must reconsider
> his position in nature and his social relations

said Professor Eugene Delporte, Director of Belgium's Royal
Observatory.[15] Dr Coutre, also an astronomer, speculated: 'The
old question of what lies on the other side of the moon may finally
be solved ... These are questions which atomic energy driving
rockets into space areas hitherto veiled even to the telescope, may
answer'.[16] But most pervasive of all was the emphasis on the use of
the energy back on earth. Typical of this was the comment of
Professor Bourgeois, a colleague of Dr Coutre:

> This may well mean revolutionary changes in industrial
> power. If atomic energy can be controlled as well as released,
> all other forms of power – coal, petrol, water power – may
> become obsolete necessitating complete changes in the indus-
> trial pattern. This is the key question: How long will it be
> before we can release the energy so that we can guide it?[17]

Behind these optimistic speculations lay the more sombre implication of the bomb itself: the unthinkable horrific role that it might
play on some future day of catastrophe. For some scientists it was
sufficient to place this consideration aside. Here it was helpful to
view one's work in a framework of technical neutrality. The power
of the bomb could be used for good or evil, for peace or war. Seen
this way, the responsibility for the holocaust at Hiroshima, and its
threat for the future, lay with politicians who made the decision to
drop the bomb. The credit for its development and any future
benefits for humanity lay with the technical community. Such a
framework can be both comfortable and comforting. It enabled
many of those associated with nuclear research to continue their
lives in the post-war years with a dedication untroubled by consideration of consequences. It even enabled scientists who had
worked on the Manhattan Project to continue working with the
military and to go on and help build the hydrogen bomb.

For many scientists involved with nuclear-related research or
development, such justifications were insufficient. Nevertheless,
for them there was often a powerful tendency to interpret the
potential of what had been begun in the very best possible light.
Later, when other problems associated with their work began to
emerge, that tendency would persist. It would not be restricted to
scientists. It would also be nourished by a community – euphoric

18 GLOBAL FISSION

first with the end of the war, and then with the stream of technological developments and consumer products that its end released –
which was happy to accept a general perspective of technological
optimism.

In the aftermath of the bombing of Hiroshima this was aided
both by the articles and speculations of scientists, and by a process
of diffusion through journalists reporting and interpreting the
comments of scientists. Books and articles appearing in the following months helped to produce images of the boundless benefits that
would be realized from the power of the atom.

In his book *Atomic Energy in the Coming Era*, which appeared in
late 1945, David Dietz, Pulitzer Prize journalist and lecturer in
Science at Western Reserve University in the USA, summed up
much of this optimism.

> Planes carrying several thousand passengers, with as much
> cabin space as a luxury liner will make non-stop flights from
> New York to India or Australia ... Instead of filling the
> gasoline tank of your automobile two or three times a week,
> you will travel for a year on a pellet of atomic energy the size
> of a vitamin pill ...
>
> The same pill will be enough to heat your house for the
> winter ... The day is gone when nations will fight for oil ...
> No baseball game will be called off on account of rain ... No
> airplane will by-pass an airport because of fog. No city will
> experience a winter traffic jam because of heavy snow ...
>
> Universal and perpetual peace will reign in the Era of
> Atomic Energy for three reasons ... First: with energy as
> abundant as the air we breathe, there will be no longer any
> reason to fight for oil or coal. Second: By using atomic energy
> to mine the ocean for its vast mineral content, every nation
> will be able to obtain easily all the raw materials that it needs.
> Third: With even more powerful atomic bombs than those
> dropped on Japan, war will become so destructive that no
> nation will dare begin one ...
>
> Changes fully as great as these will be ushered in with the
> widespread use of atomic energy. To begin with it will com-
> pletely change the costs of production. Cheaper costs of
> production will mean more plentiful production. It will also
> mean new types of production. Processes now commercially

FROM BOOM TO CRASH 19

impossible because the cost in electric power or heat would be too great will become possible at once.[18]

In the event, the prospect suggested did not occur, and shows little sign of doing so in the future. The optimistic tone was founded on a mixture of ignorance and self-deception which obscured many of the hazards and problems that were later found to be associated with nuclear power. This was combined with a confident assurance that any problems then known would easily crumble before the coming scientific assault. This view was persuasively promulgated, and the ideology that underlay it steadily began to be reflected in community attitudes.

By September 1945 the National Opinion Research Centre found 56 per cent of Americans agreeing that 'splitting the atom will prove the greatest invention in 1 000 years', and 52 per cent believing that 'people everywhere will be better off . . . because somebody learned how to split the atom'.[19] The ideology was also embraced by a section of the original Manhattan Project scientists who began to lobby for the development of nuclear energy for industrial use. It was reflected in the statement of President Eisenhower some eight years later, in December 1953, when in a speech to the United Nations he introduced his 'Atoms for Peace' programme. It was reinforced over the subsequent years by a flood of remarkable technical achievements which flowered from the military research of the war years. It is still to be found underpinning the arguments and attitudes of many of the promoters of nuclear power today.

THE 'PEACEFUL ATOM'

If civilian nuclear power was nourished by an ideology of technical optimism it was spawned almost everywhere as an offshoot of a nuclear weapons programme.

In the USA first priority was placed on establishing a mass production programme for manufacturing nuclear weapons. For this purpose a civilian Atomic Energy Commission (AEC) was established in 1946 supervised by a congressional Joint Committee on Atomic Energy. It was not until 1954 that the Atomic Energy Act was amended to give the AEC the responsibility of fostering a private nuclear energy industry.[20]

The first plant was commercial only in the loosest sense. It

20   GLOBAL FISSION

consisted of a larger version of the reactor developed by the US
Navy, under Admiral Rickover, to power submarines. The 60
MWe* reactor was built at Shippingport, Pennsylvania, and the
Duquesne Company contributed $5 million† for patriotic reasons
although they expected to make a loss.

In the UK the first nuclear power plant was established by the
Atomic Energy Agency in 1956. It was the Calder Hall plant,
Cumbria, built primarily to produce plutonium for the British
nuclear weapons programme, but with a steam generator and
turbo-generator added to provide electricity. On 17 October 1956
its output of 50 MWe was switched into the electricity grid.

The first Soviet plant began operating in June 1954, five years
after their first successful nuclear weapons test. The plant was
acknowledged to be uneconomic, producing a tiny 5 MWe for the
grid. It too was a development of the experience gained in producing fissionable materials and was designed along the same lines as
those primarily used to produce plutonium.[21]

In France, an Atomic Energy Commission (*le Commissariat á
l'Énergie Atomique*) was created in 1945 ostensibly to research and
develop 'peaceful' nuclear applications. However, the first emphasis was on the construction of plutonium-producing reactors. In
mid 1955 the French AEC and the Ministry of Defence signed an
agreement, and in April 1958 the government formally authorized
the construction of a nuclear weapon. In 1955 the French government also handed the responsibility to develop nuclear power
plants to the national electricity authority, *Électricité de France*.[22] In
1958 the first reactors to be connected to the national grid were
those designed to produce plutonium at Marcoule near Marseille.
However, the first nuclear reactor primarily designed for electricity
production was the 60 MWt Chinon reactor switched into the grid
in 1963.[23]

Germany and Japan were restricted militarily at the end of the
war. After restoration of sovereignty in 1955 the Federal Republic
of Germany placed great emphasis on 'avoiding being left behind',
although little had yet been achieved elsewhere in nuclear power

---

* The electrical output of a nuclear reactor is measured in watts electrical:
1 million watts is called a megawatt (MWe), and 1 000 million watts is called a
gigawatt (GWe). Where the measurement is for heat developed, a 't' is substi-
tuted for the 'e' giving, for example, MWt.

† US$ are used throughout this book unless otherwise indicated.

FROM BOOM TO CRASH 21

research, and appointed the world's first minister for Atomic Energy.[24] In Japan the government set up the Japanese Atomic Energy Commission in 1955 and a research institute in 1956 to construct a British Magnox plant at Tokai Mura.[25] Sweden and Belgium also began some experiments, Sweden constructing a relatively elaborate experimental underground reactor in 1954, and Belgium a test reactor in 1962.[26]

Possibly the only attempt to build a nuclear power reactor which was based on totally non-military objectives was the Canadian CANDU design. Although Canada had participated as the third partner in nuclear weapons research during the war and had been the site for some major experimental facilities, most of the British scientists who had helped construct these were recalled at the end of the war. Abandoned, Canada made a policy decision not to develop nuclear weapons and instead to concentrate on the peaceful aspect of the power of the atom. Canada's precursor for the CANDU reactor was the NRX research reactor at Chalk River which produced its full power of 40 MWt in May 1948.[27] It is ironical that the first nuclear explosion to be produced from plutonium taken from a 'peaceful' nuclear reactor was triggered by India in the Rajasthan desert on 18 May 1974, using materials taken from a Canadian research reactor at Trombay.[28]

HEADY EXPECTATIONS

Despite the flurry of research and the bewildering array of reactor prototypes that were constructed during the 1950s, it was not until the mid 1960s that nuclear power plant construction began in earnest.

In the USA, power companies ordered twenty-six nuclear plants in 1965, compared with the five that had been ordered between 1955 and 1964. During this period expectations sky-rocketed. The most optimistic predicted reactor construction would grow at 20 per cent each year until the end of the century. By the year 2000, the AEC expected between 1 200 and 1 500 1 GWe reactors would be operating in the USA.[29] Between 1963 and 1965, 7 GWe of nuclear power capacity was ordered in the USA. This grew to 54 GWe over 1966–68, dropped slightly to 40 GWe over 1969–71, and catapulted to 110 GWe during 1972–74.[30]

The eruption of orders in the USA was released by the increasing competition between two giant corporations, General Electric and

22   GLOBAL FISSION

Westinghouse, each vying to establish itself as the prime nuclear
vendor. During 1963 this competition produced a series of 'turnkey' offers. These were undertakings that the companies would
contract to deliver complete nuclear reactors at a firm price except
for adjustments due to inflation. All that would be necessary for the
purchasing utility would be to turn the key, enter, and start the
generating equipment. Both the price of the reactor and the advertised price of electricity expected to result was held at an unrealistically low level by the competing companies in an effort to gain a
foothold. The first utility to accept the offer was Jersey Central
Power and Light Company for its Oyster Creek nuclear plant, and
others quickly followed.[31]

This spurt in orders seemed to reflect an equally extraordinary
confidence. The first orders, between 1965 and 1967, were for
reactors with a maximum output between 0.87 and 1.12 GWe.
Was this based on a careful assessment of the performance of
similar reactors to check they were economic? Hardly. The largest
reactor to have operated at that time could produce at maximum
0.2 GWe. By 1969 reactors which could develop 1.18 GWe were
being ordered, while the largest reactor to have actually operated
for over one year was producing only 0.58 GWe, less than half the
ordered capacity. Even up to 1972, when large numbers of reactors
had been ordered with capacities as big as 1.25 GWe, the largest to
have operated could turn out only 0.81 GWe.[32]

The technological audacity in contracting to purchase reactors of
a size twice as large as any which had operated for even one year
was matched only by the boldness of the reactor constructors in
selling reactors twice as large as those for which they had any
construction experience Added to this boldness was an apparently
firm belief that there would be a market. So firm was this belief that
they were prepared to sell their first reactors at a loss in order to set
the market moving. This behaviour cannot be explained simply in
terms of the technological optimism of scientists and the community. In chapter 6 some of the other relevant factors which pushed
hopes so high will be examined in some detail. Here it is sufficient
to note that the optimism with which predictions were couched,
and the behaviour of even the hard-headed business community
seemed, in the early 1970s, to be quite compatible.

As sales increased in the USA, a small but steady trickle of
reactor sales commenced overseas. A few medium sized plants

FROM BOOM TO CRASH

were ordered in the late 1960s and a growing flow of larger orders began in 1970.[33] The pitch of optimism that underpinned these orders was most evident in the assessments that were made of the role that nuclear power would be playing in global energy production by the end of the century. However, this bubble of heady expectations was to burst in a spectacular way.

THE 1975 CRASH

Those who attempt to predict the future have a habit of being revealed as monumentally in error. As with the classic failure of Thomas Malthus's eighteenth century predictions of imminent starvation in England, when naive extrapolation is mixed with the intoxication of optimism startling discrepancies can result.[34] On the global scale, few expectations can have been more dramatically dashed than those that were briefly held for nuclear power at the beginning of the 1970s.

In his speech introducing Project Independence in 1974, President Richard Nixon set as a realizable goal that more than 50 per cent of the USA's electricity should be produced by nuclear power by the end of the century.[35] President Nixon was not alone or extreme in terms of the contemporary expectations. For example, Robert Kirsche and his fellow analysts employed by the US nuclear engineering company Commonwealth Associates, were predicting that the proportion should be 60 per cent by 1990.[36] Only three years later, in announcing his 1977 energy plan, President Jimmy Carter stripped the estimate for the year 2000 to 25 per cent. Official estimates of the number of 1 GWe plants that would be operating in the USA fell from 1 200–1 500 GWe to 300–400 GWe.[37] By mid 1980 the US Energy Information Administration was predicting 160–200 GWe.[38]

In Japan estimates of nuclear capacity for 1985 reached as high as 70 GWe.[39] The official Japanese Atomic Energy Commission goal announced in 1972 was 60 GWe.[40] Late in 1975 the government scaled down its forecast to 49 GWe.[41] The official Institute of Energy Economics was sceptical of even this revised target, having already in mid 1975 forecast a nuclear capacity of somewhere between 30–35 GWe.[42] By 1977 the Institute had dropped its forecast to 27 GWe,[43] and in June 1977 the government followed by dropping its target to 26–33 GWe.[44] From then to the end of

24 GLOBAL FISSION

Figure 1 **Variation of official estimates of non-communist world nuclear capacity in AD 2000 against year of forecast**

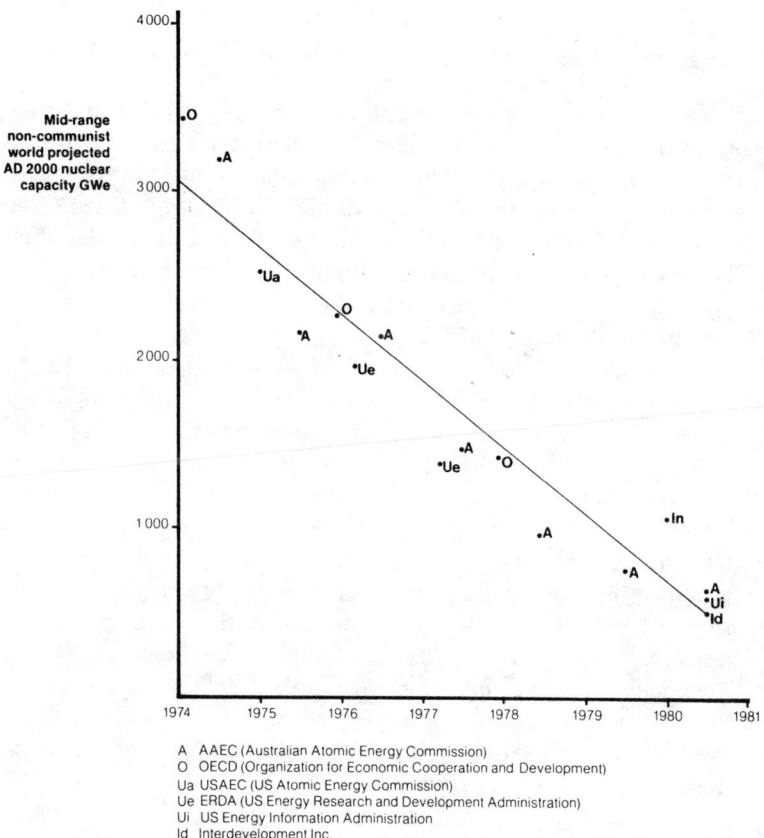

A AAEC (Australian Atomic Energy Commission)
O OECD (Organization for Economic Cooperation and Development)
Ua USAEC (US Atomic Energy Commission)
Ue ERDA (US Energy Research and Development Administration)
Ui US Energy Information Administration
Id Interdevelopment Inc.
In INFCE (International Nuclear Fuel-Cycle Evaluation)

Source: Derived from: AAEC Annual Reports 1973-4, 1974-5, 1975-6, 1976-7, 1977-8, 1978-9, 1979-80; USAEC November 1974; USERDA, *Demand for Uranium and Separative Work,* March 1976; USERDA/FEA April 1977; OECD *Energy Prospects to 1985,* 1974; OECD (NEA)/IAEA, *Uranium: Resources Production and Demand,* December 1975 and December 1977; US Energy Information Administration, *Annual Report to Congress,* July 1980; INFCF estimate quoted in OECD (NEA) *Eighth Activity Report 1979,* 1980; and Interdevelopment Inc.'s independent forecast reported in 'Official Projections Unrealistically High, Consultant Says', *Nucleonics Week,* 22 May 1980, p. 1.

FROM BOOM TO CRASH 25

the decade observers' estimates of the achievable nuclear capacity in Japan by 1985 were still lower, placing the figure at between 21–25 GWe.[45]

In 1975 *Nuclear News* estimated that West Germany would have between 45 GWe and 50 GWe of nuclear power by 1985.[46] The official government figure was still 45 GWe in December 1976 when the Association of Industrial Power Producers (*Vereinigung Industrieller Kraftwirtschaft*) stated that it expected to reach 34 GWe. By the end of 1977 the government figure had dropped further to 30 GWe.[47] From 1978 to the end of the decade a de facto moratorium on construction of new reactors set in.[48] Some commentators estimated that this could lower the figure for West Germany's nuclear capacity in 1985 to as little as 18 GWe.[49]

For the whole of the Organization for Economic Co-operation and Development (OECD), official projections for 1985 dropped from the International Atomic Energy Agency's (IAEA) 1970 figure of 555 GWe to their 1976 figure of 400 GWe,[50] and by the beginning of 1980 to 259 GWe.[51] Even this figure looked high, as it was based on the estimates of the International Fuel Cycle Evaluation (INFCE), which some other analysts considered to be over optimistic.[52] For example, a little earlier in 1979, the US Department of Energy had been more pessimistic, placing the figure at 205 GWe.[53]

The world-wide decline in nuclear expectations that occurred in the 1970s is shown graphically in figure 1. (For simplicity, where estimates were given as a pair of high and low figures the numbers have been averaged to give the middle of the range.) From this graph it can be seen that nuclear expectations declined almost linearly with time, from about 3 500 GWe in 1974, to about 500 GWe in mid 1980.

A STAGNATED INDUSTRY

The precipitate crash in expectations reflected a drastic slump in reactor sales and a slowing down in reactor construction programmes. At the beginning of the 1980s a perilous stagnation had gripped much of the global nuclear industry. An article which appeared in the conservative economic journal *Business Week* as early as December 1978 surveyed the scene, revealing a gloomy picture.[54] In the UK the six boilermaker companies involved with reactor construction had been reduced to two and even these two

might have to merge. Said David F. Hutton, managing director of Robert Morton (D.G.) Ltd, one company that has retired from reactor construction:

> We can't live on air for three or four years. The nuclear industry as a long-term prospect does not impress us at all . . . We have to live.[55]

In Italy, thirty companies faced serious problems due to the stalled nuclear programme. According to the journal, roughly $70 million worth of nuclear-related plant was standing idle.

In West Germany, an effective moratorium on the construction of nuclear plants had set in and the construction of ordered plants had been brought to almost a standstill. Kraftwerk Union (KWU), Germany's biggest reactor manufacturer, had not won a domestic order since June 1975 and more than half of its $7 billion total of orders was blocked by court rulings or lack of construction permits.

Globally, the nuclear industry had entered a period of severe market stagnation. In mid 1975, the total number of reactors on order was one hundred and fifty-six. By 1979 it had fallen to 102, and by mid 1980 to sixty.[56]

As at 1978, Westinghouse had not expanded its nuclear division at all for two years. It had received only four domestic orders in 1975 and none since. General Electric faced losses of $500 million, Babcock and Wilcox of $200 million, and Atomic Energy of Canada Ltd was expecting losses of $200 million.[57] KWU had not made a profit from the time it was set up until 1976. In 1976 it received two big orders from Iran and Brazil. However, the major part of these was later cancelled. The companies fared no better over the rest of the 1970s.

Nowhere was the stagnation felt more than in the USA, the heartland of the nuclear power industry. As *Business Week* put it, 'Not soon, but within ten years, the US nuclear industry is apt to contract dramatically, and it may collapse altogether'.[58] A report titled *Nuclear Power Costs* by the US House Committee on Government Operations in 1978 explained why:

> New orders for nuclear reactors fell off from 41 in 1973, to 26 in 1974, to 7 in 1975 and 1976 combined. There were four orders in 1977.
>
> Cancellations and deferrals have been even more dramatic.

FROM BOOM TO CRASH   27

In 1974, nine units were cancelled and 91 deferred. In 1975, 12 were cancelled and 86 deferred. Through 1976 and the first 6 months of 1977, there were 8 cancellations and 106 units deferred.[59]

In 1978 only two units were ordered, and five were cancelled. This reversal from a state of rapid expansion to a situation of a negative rate of ordering was reflected by the collapse in the total number firmly planned for construction. For the USA, the total number of reactors on order plummeted from 104 in 1975, to forty-seven in 1979, and to twenty-six in 1980.[60]

Understandably the tone of comments from the industry from 1975 to the end of the decade reflected a progressively deepening concern. 'The front line vendors have yet to make a dollar with any certainty after twenty years of effort', said Richard McCormack, a senior executive of General Atomic, in 1975.[61] Before the end of the year his company, a subsidiary of Gulf Oil, had abandoned the nuclear reactor construction business. Two years later, in 1977, John O'Leary, US Deputy Energy Secretary, was worrying that unless something changed in the foreseeable future, the nuclear option in the USA 'has essentially disappeared'.[62] Among similar gloomy comments the following year was one by a senior non-nuclear executive for one of the USA's four remaining reactor makers. He bluntly concluded, 'The existing nuclear industry can't survive. Period'.[63]

As the decade closed, the industry's concern did not seem to have abated. In 1980, A. Philip Bray, the general manager of General Electric's large nuclear power systems division, summed up the prospects of the industry equally pessimistically by stating that 'The nuclear industry is in grave danger of atrophy'.[64]

The pessimistic comments were hardly surprising given the plummeting orders and expectations of the nuclear industry. But while the cries of pain seemed justified, a question remained. Could not the present forebodings of doom prove to be just as erroneous as the earlier predictions of snowballing success? To see if an answer to this can be found requires an examination of both the obstacles which have prevented the industry's earlier expectations from coming to fruition, and an analysis of the significance of the crash in orders for the industry's future. These will be examined in much more detail later. Here it is sufficient to note that the crash has been so severe that if the industry persists in its present state of

stagnation, it may eventually face partial or complete collapse. But this will depend crucially on the outcome of a battle now raging between the nuclear industry and sections of the community.

Although there have been other contributing factors there can be little doubt that in many countries the opposition to nuclear power has been a powerful factor in the industry's present serious problems – a factor which may prove decisive. For some countries, including important potential purchasers of nuclear reactors such as Sweden, Austria, Denmark, the Netherlands, and Switzerland, government decisions have been forced which will directly halt, limit, or dramatically slow their nuclear programmes.

In other countries, particularly West Germany, court orders have been obtained which have prevented numerous proposals from coming to fruition. Elsewhere, community opposition has directly obstructed nuclear projects with such tenacity and vehemence that governments or electricity companies have cancelled nuclear plans in favour of alternative methods of generating energy. But whether directly in this way, or indirectly through court action, pressure on government, or by other means, concerns about the hazards of nuclear power will be seen to have played a vital role in the plunging expectations of the nuclear industry. To see this it is logical to begin with the USA where the nuclear industry is most massive and the greatest number of nuclear reactors are already operating. There, the intimate relationship between the pressing problems of the industry, the hazards of the nuclear enterprise, and the opposition that has risen up to meet them, is graphically illustrated by a powerful drama which began on 28 March 1979 in the state of Pennsylvania.

2
MESSAGE FROM PENNSYLVANIA

On an island in the middle of the Susquehanna River, Pennsylvania, USA, on an 113 hectare site, stands the imposing nuclear reactor complex known as Three Mile Island. The large volume of water flowing past it provides the necessary water for its cooling system and its convenient situation allows the electricity it produces to be carried easily to the state capital, Harrisburg, 26 kilometres to the south. The nearest town is Middletown, its fictional namesake 'Middletown USA', the archetype of sleepy middle America. But the sleepiness in the control room of Three Mile Island in the early morning of Wednesday, 28 March 1979, was abruptly shattered when the system's alarm bells began to ring.

At 4.00 a.m., during the 'lobster shift' in the control room of Unit Two, lights suddenly blinked to red on the instrument panel. A siren whooped a warning, the first of over 200 alarms that were to sound within the next five minutes.[1] Over the public address system an operator announced: 'Turbine trip in Unit Two'.[2] In the understated language used by the nuclear industry, a 'transient' was occurring at Three Mile Island.

The two nuclear reactors or 'units' at Three Mile Island are light-water reactors (LWRs), a design almost uniformly adopted by US nuclear manufacturers, and thus now most common throughout the world. Although the actual hardware is complex and sophisticated, the basic principles on which the reactors are constructed are easy to understand.

FUNDAMENTALS

In the central 'core' of such a reactor, sufficient rods of uranium

30   *GLOBAL FISSION*

Figure 2 **Diagram of Three Mile Island nuclear reactor**

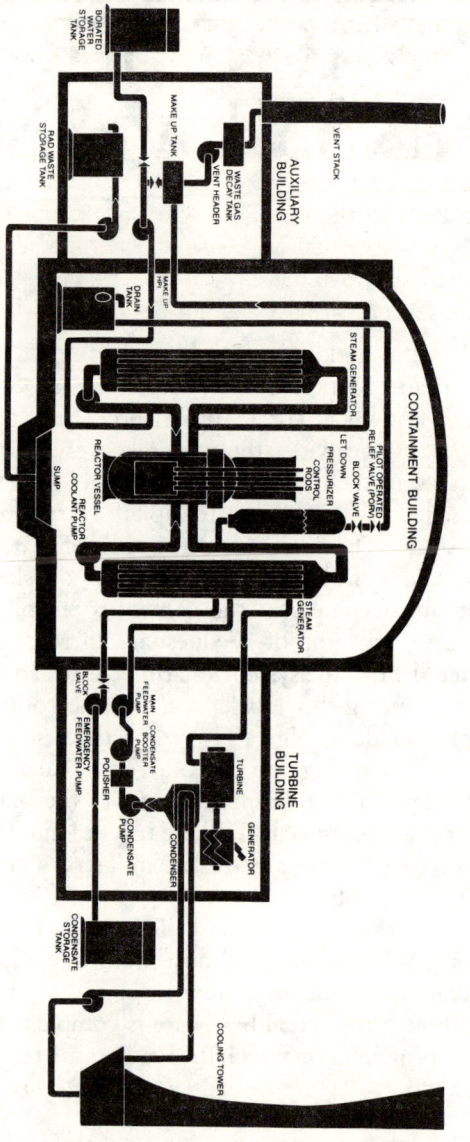

Source: J. G. Kemeny (chairman), *Report of The President's Commission on the Accident at Three Mile Island,* US Government Printing Office, Washington, October 1979, p. 86.

MESSAGE FROM PENNSYLVANIA

fuel are gathered together to yield enough neutrons to create a self-sustaining chain reaction of atomic fissions. Each fission of a uranium atom's nucleus releases heat, together with neutrons and other fragments of the original nucleus. Some of these fragments are themselves highly radioactive nuclei of other elements, such as strontium and caesium. These are nuclear wastes, and they release further heat as they too radioactively decay.

In order to control the rate of the reaction and the amount of heat produced, special 'control rods' are inserted between the uranium fuel rods in the core. These rods are constructed of a material such as boron which absorbs neutrons like a sponge. If enough of these rods are in place no chain reaction occurs. By withdrawing the rods very slowly, a point is reached when the chain reaction commences. If too many neutrons are produced, and too many fissions are taking place, then the reaction can easily be slowed by returning more of the control rods into the core.

After the reactor has been operating for even a relatively short time, the fuel rods become extraordinarily radioactive due to the radioactive waste products that are being created within them. Once this has occurred, even if all the control rods are inserted into the core so that no chain reaction takes place, the heat generated by the natural decay of the radioactive atoms in the fuel rods is so intense that they would quickly become very hot. Thus, even with all the control rods in place, the core must be continually cooled.

The Three Mile Island reactor was the Pressurized Water Reactor (PWR) variant of the design. In this design, water under high pressure is continually pumped into the core to cool the fuel rods, and the resulting scalding hot water, heated to a temperature of around 320°C, is then pumped out to a heat exchanger where it is cooled to about 290°C before being returned to the core. This self-contained circuit of water is called the primary cooling system and the water in it itself soon becomes intensely radioactive.

At the heat exchanger the steel pipes containing the primary coolant are immersed in a water jacket through which the water of the secondary cooling system is pumped. Water is pumped past the pipes of the primary cooling system to cool the primary water. The secondary water is pumped at a lower pressure which allows it to flash to steam. This steam is then piped to turbines which turn electricity generators. After use the steam is cooled and condensed by yet another stream of cold water before being pumped back to the heat exchanger. Because the heat exchanger is the source of the

32   GLOBAL FISSION

steam routed to the other lines it is sometimes called the steam
generator. The layout of the Three Mile Island reactor is shown
schematically in figure 2.

Although the core is contained within a thick steel reactor vessel
to which are welded the pipes of the primary cooling system there is
a danger that an accident could cause some part of this 'shell' to
rupture, allowing deadly radioactive materials to escape to the
atmosphere. Accordingly, the entire reactor vessel, the primary
cooling system, the steam generators, together with some other
parts of the system, are entirely surrounded by an outer 'containment building'. At Three Mile Island this consisted of a 1.2 metre
thick reinforced concrete shell.[3]

The containment is an enormously strong structure. Even so, it
is vulnerable to at least one possible accident. This is the so called
Loss of Coolant Accident (LOCA). The heat generated in the
40 000 or so fuel rods is continually carried away by the primary
cooling system. With pipes a metre across, 16 tonnes of water per
second are pumped continuously through the core at a speed of
around 200 kilometres an hour.[4] If for some reason this torrent of
water were to falter in its relentless progress through the reactor
core, the heat generated by rods could cause the temperature to rise
rapidly.

Even with the control rods in place, the heat generated by radioactive decay in the rods is considerable. In a typical modern reactor
the output of decay heat could be as high as 200 MWe, about the
output of 200 000 radiator bars, at the time of the accident. The
output would drop to approximately 60 MWe after about fifteen
minutes, but would decrease much more slowly to about 6 MWe
over the next couple of weeks.[5] The heat output would have been
lower at Three Mile Island because the Unit Two reactor had only
been in operation for three months, but the temperatures developed would still have been extremely high.

The ferocious temperatures would cause the rods first to twist
and buckle. According to an AEC study, the zirconium alloy
cladding of the fuel rods would first start to oxidize and at 1 800°C
would start to melt with the possibility of molten metal water
explosions. Once melting starts somewhere in the core it would
propagate and the core would begin to collapse. After between half,
and one and a half hours, depending on the accident, the accumu-
lating mass of molten fuel, now at a temperature of 2 700°C, would
have melted through the reactor pressure vessel, and within a few

MESSAGE FROM PENNSYLVANIA

additional hours through the concrete floor of the containment building.[6]

Once through the floor of the building it is unlikely that the earth would contain the hot radioactive gases and other materials and these would be likely to escape, forming a deadly radioactive plume. The white hot mass of fuel would continue to melt through the earth for a long but indeterminate time, a situation christened 'The China Syndrome' by those who once suggested (not totally seriously) that it might never stop! It would be almost certain to pass through the water table causing an extremely powerful steam explosion venting further large quantities of radioactive material to the atmosphere.

The final backup measure built into the hardware of the reactor to prevent this core meltdown from occurring is the Emergency Core Cooling System (ECCS). This system is designed to automatically dump hundreds of thousands of litres of water onto the core to cool it should the primary cooling system fail. Clearly, should the ECCS prove inadequate for the task, or fail to operate properly or at all, the result could be disastrous.

A study carried out under the auspices of the American Physical Society and published in 1975, estimated that such an accident could cause 10 000 cancer deaths, between 200 000 and 300 000 thyroid nodule cases, and 3 000 to 20 000 genetic defects.[7] A revision of a 1957 study known as WASH-740 carried out for the AEC by the Brookhaven National Laboratories in 1965 but only released in 1973, showed sample calculations indicating 45 000 fatalities and an 'area of disaster . . . equal to that of the State of Pennsylvania'.[8]

IN THE CONTROL ROOM

At Three Mile Island, on Wednesday, 28 March, the theoretical possibility of a core meltdown disaster became separated from actuality by only a thin tissue of lucky circumstance.

In the control room on that morning the first indication of difficulties was the alarm sounding to indicate the automatic closure of the pump in the secondary cooling system of the Unit Two reactor. This caused little concern. The operators had encountered similar 'trips' before, and the plant had developed eight such events since the previous October.[9]

Throughout the plant, maintenance crews had toiled for weeks

GLOBAL FISSION

to refuel Unit One and bring the Unit Two on line. Unit Two had been operational for less than three months and had been plagued with start-up problems. In January it had been shut down for nearly two weeks so leaky valves could be replaced and in February still other valves and pumps had had to be repaired. The announcement of the fault in the giant turbine was exasperating. 'Aw damn', muttered one engineer thinking of the additional work in finding and correcting yet another minor technical 'glitch'.[10]

Panel lights revealed the cause of the turbine trip. Half an hour earlier, at 3.30 a.m., workers began cleaning a polishing unit that traps minerals in the secondary cooling system between the two steam generators and the turbine. During the cleaning, the operation of a condensate booster pump in the secondary system had become impaired. At thirty-seven seconds past 4.00 a.m., at 00 sec., this caused the associated feedwater pumps to trip.[11] The result was that the two steam generators were now losing their supply of secondary cooling water.

The steam generators are designed to operate like giant tea kettles, powered from the core heat transmitted by the primary cooling system. But the accident that had closed off the secondary water would leave the steam generators completely dry within two minutes and unable to remove heat from the core. As the steam generators began to overheat, the temperature and pressure in the primary system also began to rise.

Six seconds after the initial pump failure, at 06 sec., the pressure had jumped to 153 times atmospheric pressure (1 atmosphere = 101.3 kilonewtons per square metre). At this pressure, 6.8 atmospheres above normal, a special relief valve automatically opened to allow the excess pressure to vent to a 'quench tank' in the containment building. The pressure continued to rise but the rate of build-up slowed steadily.

Two seconds later, at 08 sec., the pressure reached 159.9 atmospheres. Automatically, the reactor rammed its control rods into its core. The reactor had 'scrammed'. The chain reaction was quenched. But still decay heat continued to be generated relentlessly in the core.

Five seconds later, at 13 sec., the steam venting from the core vessel through the emergency relief valve had, at last, brought the pressure down to normal. At this point the valve should have closed automatically. But this time the valve stuck open.

No lights were provided to indicate to the operators that the

MESSAGE FROM PENNSYLVANIA 35

relief valve was incorrectly open. As primary coolant continued to blow out into the quench tank, the pressure began to fall and the core temperature began to rise. At 14 sec., the pressure had fallen to 143.2 atmospheres and the temperature had risen above 322°C.

In the secondary cooling system, pressure dropped sharply due to the failed feedwater pump. The designers of Three Mile Island's cooling system had anticipated that the feedwater pump would trip several times in the reactor's lifetime. In fact, a similar trip had occurred four months before, on 28 November 1978.[12] Therefore the reactor was provided with auxiliary pumps to take over from the failed feedwater pump and keep the secondary water flowing. However, at 38 sec., as the three auxiliary pumps automatically sprang into action, there was an unexpected difference.

Several days before, during maintenance, operators had shut two valves to isolate the auxiliary feedwater system from the secondary system. Contrary to regulations these had not been re-opened before the reactor resumed operation. Therefore, although the auxiliary pumps whirled at full power, the closed valves prevented them from pushing water through to the steam generators.[13]

In the control room the rapidly approaching crisis was obscured. For the primary cooling system, no light indicated that coolant was blowing out through the stuck relief valve. Operators testified later that the relief valve had leaked on so many occasions in the past that they paid no attention to gauges showing that the system was overheating. In the secondary system the panel lights showed that the auxiliary pumps had commenced operation. A tag noting that feedwater isolation valves were on had been attached to the panel, but coincidentally obscured a light that would have warned the operators that the auxiliary pumps were whirling uselessly.[14]

At 1 min. 45 sec. the steam generator boiled dry.[15] Although the pressure in the primary system was dropping, the meters in the control room, due to design inadequacy, erroneously suggested that the pressure was rising.

The pressure in the primary cooling system is maintained by adjusting the gas pressure in a tank fixed to the side of the reactor vessel called the pressurizer. Under normal circumstances this can act like the glass tube on the side of a coffee urn, the level of liquid in the pressurizer indicating the level of coolant in the reactor vessel. For economic reasons the designers of the reactor had used this feature to measure the level of coolant in the reactor vessel rather than to place a level sensor inside the vessel. Unfortunately,

36   GLOBAL FISSION

the steam 'voids' in the coolant in the reactor vessel this time forced water into the pressurizer and filled it. The gauge in the control room therefore showed a dramatically rising level of coolant, while in fact the vessel was emptying.

At 2 min. 02 sec. the reactor primary pressure sank to 108.8 atmospheres and the final backup system, the ECCS, snapped on flooding the primary system with pre-pressurized water.

The erroneous pressure readings rose dramatically, sweeping off scale. Stunned operators stared in confusion at the computer terminal which began printing out one and a half metres of question marks. Misled by the erroneous indications of the coolant level, and believing that the reactor was dangerously full of water, an operator at 4 min. 30 sec. turned off one of the two ECCS pumps. Ninety seconds later the primary cooling system flashed to steam as the pressure levelled out at 91.8 atmospheres.[16] It was 4.06 a.m., six minutes since the original pump failure, and the tapestry of actions and malfunctions seemed to be moving steadily towards a core meltdown.

The primary cooling system and the core filled with steam voids and the whole system began shaking as the massive 7 MW primary pumps, designed to pump water, not steam, sputtered and shook.[17] The primary coolant level continued to drop and the temperature of the core soared. What happened inside the core from then on is the subject of speculation. The Nuclear Regulatory Commission (NRC) believes that the core temperature may have reached 1 480°C. This is hot enough to rupture the zirconium cladding of the fuel rods, but not hot enough to initiate the collapse of the core.[18]

In the control room confusion was mixed with anxiety. At 4.07 a.m. a light on the control panel winked on indicating that a sump pump on the floor of the containment building had turned on. The primary water that had gushed into the quench tank had now overflowed after rupturing a safety disk in the tank. The sump pump was transferring this liquid to an auxiliary building outside the containment. Unfortunately, this building was not designed to hold the highly radioactive coolant.

Thirty seconds later, at 4.08 a.m. an operator finally noticed that the auxiliary feedwater pumps in the secondary cooling system were isolated. As he switched on the pumps, a torrent of water gushed into the dry, hot steam generators possibly fracturing some of the pipes of the heat exchanger.[19]

MESSAGE FROM PENNSYLVANIA

That was the first eight minutes. It was only the beginning. Over the next few hours the confusion continued. At 4.10 a.m., even as the secondary system began to regain pressure, an operator turned off the remaining ECCS pump, quite possibly leaving the core partially uncovered. After ninety seconds he apparently changed his mind and restarted both pumps.

At 5.15 a.m. steam in the primary cooling pumps threatened to break them, causing the operators to turn them off. At 5.40 a.m., following what they understood to be standard procedure in such a situation, they shut down the auxiliary feedwater pumps for the same reason.[20] Once again the core temperature soared off scale. The NRC estimates that this time it may have reached 1 980°C, the temperature region in which theoretical calculations predict a core meltdown would start. As luck would have it, it did not.

OUTSIDE

Outside the containment building the first sign that anything was wrong was the sounding of a radiation alarm at 6.50 a.m. In the auxiliary building, to which hot radioactive primary coolant had been pumped, steam relief valves had opened automatically. Extractor fans spun, taking the radioactive steam to the atmosphere in a hot plume which drifted towards Harrisburg.

At 6.50 a.m., an on-site emergency was declared and the NRC and some local officials were notified. At 9.00 a.m. the mayor of neighbouring Middletown was finally informed of the emergency, and at 11.00 a.m., seven hours after the first crucial moments, a general emergency was declared. It was the first ever at a commercial nuclear installation.

Despite these actions there was little atmosphere of crisis outside the plant. When the early morning shift of workers arrived they were turned back by locked gates. As they waited they tossed frisbees back and forth. Commented one 'It was all fun and games at first'.[21] Jack Herbein, vice-president of Metropolitan Edison, the company which operates Three Mile Island, told reporters: 'When we say general emergency, it does not mean that an emergency exists. There was nothing that was catastrophic or unplanned for'.[22]

Metropolitan Edison continued to make soothing if inconsistent comments on the developing situation throughout the day. The governor of Pennsylvania, Dick Thornburgh, is reported to have

38   GLOBAL FISSION

told aides: 'I can't make much sense out of what Met Ed is reporting. You can't make decisions about people's lives without solid facts'.[23]

By mid afternoon, twelve hours after the initial failure, Metropolitan Edison was reluctantly acknowledging that, 'some minor fuel damage', to perhaps 1 per cent of the 37 000 fuel rods had occurred.[24] They warned that more radiation might escape. Said Deputy-Governor William Scranton, deeply disturbed: 'The situation is more complex than the company first led us to believe. Metropolitan Edison has given us conflicting information'.[25]

Some families were unprepared to sit and wait. They began to leave for the homes of friends and relatives well outside the 26 kilometre circle in which radiation had now been detected.

By supper-time the NRC, who had arrived at mid morning, believed that things might have settled down. Said Charles Gallina, an NRC investigator, at a 10.00 p.m. press conference: 'The reactor is stable. They are now bringing it to a cold shutdown condition. It is in a safe condition'.[26] As chance would have it, he was wrong.

THE HYDROGEN BUBBLE

It was not until two days later, on Friday, 30 March, that the true extent of the danger was publicly revealed. At a press conference at 9.00 a.m. the Pennsylvania Emergency Management Agency reported that large quantities of radioactive Xenon-133 gas had been vented into the atmosphere a few hours earlier. A number of other radioactive isotopes had also probably been vented during this gas 'burp'. Additionally, 182 000 litres of low level radioactive water had been discharged into the Susquehanna River. Finally, there would be no cold shutdown in the near future, because of undisclosed 'problems'. Governor Thornburgh told reporters: 'We share your frustration pinning these facts down. We're getting conflicting reports too'.[27]

A key 'problem' was disclosed later in the day. At a briefing convened by NRC's Dennis Crutchfield, the presence of a major gas bubble inside the reactor vessel was revealed. Any attempt to reduce primary coolant pressure would allow the bubble to expand downwards and expose the core. The danger presented by the bubble was massively compounded by the presence of hydrogen in it. According to the NRC, when it was first discovered on Thursday, the 51 cubic metre bubble contained only 1.9 per cent hydro-

gen. By late Friday the concentration had reached 2.6 per cent. A concentration of 4 per cent to 10 per cent was thought to be likely to lead to a violent explosion.

In fact at least one hydrogen explosion had already happened, as was admitted a day later. At 1.50 p.m. on Wednesday, 28 March, only ten and a half hours after the initial pump failure, hydrogen which had presumably leaked into the containment building exploded, creating a pressure spike of 1.9 atmospheres, and stressing the containment to half its design strength of 3.7 atmospheres.[28] The pressure spike was clearly traced on a recorder sheet. But according to NRC statements, no-one took any notice of the sheet until Friday, two days later.[29]

The consequences of a hydrogen explosion that would rupture the reactor vessel and containment building would be similar to those of a meltdown. Inside the reactor vessel, a lethal cloud of highly radioactive gases had accumulated under pressure.

The impact of radiation on a human being is often measured in units of 'rem'. A dose of 0.5 rem per year is the internationally recommended maximum for members of the public. A dose of 500 rem will cause death within a few weeks.[30] Even very small doses will increase the chance of persons exposed contracting cancer or sustaining damage to their genetic material. The extent of the damage that may stem from existing allowed doses is presently the subject of a major scientific controversy.

Even outside the reactor vessel, in the containment building, high levels of radiation were being detected. Early in the accident, a gauge was reported to have shown a reading of 30 000 rem per hour inside the containment building.[31] Suspicion was expressed that this astronomical reading was due to the sensor having been hit by a jet of steam. Later it was revised down to 6 000 rem per hour, with a utility spokesperson considering that the true value might be closer to 80 rem per hour.[32]

EVACUATION

The danger from the gases inside the reactor threatening to be released by loss of coolant, or from a hydrogen explosion, was paralleled by another danger. The radiation which had been released into the atmosphere above the crippled plant that Friday morning was substantial. Would it grow? What was the danger to citizens? Should Middletown be evacuated, and beyond that,

Harrisburg? The atmosphere in the NRC conferences supervising the emergency management was tense.

In addition, there was confusion. The radioactive plume appears to have been created on the deliberate order of the utility's Supervisor of Operations, James Floyd, in order to save a primary coolant pump. 'I had that one source of water between me and a LOCA' he was later to say in justification.[33] Floyd continued to vent the gas for days. Unfortunately, this key information was apparently not transmitted to the NRC director of operations, Harold Denton, who only heard of it months later at a Commission of Inquiry.[34] The uncertainty as to the cause of the radioactive releases and the prevailing atmosphere of confusion can be seen in the transcripts of the NRC meetings published by the investigating Commission:

> *Friday 30 March. About 9.00 a.m.*
>
> Denton [Director of NRC Reactor Regulation]: – they are getting 63 curies per second . . . which would put us somewhere in the 1,200 millirem [1.2 rem] per hour, the number that I gave somebody a little while ago.
>
> Kennedy [Commissioner]: For where?
>
> Denton: Well you know wind, and it will fall off with distance, of course, 1,200 –
>
> Kennedy: It would be 1,200 at the tower?
>
> Denton: Yes, sir, correct. Then it would fall off, of course, as it went out.
>
> What we are trying to do is to figure out what to tell the Governor, who is insisting on accurate information from the NRC about what he does about evacuation.
>
> . . . [later]
>
> Denton: We calculate doses of 170 millirem per hour at one mile, about half that at two miles, and at five miles about 17. Apparently, it is stopped now, though I'd say there is a puff release cloud going in the north-east direction, and we'll just have to see. We did advise the state police to evacuate out to five miles, but whether that has really gotten pulled off, we'll just have to –
>
> Fouchard [Public Affairs]: Well, the Governor has to authorize that, and he is waiting for a recommendation from us.
>
> . . . [later]

MESSAGE FROM PENNSYLVANIA 41

Denton: Yes, I think the important thing for evacuation to get ahead of the plume is to get a start rather than sitting here waiting to die. Even if we can't minimize the individual dose, there might still be a chance to limit the population dose.

. . . [later]

Gilinsky [Commissioner]: Well one thing we have got to do is get better data. Get a link established with that helicopter [monitoring radiation above the plant] to make sure that from now on we get reasonable data quickly.

Fouchard: But it does seem to me you have to make a judgement promptly.

. . . [later]

Hendrie [Chairman]: . . . Now, Joe, it seems to me I have got to call the Governor –

Fouchard: I do, I think you have got to talk to him immediately.

Hendrie: To do it immediately. We are operating almost totally in the blind. His information is ambiguous, mine is non-existent and – I don't know, it's like a couple of blind men staggering around making decisions.[35]

A little later, after much further discussion, Governor Thornburg made a decision. He advised all persons living within a 16 kilometre radius of the plant to stay indoors with windows and doors shut. At 11.30 a.m. a further major radiation burst was released. At the mid morning press conference Governor Thornburg acted on advice from Hendrie and advised all pregnant women and pre-school children living within 8 kilometres of the plant to evacuate. He pointed out that the young and unborn are most susceptible to fallout.[36]

The discussion on evacuation was confused further by the urgent necessity to understand and do something about the hydrogen bubble. Information was desperately needed. What was the percentage of hydrogen and oxygen in the gas? At what mixture would it catch fire or explode? What rate was the growth in the bubble? Was it threatening to expose the core and create a meltdown? Most crucially of all, what options were available to control its growth and reduce its volume?

Debate over these questions continued over Friday and Saturday. Officials had to consult with scientists for two days before they

were able to determine whether oxygen was present in the bubble in a potentially flammable mixture.[37] A misread conversion table added to the confusion. The hydrogen had been unanticipated. Water had decomposed into oxygen and hydrogen. Initially the oxygen had been absorbed by the hot cladding of the fuel rods to form zirconium oxide, releasing only hydrogen. But now, with the rods cooler, radiation was also splitting up the water, releasing both hydrogen and oxygen into the bubble. The initially pure hydrogen bubble therefore seemed to be moving steadily towards an explosive mixture. Fortunately, it never got there.

By Sunday a variety of plans were ready to try to deal with the bubble. They proved unnecessary. A mixture of successful tactics and a measure of luck had produced shrinkage in the bubble. Making it easier to eliminate, the gas had first distributed itself throughout the system as smaller bubbles, although as a Commission of Inquiry would later report, 'why this occurred, no one knows'.[38] Then, quietly, the circulation of the coolant had proved sufficient to draw gas out of the core and vent it, like fizzing lemonade, in the pressurizer. Steadily the bubble continued to shrink, and by Tuesday it had completely disappeared.[39] The emergency was over. The political fallout was about to begin.

3

THE
US RESPONSE

Once the emergency period is over the chain of events that constitute an accident are fixed for all time. They are a matter of record.
But an accident is not just a chain of technical incidents. Its significance lies in the fact that it affects people, and its ramifications are a
product of the history and state of awareness of the society in which
the accident's tremors reverberate.

Before Three Mile Island there had been other serious reactor
accidents in the USA. One was the partial core meltdown of the
Enrico Fermi 1 prototype fast-breeder reactor at Lagoona Beach,
Michigan, in 1963. Another was the fire which burned out of
control at the Brown's Ferry reactor near Decatur, Alabama, in
1975, disabling most of its control system.[1] This time, however,
the accident at Three Mile Island occurred after a decade of mounting citizen concern over the hazards of nuclear power, and its
effects resonated with the debate over nuclear power surging
through many parts of the US community.

Among the people to be profoundly affected were those who had
previously placed their faith in government and the nuclear
industry to wisely and honestly look after nuclear decisions on their
behalf. On the first morning of the accident, the Wednesday, these
people were reassured to hear from Jack Herbein, Metropolitan
Edison's vice-president, that the accident was a 'normal aberration'.[2] Said Don Curry, the company's public relations officer:

> There have been no recordings of any significant radiation
> and none are expected outside the plant. The reactor is being
> cooled according to design, and should be cooled by the end
> of the day. There is no danger of a meltdown.[3]

But five days later the reactor had not cooled and no-one knew when it would. Later the director of the NRC Division of Operating Reactors, Vic Costello, would testify that radiation readings on that first morning had clearly indicated substantial damage to the core and therefore serious danger of a meltdown.[4]

On the second day of the accident, Jack Herbein was still saying 'I wouldn't call it at this point a very serious accident'. His company was only conceding that 'less than 1 per cent' of the core rods might be damaged.[5] Later in the day the NRC was calling it the worst accident in the history of the USA. Later still, the NRC would announce that 60 per cent of the rods were probably damaged.[6]

The same afternoon of the second day, the deputy governor of Pennsylvania, William Scranton, told the press that 'the utility has ended venting contaminated steam into the air'.[7] Four days later venting was still continuing and pregnant women and children had been evacuated. Also on the second day, Herbein said 'We didn't expose anybody'.[8] Two days later he would announce that four workers had been overexposed to radiation.[9]

For the people living near the plant the effect of the early contradictions was devastating. Said Austin Buser, the postmaster of Goldsboro, a town close to the plant: 'I'm sick of it. We hear lies and then truth, more lies and then more truth. When can we start believing the people who have our lives in their hands?'[10] Within five days, a CBS poll showed that 66 per cent of the people in Philadelphia believed that the danger from the accident was greater than the public officials were disclosing. Only 16 per cent believed that the officials were being honest.[11]

But the effects of the accident went beyond disquiet or even disillusionment over contradictory statements. More significantly, in increasing numbers, people were losing faith in the safety of nuclear power itself. The catalytic effect of the accident on public opinion was reflected in opinion polls. Two successive polls, taken by Harris, one before the accident in October, and the other after, in mid 1979, showed a dramatic decline in support for nuclear power. In the earlier poll people had supported the building of nuclear plants by an overwhelming 2 to 1 margin. After the accident, people were evenly divided on the same question, opposition to nuclear construction having risen from 31 to 45 per cent.[12] A later poll by Associated Press–NBC showed a dominating 65 per cent opposed to further nuclear construction, at least until the safety issues were resolved.[13] The message was clear enough.

THE US RESPONSE

Beyond the opponents and proponents of nuclear power, beyond the scientific and technical community, many members of the general public were questioning their previously held opinions on the hazards posed by nuclear power. In doing so they were embracing issues which had been the subject of technical controversy for a much longer time.

THE TECHNICAL DEBATE

Over the previous decade the problems posed by nuclear power had, within the technical community and the nuclear industry, seemed to loom ever larger. Early in the development of the nuclear programme, most people associated with it had believed that the list of unsolved problems – environmental and health hazards, disposal of wastes, the dangers of a reactor core meltdown, the spread of nuclear weapons – would succumb easily to a concerted technical assault. However, each attempted solution seemed to bring with it further problems, while continued scientific research and experience kept bringing to the surface problems that were either unsuspected, or had been considered remote possibilities. For many of the non-technical people, however, much of the debate had been obscured by a curtain of technical jargon. Only recently had the debate carried sufficient power for real efforts to be made to interpret its meaning for them. One important consequence of the Three Mile Island accident was that it acted as a searchlight illuminating for the general public some of the more contentious areas of uncertainty in the technical debate over nuclear power.

Within eight days of the accident President Carter had announced the appointment of a Commission of Inquiry to investigate the causes of the accident and to recommend safety measures that might be necessary. 'You deserve a full accounting, and you will get it', he said in his nationally televised address to the American people.[14] Soon it was clear that it was to be one of the most investigated technical mishaps in history. By May 1979 fourteen separate inquiries had been launched by congress, the US administration, and the US nuclear industry.[15]

It is not the purpose of this book to dwell on the detail of the technical debate over nuclear power. These issues are already canvassed in a wide range of excellent books.[16] Here it suffices to examine just several technical issues which, aided by the accident at

46 GLOBAL FISSION

Three Mile Island, have played an increasingly important role in the political arena. Before the inquiries had begun to hear their first testimony, different groupings in the community were already making the first moves to sway the public's verdict. At stake was a crucial public judgement: did the accident show how dangerous the nuclear industry is or, conversely, did it by the fact that nothing worse happened show how safe it is? It was more than a quibble over whether a glass is half full or half empty. The fate of the nuclear industry might well depend on it.

To answer the question it is necessary first to gauge how much damage was actually caused. As with many nuclear accidents no-one died directly and immediately as a result. The nuclear industry was quick to use this to its advantage. When Senator Edward Kennedy called for the industry to be further regulated, supporters of the industry replied with the slogan 'More people died in Kennedy's car than at Three Mile Island'.[17] This answer is too glib however. As often happens in nuclear accidents, radiation had been released through the containment building. Radioactive clouds had drifted over Pennsylvania. How many, if any, deaths would this radiation cause? The question illuminated an important controversy burgeoning within the US and global technical communities.

THE RADIATION CONTROVERSY

The effect of radiation on human cells is only partially understood, and the level of damage caused by small doses of radiation is the subject of fierce scientific controversy. Since the nuclear fuel cycle continually emits low doses of radiation a knowledge of the magnitude of its effects is vital. To date, however, the calculation of these effects relies on the use of crude assumptions embodied in simple mathematical models.

Initially, the levels of radiation emitted by the industry were justified in terms of the 'threshold model'. Below a certain level of radiation exposure (the threshold), cell repair mechanisms were assumed to be capable of removing damage as fast as it was created. Thus workers or others exposed to radiation were assumed to be completely safe provided that the rate of exposure was kept below the threshold. It was a convenient model for the industry but, unfortunately for them, emerging evidence gave little justification for it.

In place of the threshold model, regulatory agencies began to use

THE US RESPONSE

a new 'linear dose model'. In this model damage from radiation is assumed to be directly proportional to the dose received. That is, if a dose of one unit of radiation produces one unit of damage, then a half unit of radiation is assumed to produce a half unit of damage. Here, however small the radiation dose, there is always some damage. The problem then becomes to set the allowable dose sufficiently low to give no more than an 'acceptable' level of damage. Until recently it was widely accepted that this model would overestimate the damage from a given dose of radiation. Using the model, the allowed radiation exposures for workers (5 rem per year) and members of the general public (0.5 rem per year) were considered to lead to 'negligible' increases in cancers and genetic defects. Important evidence has now emerged that the damage per unit dose at low exposures to radiation may be much greater than is deduced from the linear dose model.

Some of this evidence is still circumstantial. A Medical Research Council study in Edinburgh, Scotland, has found chromosome damage in the blood of workers at a nuclear submarine maintenance yard. Their monitored exposures over ten years were less than the permitted dose.[18] Similarly, at another submarine yard at Portsmouth in the USA, the cancer death rate of exposed workers was reported to have been more than twice that for other workers at the same place.[19] However, after carrying out their own study of the data, the National Institute of Occupational Safety and Health disagree that there is any significant increase. But the institute's methods and results have also been attacked by other prestigious scientists.[20]

Perhaps one of the most important studies has been the $6 million investigation carried out by Dr Thomas Mancuso of the University of Pittsburg, and Dr Alice Stewart and Dr George Kneale, both of the University of Birmingham.[21] By analyzing the radiation exposure and death records of workers employed at the Hanford nuclear complex in Washington, they concluded that radiation doses below the present allowed levels had caused a significant excess of cancers. Their analysis was immediately subject to the most detailed scrutiny.[22] Some of their other conclusions were later abandoned.[23] The key conclusion, that the 'conservative' linear dose model (which is used to set acceptable radiation levels) underestimates the number of cancers produced by low levels of radiation, remains the subject of bitter scientific controversy.[24]

Equally important was Mancuso's finding that people are more

48   *GLOBAL FISSION*

sensitive to radiation in early and late adult life than in intervening years. In another study, Dr Irvin Bross, director of Biostatistics at the Roswell Clark Memorial Institute in Buffalo, New York, had already found cancer increases of 5 000 per cent in unborn children exposed to low levels of radiation.[25] The strong implication from these studies is that sensitivity to radiation varies from person to person. If this is so, a powerful argument can be mounted for reducing allowed levels to the point where the risk to the most sensitive person is reduced to an acceptable level.

The potent implications for the nuclear industry seem not to have been lost on the US government's Energy Research and Development Administration (ERDA). In January 1976, as the first significant results began to emerge, it removed Mancuso's funding and gave it to Oak Ridge National Laboratory, a company with many nuclear contracts. As representative Paul Rogers later charged, the termination of the contract seemed to bear all the signs of deliberate concealment.[26] If so, it failed. The subsequent uproar helped place Mancuso's results under a spotlight and sparked numerous inquiries into the effects of radiation.[27] As these proceeded, evidence of more death and illness from radiation suffered by different groups in the community was uncovered. Public interest was also drawn to 'an epidemic of monumental proportions' among uranium miners,[28] to leukemias suffered by soldiers ordered to take part in exercises in close proximity to nuclear bomb tests,[29] and to excessive cancers among residents of Utah, apparently caused by their exposure to radioactive fallout from US nuclear test explosions.[30]

Concern over low level radiation had already reached an all-time high when on 27 February 1979 a special Interagency Task Force on Ionizing Radiation released its 800-page report. It reported that 'the results of recent studies of populations exposed to very low levels of radiation raise serious questions . . . and suggest that risks may be higher than earlier predictions'.[31] Thirty days later the message was dramatically highlighted as radiation pierced the containment building and emanated invisibly from a cloud of gas drifting from the stricken reactor at Three Mile Island.

The technical confusion was graphically illustrated by the differing announcements that followed the accident. Most sanguine was Jack Herbein of Metropolitan Edison, who said, 'We didn't injure anybody, we didn't overexpose anybody and we certainly didn't kill

THE US RESPONSE 49

anybody'.[32] Secretary of Health, Joseph Califano Jnr testified that the 2 million people living near the plant had received a dose of 1.8 rem. He said that this could mean that no additional cancers would result.[33] Respected health physicist Karl Morgan, however, calculated that this dose would cause fifty excess cancers,[34] and cancer specialist Sister Rosalie Bertell calculated that higher absorption by vital organs could have rendered the NRC dose estimates too small by a factor of ten.[35] One month later Califano increased his dose estimate to 3.5 rem, saying that one to ten additional cancers could be expected. However he conceded that the figures did not include the doses received by workers at the plant, and that the extent of public exposure remained uncertain.[36]

The array of claims, counter claims, and contradictions which appeared in the wake of the accident illuminate the primitive and confused state of the technical community's present understanding of the effects of low level radiation. By illustrating the uncertainty over the hazards associated with low doses of radiation it adds powerfully to the momentum already in the community to see permitted doses reduced substantially, and for situations where permitted doses can be exceeded to be avoided. As will be seen in the next chapter, the effects of this could be ruinous for an industry already in financial difficulty. But beyond this, it raises a more profound question. If, despite past public assurances, the experts are so uncertain about this hazard associated with nuclear power, how much faith can be placed in their past assurances about the safety of other aspects of the nuclear fuel cycle?

THE POTENTIAL FOR CATASTROPHE

Of all the assurances given to the public by the experts in the past, the confident assertion that nuclear reactors are to all intents and purposes completely safe has probably been the most important. People have been assured for years by industry and government experts that the chance of a catastrophic reactor accident was less than 'the chance of a meteorite falling on your head'. The basis of this claim has been a study commissioned by the AEC in 1972 and carried out for them under the direction of the Massachusetts Institute of Technology's Professor Norman Rasmussen.

During the early part of the 1970s there was a growth of concern about the possibility of catastrophic reactor accidents. This was

50   GLOBAL FISSION

nourished by the publication of various estimates of the substantial
devastation that could be precipitated by a core meltdown. Professor Rasmussen's study was used to counter this concern by
confronting critics with a seemingly impenetrable set of calculations which concluded that the possibility of a core meltdown was
extraordinarily remote, occurring less than once in a billion years of
reactor operation.[37]

Because the commercial reactor industry developed rapidly
during the early 1970s the overall operating experience is still
relatively small. In addition, new reactors are much bigger and
more complicated than earlier reactors. If the industry's more
ambitious plans go ahead up to the turn of the century, the entire
accumulated reactor experience to date would represent less
reactor hours than would be clocked up in a single year in about
AD 2000. The present history of malfunctions among the few
reactors that have operated to date also give little indication of how
the many intended reactors will operate in the future. The
Rasmussen study therefore had to rely on a mathematical model.

Any mathematical model is only as reliable as the degree to
which its equations and data mimic the real world. Rasmussen's
model was criticized for having glaring theoretical limitations and
using data, some of which was apparently plucked from the air.[38]
In particular, the critics argued that since it is impossible to know
all the possible sequences of events that can lead to an accident it is
impossible to calculate the likelihood with which these unforeseen
sequences will occur.

By the time of the accident at Three Mile Island, the Rasmussen
Reactor Safety Study had been largely discredited within the technical community. The *coup de grâce* was provided by the same body
that originally released the study in 1975. On 19 January 1979 the
NRC released the report of a review group headed by Harold Lewis
of the University of California. Although the group still considered
the study the best calculation available, its overall conclusion was
damning: 'The commission does not regard as reliable the Reactor
Safety (Rasmussen) Study's numerical estimate of the overall risk of
reactor accident'.[39]

For the technical community, the accident at Three Mile Island
two months later merely reinforced their now acknowledged
inability to compute the likelihood of reactor accidents. The
Rasmussen Study had included a very similar sequence of system
failures, but had calculated the chance of them occurring as less

THE US RESPONSE 51

than one in 10 million reactor years. Yet the unexpected had happened.

A soaring barrier divides the opinion within the scientific community from the response within the general public. It obscures technical uncertainty and reinforces the authority exercised by the technically trained. Transmission through the barrier is mediated by interpretation, by what scientists choose to translate, and what media and industry choose to present. For the public, the accident at Three Mile Island not only helped shake the ideology of reactor safety which had been established by years of authoritative use of Rasmussen's conclusions. It also helped highlight the uncertainty which surrounds many problems posed by the nuclear fuel cycle.

Problems connected with the nuclear fuel cycle, from waste disposal, through to the risk of meltdowns, and the problems of weapons proliferation, all tend to be answered by pro-nuclear experts in a similar vein. In general the answer is of the form: 'to every problem that can arise with a machine we can and will engineer a solution'. This assertion, which could be called the 'claim of technical certainty' was once again presented in the wake of the accident. As one nuclear utility executive put it:

> It appears from what we know that the things that happened there can be dealt with by engineering, training and operations . . . As far as this company is concerned we see nothing to stop us going ahead with our program of expanding our nuclear power commitment.[40]

The claim had been convincing in the past, but now the groundwork developed over years by an expanding technical and broader citizen opposition to nuclear power combined with the graphic illustration provided by the accident to help reduce the claim's effectiveness. In particular, the accident showed that the unexpected should be expected. It challenged the validity of a crucial component of the claim: that we can design for the unknown.

DESIGN AND THE UNKNOWN

As hearings commenced into the causes of the accident it became clear that some of the key features of the accident had come as a surprise not only to the owners of the reactor but also to its designers. One of these was the turning off of the emergency core cooling

system.[41] Another was the bubble of hydrogen that had formed at the top of the vessel threatening to blow it apart. As one executive of Babcock and Wilcox, the designers of the relevant sections of the plant, acknowledged, he knew of no analysis by the company which predicted the formation of such a bubble.[42] Similarly, the NRC's principal document on reactor safety had made no mention of such a problem.[43] Yet no sophisticated science was necessary to predict that hydrogen would be produced. As Sir Martin Ryle of the Cavendish Laboratory, Cambridge, put it, the hydrogen should have been predicted as a matter of elementary 'textbook knowledge'.[44] Five years earlier the Union of Concerned Scientists had used their knowledge to describe the likely progress of a loss of coolant accident. They had noted:

> heat is provided by chemical reaction of the zirconium fuel-rod cladding with steam or liquid water. The reaction rate increases strongly with temperature liberating hydrogen ... Irrespective of the occurrence of explosions, hydrogen generation would continue and would put the integrity of the containment at risk.[45]

Perhaps, as one article suggests, the problems of the zirconium were overlooked by the designers because there is no readily available alternative for cladding the fuel for these sort of reactors.[46] Whether or not this is so, the impossibility of being sure that every contingency has been designed for, was graphically illustrated by the accident. Here, even problems which were known to exist were inadequately designed for and could have precipitated a disaster. As Tennessee Valley Authority safety expert Carl Michelson testified later, the hydrogen was not the only feature of the accident which had been predicted but ignored by the designers.[47]

The demonstrated inability to cover every contingency in advance reveals another design dilemma for the industry. To what extent should the reactor be designed to allow the operators to intervene? In previous accidents, such as that at Brown's Ferry in March 1975, operator intervention may well have saved the reactor from meltdown. But at Three Mile Island various steps taken by the operators had contributed to the danger. Since the machine cannot be guaranteed to be designed 'safe' the operators are no irrelevance. At Three Mile Island they were a crucial part of the safety system, and were provided with lights, displays, and controls linked to almost every part of the reactor. The operators were not inexperi-

THE US RESPONSE

enced. Based on NRC exams, the staff ranked ninth out of those of thirty similar facilities. Each of the licensed operators had had five years experience of commercial reactors.[48] What then had gone wrong?

One conclusion was that of the Tennessee Valley Authority, that the instrumentation was inadequate. It ordered more displays for its reactor designs.[49] A House task force into the accident came to a similar recommendation.[50] Yet would such a step increase the safety? A study carried out by the Electric Power Research Institute suggests that it was the very provision of the safety systems that may have caused the accident: 'The idea that the Three Mile Island operators were incompetent or insufficiently trained is wrong. In fact they were trapped in a system that had successive layers of safety equipment added until the safety systems became mutually destructive'.[51] Some 200 alarms sounded in the first two minutes of the accident. Thousands of indicator lights were available and at times 'lit up like a christmas tree'. So much information poured out of the computer typewriter that it was running more than an hour behind. The paper jammed for almost an hour and a half but went unnoticed in the excitement.[52] In order to handle the situation there were, at peak, as many as sixty people in the control room. Communication was made more difficult after they donned their protective masks.[53]

It was not the first time that a 'safety' system had almost precipitated a disaster. The partial core meltdown for the Fermi 1 prototype fast-breeder reactor had also been caused by a safety system, a piece of which had lodged in the cooling system. This time not only had the mechanical systems contributed, but so too had the flexibility of the operators, so crucial to safety. As *Nucleonics Week* put it: 'The operators had become accustomed to coping with faulty components and instruments that they didn't believe in'.[54]

Beyond each of the particular problems and uncertainties associated with nuclear power illuminated by the accident at Three Mile Island there was also a general message. It was a message with sufficient force to penetrate to many outside the technical community, and with the potential to transform vague doubts into conscious concern. The project supervisor for the Diablo Canyon reactor in California summed up this message succinctly: 'What couldn't happen, did happen'.[55] For many it was not difficult to make the inductive leap to the more ominous conclusion: what can't happen could happen.

54   GLOBAL FISSION

The conclusion was reflected in agonizing official statements. Said Rep. James Weaver, head of the House investigatory committee task force on the accident, when announcing his findings to the press: 'an accident such as occurred at TMI not only could happen again, but is likely to at any time. It was not a next-to-impossible fluke'.[56] As the president's commission on the accident warned later, in October 1979, 'We have not found a magic formula that would guarantee that there will be no serious future nuclear accidents. Nor have we come up with a blueprint for nuclear safety'.[57]

It is right to be wary of the statements of public officials and politicians. Public postures can be assumed and changed in almost perfect synchrony with the shifting winds of political opportunity. Many bold statements were made by key political figures, right up to the president himself, in the wake of the accident. But the accident could still have either limited or profound long-term political consequences. It would depend on the state of the ground on which the political fallout descended. As it happened, the accident occurred at a time when opposition to nuclear power was already widely planted, and the ground well prepared for rapid growth.

THE POLITICAL FALLOUT

Spanning the entire USA, from coast to coast, sprawl a tangled network of hundreds of increasingly active citizen groups often deeply locked in struggle against a particular local nuclear project.[58] Emerging during the 1970s in response to the rapid development of the nuclear industry and its ambitious plans, these groups have been the motive force behind the debate over nuclear power that has contributed to the steady paralysis of the nuclear industry during the latter half of the last decade.

It is not possible here to detail the structure and history of the diverse groups which make up the US citizens movement against nuclear power. This could be the subject of a book in itself. Nevertheless, it is important to have a picture of the ground over which the accident at Three Mile Island cast its shadow in order to understand the expressions of opposition which erupted as a response. The US movement is the best documented in the world, and excellent descriptions of key events, its history, and contemporary reports, are contained in books such as *The Nuclear Power Rebellion*,

The Electric War, *No Nukes*, and *Energy War*.[59] Here only a broad outline is given, but important trends in this movement will be discussed later as part of an overview of the history and significance of the global battle over nuclear power.

Much of the 1960s and early 1970s anti-nuclear activity in the USA came from concerned lawyers and scientists who, with the support of local citizens, began to oppose nuclear projects at licensing hearings. Intervention in the early 1970s against the licensing of the Calvert Cliffs, Vermont Yankee, Indian Point, and Midland reactors (in Maryland, Vermont, New York, and Michigan respectively), laid the legal groundwork for many similar actions later. Scientists came together in the Union of Concerned Scientists or collaborated with citizen groups or conservation organizations such as the Natural Resources Defence Council. One such organization, the Sierra Club, split over the nuclear issue in 1969 when the then president, David Brower, tried to get them to oppose the siting of the Diablo Canyon reactor at Avila Beach, California.[60] Brower went on to form the strongly anti-nuclear organization Friends of the Earth (FOE). It set up numerous local chapters across the country. Eight years later the public were alarmed to hear that the US Geological Survey had disclosed the existence of an offshore geological fault capable of causing an earthquake at the Diablo Canyon plant of a force considerably greater than the reactor's design strength.[61]

FOE was not the only coalition of groups to spread the anti-nuclear message. First Ralph Nader's citizens organization Critical Mass, and then other diverse coalitions were formed across the country to take up the fight against nuclear power. Some centred around particular concerns: the church, conservation, Indian land rights, peace, and pacifism. Others brought together people from particular social groupings. Parents, women, technical workers, students, and farmers became involved. One important influence on a number of the groups was the idea of non violent action, a strategy pushed forward by the Quakers.

The power of this strategy was first demonstrated in the USA by the Clamshell Alliance, a coalition of local citizens groups which sprang up in opposition to a proposed nuclear reactor at Seabrook, New Hampshire.[62] The concern around which it initially organized was the effect that the several billion litres of hot water the reactor would pump out each day would have on the clams and other sensitive ecology in the sea near the plant. On top of this, earth-

56 GLOBAL FISSION

quake, radiation, and other hazards began to trouble increasing numbers of local people. On 1 August 1976, eighteen local 'Clams' walked onto the plant site. It was the first 'occupation' of a plant site in the USA and it brought 600 people out in support. Three weeks later the Clams returned with ten times their previous numbers. A support rally attracted 1 200. A year later 2 000 people occupied the site for twenty-four hours. This time however the occupation was no longer merely symbolic.

On the following day the arrests began. The Clams had organized themselves into 'affinity groups' which had carefully trained themselves how to resist provocation and arrest by totally non violent means. As the police moved in the demonstrators lay down. It took many hours to carry the 1 414 Clams to waiting army trucks and when they were finally imprisoned in five National Guard armouries they were far from disconsolate. Said one Clam, 'It was like the state had given us five free conference centres'.[63] The incarceration posed enormous problems for the state, not least of which was how to feed them. Eventually they were released pending a mass trial.

A year later, in June 1978, more than 18 000 people returned to the site. This time the state offered them land on the site. To an extent this successfully defused the action. The saga of the Seabrook reactor continues. The success has by no means been all on the side of the demonstrators. Whatever the final outcome, the substantial citizens movement built up against the Seabrook reactor in New Hampshire has already left its mark. At the beginning of the 1980s, some twelve years since the plant was first proposed, and with a cost that has escalated from $900 million to more than $2.5 billion, the plant was still not completed. Perhaps more importantly, the influence of the tactics and successes of the Clamshell Alliance were not restricted to New Hampshire. Across the country other coalitions – the Abalone, Catfish, Potomac, Bailly, Paddlewheel, Cactus, and Armadillo alliances, to name just a few – were formed and had taken up the message and tactical lessons of the Clamshell Alliance.

One of the issues which contributed first to the successes of the Clamshell Alliance, and later to some debilitating strategic disagreements, was the importance of the strategy of non violent action. The idea of taking action in which enormous emphasis is placed on ensuring that no violence is initiated by any of the opposition movement's participants has been in part a reflection of a strong Quaker influence within some of the anti-nuclear groups.

THE US RESPONSE

By actions such as non violent demonstrations, sit-ins, occupations, shackling of demonstrators to gates, and the like, citizens in the opposition seemed to have found a vehicle to oppose nuclear projects which could often prevent a counter-productive opinion backlash within the generally conservative climate of the USA.

Although a multitude of alliances against nuclear power formed across the USA, no truly effective national coalition of all the forces developed to routinely coordinate their activities. In part this was due to the way the movement grew around local groups whose prime concern was a particular nuclear project. But it was also due to a frequent desire to move away from old forms of organization. Decentralized networks of groups tended to be developed rather than have highly centralized coordination, and many groups chose to adopt lowest common denominator decision-making procedures rather than the sometimes more divisive, but occasionally more decisive, practice of taking votes. The result was the growth of a broad undertow of opposition to nuclear power unmarked by signs of substantial formal organization which would make it readily identifiable as a national political force.

Occasionally, however, the movement did produce nationally significant symbols which helped further strengthen the organic growth of the movement's decentralized parts. One was the action taken by Sam Lovejoy, a member of Nuclear Objectors for a Pure Environment (NOPE), a Montague residents' action group opposed to the construction of a twin 1 150 MWe nuclear plant in western Massachusetts. On 24 February 1974 Sam Lovejoy used a crowbar to undo the steel cables holding up a meteorological monitoring tower placed on the site by the nuclear company. Later he gave himself up. During the subsequent trial he eloquently defended his actions on the grounds that the destruction of property had been necessary to prevent the far greater evil of the hazards posed by the proposed reactor. Both judge and jury seemed to be impressed and he was acquitted on a technicality. His action was the beginning of delays still preventing the first sod from being turned for the plant's construction six years later at the beginning of the 1980s.[64]

Dramatic actions such as this have been important. But more generally the nuclear programmes in the USA have ground to a virtual halt through the increasing costs of nuclear power, the incremental obstruction of the licensing process, and the uncertainty accompanying the steady closure of paths for nuclear fuel

cycle development through the passage of anti-nuclear regulations by the states. In 1977, sixteen states had passed laws related to nuclear power: seven had banned disposal of nuclear waste within their borders, eight had regulated the transport of waste, and six had limited the storage of radioactive materials. By 1978 the pace had slowed, but a further four had passed such legislation. Nevertheless, by mid 1978 Maine, New Hampshire, New Mexico, and West Virginia were considering legislation prohibiting radioactive waste dumping, and Louisiana, Minnesota, Montana, Oregon, South Dakota, Texas, and Vermont had already passed such legislation. Colorado, Hawaii, and Michigan had adopted resolutions urging such legislation.[65]

Added to this, California, which in 1976 had defeated a referendum to place restrictions on nuclear reactor construction by two votes to one, underwent a complete transformation. In 1978 voters in Kern County overwhelmingly rejected a proposal for a reactor complex in the town of Wasco. Coming as it did in one of California's most conservative areas, an area that had voted strongly against the referendum only two years before, it signified a dramatic change of attitude in the state. In order to head off the opposition in 1976 the state legislature had passed a bill requiring a halt to the construction of any further nuclear plants until the waste problem is solved. With California now the USA's first avowedly anti-nuclear state, the clause seemed likely, at least for the near future, to be enforced. Similar moratoria on nuclear reactor construction existed in Iowa, Montana, New York, Vermont, and Wisconsin.[66]

The state legislative actions were merely a reflection of the strength with which concern was bubbling up within diverse sectors of the US community. These signs were significant enough, as was the overall malaise of the USA's nuclear industry. However, in March 1979, as the headlines poured out highlighting the ominous events taking place at Three Mile Island, the steadily fermenting concern accelerated, breaking above surface as an eruption of powerful, visible opposition.

ERUPTING OPPOSITION

From the day the first alarm bells sounded, a series of surges of demonstrations of opposition to nuclear power swept across the country, steadily increasing in strength. One of the first was in San

Luis Obisbo, California, only a day after the first news of the accident. There a small crowd gathered to oppose the licensing of the Diablo Canyon reactor. They carried placards warning, 'what happened in Pennsylvania can happen here'. Four days later the state's governor, Jerry Brown, cabled the NRC requesting a 'precautionary and temporary closure' of the state's Rancho Seco reactor.[67] By the end of the week small rallies of up to 5 000 people had also occurred in New York, Boston, Washington, Vermont, New Jersey, Indiana, and Plymouth. Said Irwin Becker, a speaker at a small rally on the steps of the State House in Providence, Rhode Island, 'People are angry. People are mad. They're also frightened'.[68]

A week later, San Francisco saw 25 000 people jam the Civic Plaza for six hours to protest the building of the Diablo Canyon reactor. It was, up to that time, the nation's largest demonstration against nuclear power.[69] Responding to the mounting wave of concern, San Francisco's board of supervisors, and the board of nearby Marin County, had already resolved to try to stop the Diablo Canyon reactor and close down the others in the state.

Over the next month smaller demonstrations of up to 10 000 people spread across the country. From Harrisburg, Pennsylvania[70] to Los Angeles, California,[71] Midland, Michigan, Limerick, Pennsylvania, and Denver, Colorado; and in Massachusetts, Oklahoma, Arkansas, Vermont, Arizona, New Mexico,[72] and New York,[73] people gathered in their thousands to express their concern and anger. In many places the demonstrations were not only larger than any previous anti-nuclear protests, but twice, five times, or even twenty times bigger.

It was almost a month to the day of the first traumatic beginnings of the accident. At last the plant was being brought to the relatively safe state known as 'cold shutdown'.[74] The reactor core had cooled substantially, but as the next weekend showed, the political and economic consequences were still coming to the boil. On the morning of Sunday, 6 May 1979, tens of thousands of people began to congregate on the grassy Ellipse outside the White House in Washington. During the day the crowd swelled to proportions that surprised even the organizers, one of the whom, Don Ross, had expected 50 000.[75] Police figures put the size of the demonstration at 65 000. Reporters placed it at 100 000, and the organizers at a massive 125 000. All were agreed that it was the largest demonstration in the USA since the biggest demonstrations in the 1960s

60 *GLOBAL FISSION*

during the Vietnam War. From scratch, it had taken only three weeks to organize.[76]

At noon, the sea of people rose and funnelled into Pennsylvania Avenue, solidly filling one half of the street from the Treasury Department to the Capitol building for more than an hour. 'Remember May the sixth, 1979' triumphantly shouted one speaker, biologist Barry Commoner, 'It will be remembered as the day the solar age began and the day the nuclear age died'.[77]

From the fervour of the crowd and its immense size it was clear that opposition to nuclear power in the USA had reached a new level of activity. The proposals launched at the demonstration – a moratorium on new construction, an end to corporate liability for nuclear accidents, and legislation to give states the rights to ban the location of radioactive wastes within their borders – might well gain further momentum. More importantly, the gathering indicated that grassroots agitation was growing across the USA. It was already shifting the political centre of gravity in state legislatures, and even congress. Beneath the effervescent demonstrations it was clear that public concern over nuclear hazards had hardened at a time when the nuclear industry was already badly wounded. Undoubtedly the opposition had gained considerable political momentum. The question remained however: what were the implications for the nuclear industry?

4
POLITICAL ECONOMICS

Every country has a mainstream of assumptions and language in which it is considered legitimate to express ideas and to develop analysis. In the USA, one dominant ideology contributing to this mainstream is that of the free market. The idea that the market place operates to provide an optimum allocation of goods and services pervades social commentary and analysis.

It is not surprising then that in the USA, as well as in many similar industrialized Western countries, the debate over nuclear power dwells heavily on its performance in the market place. On the one hand the nuclear industry argues that nuclear power is the cheapest and therefore best way available for generating electricity. On the other, the opposition draws attention to the recent disappointing economic performance of new nuclear power plants in relation to the cost of new coal-burning stations. As one New York consultant on nuclear economics, Charles Komanoff, after an extensive and careful study of the comparative costs of new coal and nuclear stations in the USA, concluded in testimony to a US congressional committee:

> In my judgement, no utility executive with an accurate perception of the costs of nuclear power and a sincere desire to minimize customer costs would propose ordering a new nuclear plant, with the possible exception of New England . . . On an overall lifecycle cost basis, I believe that electricity will cost 22 per cent more than electricity from new coal plants, averaged over the seven regions of the country.[1]

This conclusion of course does not go undisputed by the industry, but not only Komanoff's subsequent work[2] but also the continued slump in reactor ordering and a tendency for electricity utilities to cancel nuclear stations and instead build coal-fired plants, seems to bear testimony to the correctness of his conclusion.[3] Komanoff's conclusion is also supported by the decline in the proportion of US electricity produced by nuclear reactors. In 1978 nuclear reactors produced their maximum ever contribution of 12.5 per cent. By 1980 that contribution had declined to 11 per cent. By contrast, over the same period, the share of electricity produced from coal-burning stations rose from 44 per cent to over 50 per cent.[4]

The observation that nuclear power is appearing increasingly less competitive in the market place, raises two questions: to what extent are the economic problems of the nuclear industry and the political conflict over its plans interconnected? In particular, will the political future of the industry fundamentally be determined by the operation of underlying economic forces or, on the contrary, will the economic future of the industry be determined by politics? The answers are of considerable importance. If economics is the more fundamental factor in the fate of nuclear power then the political confrontation over the future of the industry is of little long term significance. If, on the other hand, politics is the more fundamental factor, then the outcome of the political conflict must be recognized as being crucial.

In the USA particularly, there is a marked tendency for not only the nuclear industry and media analysts to suggest that the economics are central and the conflict only peripheral, but also for the opposition to suggest the same idea from a different angle. However, in this chapter it will be argued that this apparent consensus is the result of a shared ideology, and that, at least in the case of nuclear power, it produces conclusions which are certainly misleading, and can be quite wrong. To see this, a look at the accident at Three Mile Island provides a graphic illustration of the extraordinary sensitivity of the nuclear industry's economic fortunes to growth of community concern over nuclear hazards. But first attention needs to be narrowed to the turmoil which boiled up in Pennsylvania and its effects on Metropolitan Edison, the local owners of the reactor, and on General Public Utilities (GPU), its parent company.

THE PROBLEMS OF GENERAL PUBLIC UTILITIES

The issue of nuclear hazards had slumbered in Pennsylvania, occasionally awakening to surface over the years before the accident. As early as 1973 there had been evidence of concern. In that year Pennsylvania's insurance commissioner, Herbert S. Denenberg, had issued the following statement:

It may be that nobody but God could write the insurance policy we need on nuclear power plants . . . the only adequate insurance against catastrophic loss from nuclear accidents is to stop building more nuclear power plants and to begin closing down the ones we have now. It's that simple.

His office had issued an eight point Consumer's Guide to Nuclear Non-Insurance. The final point had been:

Don't Take the Advice of the Nuclear Establishment on the Issue of Nuclear Safety. The people who make and run nuclear power plants have assured us that there will never be a major catastrophe. But some manufacturers of nuclear reactors also make toasters, dryers, washers, television sets, and other household appliances. Are nuclear reactors that much more perfect and dependable?[5]

After the accident at Three Mile Island such prescient advice suddenly seemed little more than common sense. The mood among Pennsylvanians was not just concern but anger. They were angry at the danger to which they had been subjected, at the confusing and misleading information that they had been fed, and the inconvenience caused by the evacuation. Their anger was not placated by the subsequent actions of Metropolitan Edison. Insensitively, the company had added to its already abominable image by announcing that any of its women workers who were pregnant, and had thus evacuated from work on the advice of the state's governor, would have their time off the job deducted from their vacation days allowance.[6]

The company's rapidly growing reputation for one-tracked pursuit of profit was not diminished by the news that before the Unit Two reactor accident it had been rushed into service, thereby achieving a tax rebate. By pressing the reactor into operation on 30 December 1978 the company had qualified for between $17–$28

million in tax investment credits, plus a $20 million depreciation allowance. Said one of the plant's engineers in the wake of the accident, the plant 'went commercial too quickly'. A stream of multiple failures should have been taken as an indicator that all was not well. He added, 'If the lights in your house blow out every time you turn on the toaster, you know something is wrong'.[7]

With this background, the company's announcement that it intended to recover the $10 million cost of replacing the electricity from the shut down reactors by raising the electricity rates paid by residents by a factor of 40 per cent, was not well received.[8] In the face of the subsequent uproar the governing Public Utilities Commission not only suspended the $49 million rate increase which had been granted to the company only six days before the accident, but also suspended $22 million of a $56 million increase which had been granted earlier in January.[9]

The suspensions were an indication of the strength of the community's anger, but the commission's subsequent behaviour suggested strength on the other side also. Three months later the commission responded to the company's pressure by granting a rate increase of 11 per cent.[10] Three months later again, the commission prepared to receive yet another rate increase application from the company, this time for 12 per cent.[11] Despite this minor progress, however, the prospects for the company's nuclear programme did not look good.

Soon after the accident State Governor Thornberg had gone on record endorsing the minimum one to two year suspension of operation imposed on the Unit One reactor by the NRC. He said that the duration of the suspension would be 'irrelevant' to him.[12] (As late as October 1980 the suspension was still in force.[13]) More significantly, he told a senate panel that Pennsylvania, the birthplace of the USA's commercial nuclear programme, might now begin a shift to coal as its future energy mainstay.[14]

The enhanced hostility to nuclear power made it difficult for Metropolitan Edison to even begin the clean up in what would be the first step in the anticipated four-year, $400 million plan to put the damaged Unit Two reactor back into service.[15] More than 2 million litres of radioactive water remained inside the containment building together with highly radioactive gases and the lethally radioactive core. To begin the clean up, which alone was initially estimated to cost $200 million,[16] the water must be pumped somewhere and the gases vented.

POLITICAL ECONOMICS

As a Pennsylvanian group called Antinuclear Group Representing York (ANGRY) petitioned the NRC to hold public hearings before any planned venting of gases, Metropolitan Edison once again found itself thwarted from releasing trapped radioactive krypton gas. Waste destined for the Barnwell treatment plant was turned back by South Carolina which gave notice that regardless of its characteristics no waste would be accepted from the plant.[17] Said one commentator, 'Nothing at Three Mile Island is really a technical problem any more; it's all political'.[18]

For Metropolitan Edison, the political backlash continued to exacerbate the increasingly complex technical feats that would prove necessary to clean up the plant while at the same time maintaining acceptable levels of safety. By November 1980 the estimated time before the core could be removed had extended to August 1985, and the estimated cost of the clean up to that date had risen to $750 million or, with adjustments for inflation, to a neat $1 billion.[19] Six months later the cost was being put at $1 billion to $1.3 billion.[20] Added to this would be the cost of replacing power lost, an estimated $1.2 billion to $1.9 billion.[21] In addition to the technical problems each delay, each safety inroad into its operating programme, and each restraint imposed on its attempts to pass on the costs of the accident to consumers, brought the prospect of severe financial embarrassment ominously closer. The company's owner, GPU, first slashed salaries, eliminated 600 jobs, halted all new construction, and severely reduced its quarterly dividend.[22]

Later, in December 1980, after an appeal by GPU's chairman for government funding to save the company from bankruptcy had failed,[23] the company took another seemingly desperate measure. It launched a suit against the NRC alleging that it failed to warn the company of a similar accident that had occurred at a plant in Toledo, Ohio. More significant than the action were the estimated damages. The company demanded $4 billion to cover costs accruing from the accident.[24] By early 1981 the only buffer standing between the company and bankruptcy was a $412 million credit extended to it by a consortium of forty-five banks, and the company had made clear that it would not have enough money to pay for the Three Mile Island clean up.[25]

Undoubtedly, for GPU the economics of nuclear power have been substantially affected by the political consequences of the Three Mile Island accident. But the sensitivity of nuclear economics to the accident has not been restricted to GPU or

66   *GLOBAL FISSION*

Pennsylvania. Across the country, touching both the electricity utilities and the companies involved in reactor manufacture, the political consequences of the accident have shown a powerful ability to further raise the costs of nuclear power.

OPERATING COSTS

Nuclear reactors have always cost more to operate than expected. One reason is the unexpectedly large proportion of their operating lives that they spend switched off for repairs. The amount of electricity that a reactor produces, compared with the amount of electricity that it could theoretically produce if it operated at full power all the time is measured by the 'capacity factor'. A reactor has a capacity factor of 100 per cent if it produces continuously at its maximum theoretical output. It has a capacity factor of 50 per cent if it produces on average half the electricity that it is theoretically capable of generating.

When nuclear reactors were first introduced it was confidently predicted by the industry and the US Atomic Energy Commission that the reactors would yield capacity factors of around 80 per cent.[26] However, in actual experience, the factor has been much lower. In 1977 the average factor for all US reactors was only 63.9 per cent. That was the highest for seven years.[27] In 1978 it moved a little higher to 66.1 per cent.[28]

Smaller capacity factors than expected mean less electricity produced and sold. This means lower profits than expected. In the first half of 1979, in the wake of the accident at Three Mile Island, the average capacity factor in the USA fell to 59 per cent.[29] This reflected the temporary closure of reactors of similar design to those at Three Mile Island. However it also highlighted a possible longer term implication of the accident.

Capacity factors are sensitive to changes in attitudes to safety. As one journal put it:

> there are those that worry that whereas before TMI plant operators were loath to shut down the plant every time they received a danger signal, which could be spurious, now they will be tempted to take the plant off line – at a cost of $300 000 per day – every time an alarm goes off.[30]

Another major contributor to operating costs is maintenance and

POLITICAL ECONOMICS 67

repair of faults. Often this is made much more expensive by the need to use large numbers of maintenance workers in order to keep their exposure to radiation below a safe level. However, as discussed in the last chapter, what constitutes a 'safe' dose of radiation is a contentious scientific question. Regulational agencies are forced to choose between conflicting opinions, and the safety levels that they decide to enforce may be sensitive to public concern.

THE POLITICAL ECONOMICS OF RADIATION

In 1934, 50 units of radiation impact (rem) per year was adopted by the International Committee on Radiological Protection (ICRP) as the permitted level of radiation exposure for workers. This standard was followed by most nations until 1950. In the USA, the National Committee on Radiological Protection (NCRP), also in 1934, adopted a lower level of 25 rem per year. By 1949 the NCRP and the ICRP had dropped their standard to 15 rem per year and in 1957 the maximum permissible level was dropped again to 5 rem per year. This is the standard which is presently in force.

The 1979 report of the US National Academy of Science's Biological Effects of Ionizing Radiation Committee (BEIR), has continued to support the 5 rem per year level. However the committee has been divided over the suggestions which have emanated from various scientists that the level should be reduced by a factor of between two and twenty. The Chairman of BEIR, Dr Edward Radford, has testified to a congressional hearing that, in his opinion, the level should be reduced by a factor of ten.[31]

In the final analysis, what constitutes an acceptable number of deaths from the radiation emitted from the nuclear fuel cycle is not a technically answerable question. It is a social judgement which will come about as a political outcome of a battle between competing forces. If Dr Radford is right, and if, in addition, his view is adopted, then workers will be permitted to be exposed to no more than 0.5 rem in any one year. The cost of doing this could be astronomical.

The nuclear military and research complex in the USA employs just one part of the nuclear workforce. The Department of Energy has estimated that to reduce the exposure of this group of workers to 0.5 rem per year would cost a massive $3.5 billion initially, and $440 million per year thereafter.[32]

68   GLOBAL FISSION

In the US nuclear power industry in 1976, according to a report of the NRC, 13 387 workers, more than 20 per cent of the industry's workforce, received doses greater than 0.5 rem. Thirteen per cent received more than 1 rem.[33] A study of NRC documents by the Public Citizens' Health Research Group concluded that the radiation doses received by workers in 1977 were 25 per cent greater than in 1976.[34] Seven per cent received more than 2 rem.[35] Finally, a report of BEIR concluded that 30 000 nuclear power workers received between 0.6 and 0.8 rem in 1979.[36]

Not only are nuclear power workers exposed to levels of radiation that are increasingly being challenged as unacceptable, but the number of workers so exposed is increasing rapidly. This seems to be due not merely to the increased number of reactors in operation. A study of NRC statistics published in the industry journal *Nuclear Engineering International* in 1977 showed that the average exposure of workers to radiation in each nuclear reactor complex had increased steadily from 178 rem in 1969 to 457 rem in 1975.[37] The paper concluded that 70 to 80 per cent of this dose was due to workers' exposure during maintenance to built up radiation in the reactor. This radiation builds up constantly, and even faster as reactors get older.[38] Most of the reactors now operating in the world have been built recently. As nuclear reactors become increasingly radioactive with age, the large numbers of workers who are already receiving substantial doses of radiation are therefore likely to constitute only the tip of an iceberg. Even to keep the dose within existing levels, the number of workers employed on each plant may have to be increased dramatically.

The number of workers required for major repairs on nuclear reactors can already reach an impractically large figure. In the USA the Indian Point 1 reactor near New York developed problems with its steam generator. It took 1 500 men to repair the defective welds. The repair required virtually every qualified welder in the area to be exposed up to the dose limit. It used up almost every qualified ultrasonic tester in the country. Workers received their entire year's exposure in only twenty minutes.[39]

The debate over the effects of radiation is of profound importance to the nuclear industry. It hangs like the sword of Damocles over the nuclear fuel cycle. Even by itself, the outcome of the debate, heated by the accident at Three Mile Island, could substantially reshape the economics and political feasibility of nuclear power.

THE CRUCIAL CAPITAL COSTS

The major factor in the final cost of the electricity delivered from nuclear reactors is the capital costs – the costs of building them. These account for two-thirds of the final electricity price.[40] And the economic competitiveness of nuclear reactors has suffered because over the last fifteen years these costs have soared. Over the eleven years between 1964 and 1975 nuclear construction costs increased by more than ten times the rise in the Consumer Price Index, and ten times the cost of an oil refinery.[41] By 1977, a reactor was costing $900 million, four times the cost of an equivalent plant in 1972.[42] By contrast, coal-fired power stations suffered much smaller increases. Between 1971 and 1977 coal-fired station capital costs rose by 9 per cent above the rate of inflation, due in part to vastly improved regulation of pollution through fly ash controls and scrubbers. In the same period, nuclear reactor capital costs rose by 17 per cent above inflation.

In what has by now become a classic paper,[43] Professor I. C. Bupp and J. C. Derian, two colleagues from the Harvard Business School, have reported important factors common to those reactors which experience the greatest capital cost increases. Consistently, those with the greatest capital cost rises are those which have experienced the greatest delays in being built and licensed to produce electricity. The importance of this stems from the factors which could extend this period of delay. Highly prominent among these were the delays caused by environmental hearings and the demands of strengthened safety regulations. That is, intervention by citizens causing the nuclear industry to improve its safety has been the crucial factor in the rising capital costs. In this sense the community is forcing the industry to incorporate into its cost structure costs in the form of hazards previously borne by the general public. Bupp and Derian summed it up this way: 'The issue here is not merely technical or economical, but is inherently political. Present trends in nuclear reactor costs can be interpreted as the economic result of a fundamental debate on nuclear power within the US community'.[44]

The importance of this debate is reflected in the overall doubling of the average construction lead times for nuclear reactors in the USA (from forty-five to ninety months) between 1968 and 1977.[45] Overall it now takes typically twelve to fifteen years to plan and construct a nuclear reactor in the USA. Of this, some three and a

70   GLOBAL FISSION

half to seven years are attributable to siting and regulatory problems.[46] According to one study by a US utility, the cost of one
year's delay in the construction of a 1.1 GWe reactor can be as high
as $200 million per year, giving rise to a 15 per cent rise in the cost
of electricity generated.[47]

The same effects are also evidenced in other countries, although
usually less dramatically. Thus, a study of licensing times in the
USA, Japan, West Germany, and Canada, by R. Lester (of the
International Consultative Group on Nuclear Energy) concludes
that the times have increased markedly in each country during the
1970s. He notes:

> The increasing number and stringency of safety and environ-
> mental regulations, the linkage of power plant licensing to the
> resolution of the nuclear waste disposal issue, and the continu-
> ing fluidity of the general regulatory environment have been
> directly or indirectly responsible for many of the increases . . .
> [and] is linked inextricably to the more general question of the
> degree of consensus within societies over the proper role of
> nuclear power in the energy sector.[48]

In the USA, in early 1979, Komanoff announced his conclusion
that if present capital cost increases continue, then by 1986–87 the
electricity from newly-ordered nuclear power stations will be twice
as expensive as that produced from coal stations.[49] His conclusion
depends on a sophisticated extrapolation of past trends. Concern
developed around the accident at Three Mile Island undoubtedly
increased the likelihood that those trends would not only continue,
but even accelerate.

One contribution to rising capital costs has been the cost of new
safety features required by regulatory agencies or courts. An early
demonstration of this followed in the wake of the Three Mile Island
accident as a safety committee in Illinois responded with forty-nine
recommendations for altered safety precautions. It was estimated
that implementation of these would add 1 per cent to the cost of
Commonwealth Edison's reactors.[50]

Overall, the number of codes and standards applicable to nuclear
power plants in the USA is reported to have risen from little more
than 100, in the mid 1960s, to around 1 700 in 1976.[51] At the same
time, the amount of concrete, steel, and construction labour
required to build a nuclear reactor, per kilowatt of electricity that it

POLITICAL ECONOMICS

can produce, has in each case more than doubled.[52] Safety engineering already comprises about half the total cost of nuclear reactors.[53] Small changes to safety design can cost a great deal. Additionally, costs can be further escalated in less obvious ways, such as changing reactor siting regulations.

In the past, nuclear reactors have often been built close to major population centres. This saves on the costs of transmitting electricity over long distances. Even when sited close to consumers transmission costs are substantial, amounting to about 12 per cent of the total reactor capital costs.[54] On top of this is the large operating cost due to electrical losses along the transmission lines. With the accident at Three Mile Island the safety of placing reactors close to urban centres came under increased attack. The initial spark was struck by the President's commission into the accident. A key recommendation concluded:

> In order to provide an added contribution to safety, the agency should be required, to the maximum extent feasible, to locate new power plants in areas remote from concentrations of population.[55]

The US senate took up the matter. It adopted an amendment to an NRC authorization bill which required the NRC to develop a new siting regulation within six months to take account of population density factors to 'the maximum extent possible'.[56] Subsequently, in November 1979, the NRC announced that it would have to determine whether some of the nation's seventy-two existing reactors would have to be closed down because of their proximity to major population centres such as New York and Chicago.[57] If this pressure results in reactors eventually being sited further from population centres the cost penalty could be substantial. As noted earlier, there is striking evidence of a strong correlation between the capital cost rises plaguing the nuclear industry and the lengthy delays now characteristic of the reactor licensing process. One way to alleviate this problem would be to streamline this process. The possibility of local citizens forcing improvements to reactor safety by intervention in the licensing process would then be decreased.

However, the implementation of the industry's dream of cutting back licensing times, often referred to somewhat whimsically as 'licensing stability', requires the support of congress. In the aftermath of Three Mile Island's accident the prospects, at least in the

GLOBAL FISSION

short term, seemed greatly diminished. As one prominent industry executive pointed out, the goodwill in congress, patiently built up over several years, had vanished. He went on: 'Much of our work will now be keeping our finger in the dike rather than going after some of the initiatives we wanted'.[58] Another added that the prospects for licensing stability were now a 'total loss'.[59] Indeed, as the accident has shown, the sensitivity of nuclear reactor design requirements to the forcefulness of citizen concern poses a serious dilemma for the industry's strategy to decrease licensing times. A key element has been to create a standard reactor design. After one is licensed the others might be licensed with much less scrutiny. But the temporary closing around the USA of eight Babcock and Wilcox designed reactors after the accident at Three Mile Island, points to a potentially fatal flaw in this approach. If the Three Mile Island design had been standard throughout the USA, what would it have done to the economics of nuclear power, let alone the economy of the USA, if almost every nuclear reactor had been forced to close without warning?

As if to add insult to injury, increased opposition to nuclear power reinforces the prevailing uncertainty about the long-term economic viability of nuclear power. This itself can have significant economic consequences, producing a cycle of decreasing confidence and further raising capital costs. A business community uncertain about the future of nuclear power will consider money borrowed for nuclear projects as being money lent at risk. The result is that the needed money is lent only at higher interest rates. When Virginia Electric and Power Company went to the money market for a construction bond issue soon after the accident at Three Mile Island it was charged a higher fee and had to pay a higher interest rate than anticipated.[60] Capital costs are therefore greater. This in turn further unsettles the future of the industry, contributing to even greater uncertainty. Said the chairman of GPU, the market for all US electrical utilities 'is likely to be seriously affected'.[61]

For the world's largest commercial bank, the Bank of America, the uncertainty in the wake of the Three Mile Island accident was too high. The bank's chairman announced that all new loans for nuclear power facilities in the USA and elsewhere would be frozen. The freeze would continue until the bank was satisfied with the results of the various government agency reviews of the accident presently underway.[62]

CONSEQUENCES

The political reaction in the wake of Three Mile Island contained a spectrum of powerful pressures which could increase the operating and capital costs of the nuclear industry. This contributed to the prevailing uncertainty jeopardizing the industry's cost structure. The effects of uncertainty were not limited to the lenders. With the prospect of additional nuclear diseconomies and continued growth in the political currents which underlay them the electricity utilities developed a further aversion to the purchase of nuclear reactors.

A few months after the accident *Nucleonics Week* reported that 'many plant cancellations' were forseen in its wake. It predicted that 30–53 GWe, or 30 per cent of present committed reactor orders, would be cancelled. As cancellations began to occur across the country it became clear that the expectations expressed in the journal were realistic. Between June 1979 and May 1980 the number of reactors on order in the USA was slashed from forty-seven to twenty-six.[63]

For the industry the effect has been to reinforce a queasy and mounting anxiety over its future. It would have been bad enough if the accident had occurred in a time of boom. But it occurred at a time of unprecedented market stagnation. With orders already ground to a halt since the middle of the decade, executives of the more vulnerable reactor vendors have begun to wonder whether the political reaction to the accident might not prove the final obstacle to ever extricating themselves from the morass. First to go would be Babcock and Wilcox, the designers of the cooling system of the Three Mile Island reactor. Since January 1979, when two of its backlog of orders were cancelled and four more placed under review, it has become clear that the company is likely to exhaust its remaining backlog by 1984. From then on its retreat from the market could be only a matter of time.

The process of decline which faces the nuclear industry in the USA and elsewhere if orders do not revive are examined in chapter 6. The seriousness of the situation has been emphasized by a draft report to the International Consultative Group on Nuclear Energy, a grouping of nuclear experts from fifteen countries partly funded by the Rockefeller Foundation.[64] The report concludes that unless major domestic political and economic changes occur in the next five years, two of the four largest US companies, General Electric and Babcock and Wilcox, together with the West German and

Swedish reactor vendors, will be forced to withdraw from the reactor market. If major changes do not occur by 1988 the remaining US and Canadian companies would also be forced to withdraw. Beyond that, the remaining companies might well lack the facilities to maintain a global nuclear operation. The report concludes that a real risk exists of 'further deterioration of the political and economic conditions facing utilities and reactor industries, perhaps even to the point of jeopardizing the nuclear option itself'.[65]

THE BRONTOSAURUS

It is one thing to demonstrate that the prevailing opposition is capable of exerting a profound influence on the economics of nuclear power in a situation which is far from healthy for the reactor manufacturers. It is quite another, however, to conclude from this that the industry has no hope of extricating itself from its present economic morass. Such a conclusion is made particularly tempting by the comments from the industry, which in recent times have bordered on despair. 'Many Wall St Councillors are not sanguine about a market comeback and have been steering their customers away from Utility Stocks' observes *Business Week*.[66] 'The industry will run out of business in the early 1980s, reflecting the low level of orders' adds Robert McCoy, vice-president of Kidder Peabody, a US company.[67] And, proclaims Jack Robertson, a Department of Energy deputy energy secretary, 'The nuclear option is today dead. It is a national tragedy'.[68]

This view, that the nuclear industry has been dealt a death blow, is shared with more enthusiasm by some opponents of nuclear power. For example, Anthony Roissman, a respected and adroit anti-nuclear activist in the influential Natural Resources Defence Council, refers to 'The nuclear carcass' which, he complains, 'seems to be more attractive to the anti-nuclear movement than the living body of alternatives'.[69]

Comments implying that the ultimate demise of the nuclear industry is a *fait accompli* have been reflected for years in statements from some opponents of nuclear power. Perhaps one of the earliest examples has been the often repeated comment of FOE UK's respected energy analyst, Amory Lovins. His testimony to one congressional committee in September 1977 summed up his view succinctly.

It is my considered judgment that nuclear power is dead – in

POLITICAL ECONOMICS 75

the sense of a brontosaurus that has had its spinal cord cut, but because it is so big, and has all those ganglia near the tail someplace, can keep thrashing around for years not knowing it is dead yet ... I shall therefore suggest why the death of this particular brontosaurus was inevitable.[70]

Lovins goes on to attribute the past and predicted future decline of the industry to the action of 'straightforward market forces: As Adam Smith might have said, "the invisible fist strikes again" '.[71] His conclusion is based on the dramatic collapse of nuclear expectations, and the escalating costs of nuclear reactors. If his comment is indeed the postmortem, then the death certificate of the nuclear industry will be written soon, sealed with the epitaph 'uneconomic'.

It might seem appealing to take this apparent bipartisan agreement between representatives of the industry and its opposition as the final word. If so, all that is needed is to sit back and relax as the invisible fist completes its predestined knockout bout with the nuclear brontosaurus. However, to do this is to make the dangerous error of analyzing the problem within the restricted confines of the ideology of economic determinism. It is to overlook the fact that if some political developments can increase the costs of nuclear power, then others can reduce them. It is to ignore the plastic nature of the economics of nuclear power.

THE PLASTIC ECONOMICS OF NUCLEAR POWER

The knowledge that nuclear power appears in many places to be uneconomic can be grossly misleading. We have already seen how many future increases to the costs of nuclear power are underpinned by political considerations. Added to this must be the further observation that nuclear power, in another sense, has never been economic. From the earliest days of nuclear power to the present, from one end of the nuclear fuel cycle to the other, diverse subsidies permeate every stage.

RESEARCH, DEVELOPMENT, AND SUPERVISION

The early days of nuclear power, when all research and development lay within the ambit of the military Manhattan Project, set the scene for a government-nourished industry. The state AEC gave birth to the first 'commercial' reactors and a panoply of national and international regulatory agencies were set up to supervise and

76   GLOBAL FISSION

limit the hazards of the nuclear fuel cycle. All of this was provided by the taxpayer, free of charge, to the nuclear industry.

Government participation in the necessary research and development continued over the following years. In 1977, for example, ERDA's annual budget for 'direct energy programs' requested $1 642 million (63 per cent) for nuclear research and development. It compared with $968 million (37 per cent) for all other energy programmes combined.[72] By 1975, according to one US estimate, $5 000 million, irrespective of military funding, had been provided by the federal government for the direct development of nuclear power. Not one cent of this appears in the cost of electricity delivered from nuclear reactors.

The extent of the funding provided for regulatory agencies can be gauged from the amount that the NRC alone would spend as a result of the accident at Three Mile Island. According to the agency's own estimates the accident would require it to spend $14 million.[73]

ENRICHMENT   In day to day operations the first obvious subsidy is the cost of separating fissile Uranium-235 from inert Uranium-238 to provide 'enriched' uranium fuel. In 1976 ERDA was providing this at a cost of $61.3 for each fuel unit (measured in 'separative work units'). There was no commercial competitor but industry sources estimated that if there had been it would have had to charge around $100 for an equivalent amount of enriched fuel.[74] This accords with the conclusion of the Australian Ranger Uranium Environmental Inquiry which noted that 'the real costs of enrichment per unit of electricity generated may be at least double recent ERDA prices'.[75]

WASTE AND DECOMMISSIONING   At the other end of the nuclear fuel cycle, the costs of storing the radioactive wastes and decommissioning the highly radioactive reactors at the end of their lives, are all generally omitted from the price of nuclear generated electricity. No generally accepted methods exist for disposing of high level waste, and decommissioning of reactors has been carried out only on an experimental basis. The costs of waste disposal vary dramatically with the process envisaged as being 'acceptable'. The possible magnitude of the omitted costs of decommissioning are indicated in an article by Tom Wicker, the distinguished editor of the *New York Times*:

POLITICAL ECONOMICS

When an experimental nuclear facility at Elk River, Minn., was dismantled, the cost ran to $6.7 million – although the plant had cost only about $6 million to build. At Oyster Creek, NJ, a nuclear plant could be safely dismantled for an estimated $100 million, but that's more than 150% of the original $65 million cost. As much as $600 million may be needed to decommission and decontaminate a privately owned fuel reprocessing plant at West Valley, NY.[76]

ALLOWANCES   On top of all these subsidies towards the present costs of nuclear power are a range of substantial tax allowances. Prominent among these are the 'accelerated depreciation allowances' and 'investment tax credits' granted on capital works. Coal stations too enjoy these, but the allowances are much more valuable to the nuclear industry as so much of the cost of nuclear power is due to the initial cost of constructing the plant. A research group at California's Lawrence Berkeley Laboratory concluded that:

> Effectively the federal government pays about 20 per cent of the cost of each new power plant that comes on line (10 per cent is tax credit and about 10 per cent is accelerated depreciation allowance).[77]

These tax allowances are a major incentive for utilities to buy nuclear reactors as under federal tax law they are not required to pass the savings on to consumers. According to one 1975 investigation the difference between the 'phantom taxes' charged to consumers and the taxes actually paid by the major power companies amounted to around $1.5 billion.[78]

Because utilities have a captive market to which they sell their electricity, prices must be regulated by some state controlled board. One concession made to the utilities, which substantially advantages the capital intensive nuclear plants, is to allow them to include 'Construction Work In Progress (CWIP)' in the price. This means that existing customers pay now for the construction of plants which will produce electricity in the future. This subsidy, by present customers to future plants, has become substantial. According to Federal Power Commission Vice-Chairman, Don Smith:

> From 1966 to 1975, CWIP increased 630 per cent, from $3.6 billion to $26.3 billion . . . the higher proportion of planned

78   *GLOBAL FISSION*

nuclear facilities and the continuation of carrying expenditures on deferred nuclear projects has contributed to this phenomenal growth.[79]

LOW INTEREST LOANS   Both in the USA and elsewhere nuclear companies receive substantial subsidies through the medium of government loans. Low interest loans granted to overseas customers for the purchase of nuclear reactors represent an implicit transfer of money from the taxpayer, through the government, to the company. In the USA these loans are handled by the Export–Import Bank. In the fiscal year 1975 ExIm's grants for the finance of nuclear power stations was the bank's second largest category of loans, representing 16.6 per cent of its total outlay.[80] In the following year its loans and guarantees for nuclear power totalled more than $1.2 billion.[81] The bank was also prepared to lend to countries such as the Philippines and Brazil which, as one Bank of America executive points out, commercial banks would rule out as being too great a credit risk.[82]

The implicit subsidy can be estimated by calculating the difference between the interest rates charged by ExIm and the interest it would have received by lending to US domestic borrowers. According to a Congressional Budget Office estimate, the implicit subsidy for nuclear power by ExIm amounted to $163.3 million in the fiscal year 1975.[83]

LIMITATION OF RISK   A final but important subsidy is legislation which limits the financial liability of the nuclear industry should a serious accident occur. Under the Price-Anderson Act the maximum liability of a reactor operator in the event of an accident is the maximum available private insurance. In 1975 this was $125 million. On top of this the government provides an additional $435 million in insurance.[84] The combined maximum liability of operator and government is thus $560 million, less than one-twentieth of the maximum damage predicted in the NRC's own investigation of the effects of a reactor accident.[85] Similar acts with provisions to limit insurance liability, such as the British Nuclear Installations (Licensing and Insurance) Act 1959, exist in other countries with nuclear reactors.

In 1973 Pennsylvania's Insurance Commissioner, Herbert S. Denenberg, estimated that the Price-Anderson Act had reduced the annual insurance premiums from the level necessary to cover the

POLITICAL ECONOMICS 79

maximum damage estimated in government reports by $23 million for each reactor.[86]

The economics of nuclear power are thus borne gently upon a feather-bed stuffed with a bewildering mixture of subsidies, allowances, and cost omissions. One estimate in 1978 was that if some of these subsidies were factored in nuclear-generated electricity would cost between three and seven times as much as it does at present.[87] Another, a seventy-five page report from the US government Energy Information Administration's Office of Economic Analysis, made available to the media in December 1980, concluded more modestly that it would cost twice as much as it does at present. It put the total subsidy to nuclear power (excluding tax arrangements and insurance liability limits) at $40 billion.[88]

The rich profusion of subsidies on which the economics of nuclear power rest have not materialized unaided. They have arisen through the relationship between the nuclear industry and government which, as will be seen later, has been particularly intimate, and through pressure by the industry.

The economic and political muscle of the major nuclear corporations is awesome. The two principal reactor vendors in the USA, General Electric and Westinghouse, are economic entities of global scale. In 1977 General Electric was the thirteenth largest industrial corporation in the world ranked by annual sales, and the sixth largest in the USA. It shared 85 per cent of all reactor sales within the USA with Westinghouse. In 1970, before the early boom in reactor sales, these two companies together had annual sales worth $13 000 million, greater than the gross national products of Saudi Arabia, Libya, and Portugal combined. Together with the remaining two nuclear reactor vendors, Babcock and Wilcox and Combustion Engineering, the major oil companies which have interests in nuclear power, and the many smaller companies which supply components for the nuclear fuel cycle, the US nuclear industry forms a grouping with common interests and enormous political strength.

That the power of these companies, in concert with that of those utilities which have become committed to nuclear energy, is considerable, is testified to by the mountainous structure of subsidies which underpins nuclear costs. Nevertheless, that power is anything but unlimited. Over the second half of the 1970s the economic position of the industry, though still supported by an

array of subsidies, has been steadily eroded. Costs have risen as the costs of hazards previously borne by the community have been at least partly internalized. At the same time the political climate has prevented the introduction of sufficient new subsidies to halt the decline in the economic health of the industry.

In the wake of the accident at Three Mile Island the political climate in which the industry has sought to survive seemed to be turning rapidly more inclement. Congress, state legislatures, the NRC, and other agencies were being pressed into actions restricting and regulating numerous hazards presented by the nuclear fuel cycle. Increased regulation of radiation, heightened concern over reactor safety, declining credibility of the industry and its experts, expensive new design changes, and interruptions to operation and construction were all exacerbating the already severe economic problems of the industry.

Political opposition to nuclear power, whose roots run back to the first tentative formation of a citizens movement against nuclear power years before, was now surging forward with seemingly irresistible force. Perhaps with the impetus of the accident, would it not be reasonable to declare the battle won, and pronounce the death sentence for the nuclear brontosaurus? To do so would be to underestimate the resilience of the opponent. To win a battle, even a prolonged run of battles, is not necessarily to win the war. By campaigning among the community, by intervening in the licensing of nuclear reactors, and by relentlessly attacking the safety of nuclear power, opponents opened the route for the rapidly rising costs which have dogged the industry over the past five years. They have deprived the brontosaurus of the orders so crucial to its continued health, and by doing so have slowly but successfully reduced it almost to starvation. But in that very success lies the possibility of failure. They have made the beast hungry. It would be foolish to forget, in this moment of apparent triumph, that it is when a beast is starving that it is most dangerous.

The pressures created by the industry, and those by the opposition lead to a final price for nuclear electricity which is totally artificial. The 'economic reality' in which nuclear power is perceived to be uneconomic is nothing but a chimera. And the fact that the economics of nuclear power are sensitive to political currents is for the nuclear industry a strength as well as a weakness. In the past the economics of nuclear power have been constructed by the government underpinning the costs with subsidies. Now, faced by

POLITICAL ECONOMICS   81

plummeting prospects in the market place, the nuclear industry is far from passively resigned to accepting its problems as an unalterable economic reality. Instead, it seeks to solve them by pressing for more of the help that has supported it so effectively in the past. In short, it seeks to force measures by which its present economic reality may be reconstructed.

RESTRUCTURING ECONOMIC REALITY

The various components making up the nuclear industry are continually manoeuvring to achieve a revised cost structure. In the aftermath of the accident at Three Mile Island some of these manoeuvres could be seen breaking above surface. Key among these were the attempts to pressure the legislature. On 10 May 1979, just one month after the accident, nearly 400 nuclear industry leaders assembled in Washington to lobby congressmen. Asked how congress might react, Representative Clarence Brown said: 'The position of Congress is clearly negotiable'.[89]

To an extent the lobbyists achieved their main short-term objective. Although unable to solve their economic problems they were at least able to head off what could have been a politically crippling body blow. Two months later, despite many proposals for the enactment of a moratorium on further nuclear construction, the US senate voted 55 to 35 against a six month freeze on construction permits. In a different part of the same bill, a proposal to give each state power to veto federal efforts to store nuclear waste within its boundaries was also defeated, narrowly averting the possibility that there might be nowhere left to dump even low level waste.[90]

President Carter, too, was quick to pay his respects to the power of the industry. Less than two weeks after the accident he stated: 'There is no way for us to abandon the nuclear supply of energy in our country in the foreseeable future', and although feeling unable to give them their longed for 'streamlined' licensing procedures, he added 'I think it does not contribute to safety to have a bureaucratic nightmare or maze of red tape as licensing and siting decisions are made'.[91]

Momentarily thwarted from its aim of getting the administration's draft Nuclear Siting and Licensing Bill introduced, the nuclear industry's supporters did succeed in attaching a special clause to a proposed House Energy and Water Appropriations Bill. This required the NRC to place 100 additional staff onto the job of

processing reactor licences. However, with the accident reverberating, the lobbyists' strength was insufficient to protect the bill from a further amendment which destroyed its effect.[92] For the short term, what would have been interpreted as a clear signal to the NRC to forge ahead with licensing approvals had been averted by the opposition. In the longer term, such successes could not be guaranteed.

A move which could substantially weaken future licensing procedures was later set in train. A bill sent to congress by the Administration proposed the creation of a three-member Energy Mobilisation Board (EMB). It would have the authority to designate priority energy projects and facilitate their construction.[93] Ostensibly intended to expedite the Administration's $88 billion synthetic fuel programme, it granted powers to the Board which could have important implications for the nuclear industry. Included were powers to waive procedural regulations of federal, state, and local laws. These could be used to weaken review procedures, limit public participation, shorten licensing times and, by designating a reactor project a 'critical energy facility', render it no longer subject to judicial review.

Unlike these types of proposals much of the industry's political manoeuvring is less well advertised. Like an iceberg, most of it lies beneath the surface. Several of the industry's measures have been summed up by Charles Cicchetti, who chairs Wisconsin Public Service Commission, in a speech before the Atomic Industrial Forum. Noting that the economics of nuclear power tilt his members away from nuclear plants, he recommended that the Carter Administration 'revise' the economics by establishing federal responsibility for permanent waste storage, plant decommissioning, and guaranteed fuel availability. These three factors, he said, 'leave such a great financial and political risk that the federal government's virtual silence on these matters leave decision-makers few options than to be virtually negative'.[94]

The strategy of the industry, however, reaches beyond seeking further subsidies. As will be seen later, the companies may be capable of withstanding considerable financial hardship provided that their fortunes eventually improve. This depends not only on revising the economics of nuclear power, but also on a substantial growth of energy consumption. This will be determined not only by the future course of the economy, but also by the energy path that advanced industrialized countries follow.

POLITICAL ECONOMICS   83

It is not possible in this book to dwell on the detail of the possible energy paths that are available. These are dealt with in a growing number of excellent books.[95] The difference between the arguments of the nuclear industry and its opponents is illustrated by two reports released not long after the accident at Three Mile Island. The first is by National Economic Research Associates (NERA), a New York based consulting firm. It predicts dire economic penalties, amounting to $4 to $26 billion in the year 2000, if no more nuclear plants are added on line.[96] The other, titled *The Easy Path Energy Plan*, written by Vince Taylor and published by the Union of Concerned Scientists, comes to a diametrically opposite conclusion. Noting that since 1975 the USA has gained 2.5 times as much energy from increases in the efficiency of energy use as from growth in oil imports, nuclear power, and coal production combined, Taylor proposes that this trend be continued. By developing existing energy-saving measures further and launching several new measures in specified areas, he argues that by 1985 oil and coal imports could be decreased and nuclear power generation dropped to half its present level, all without economic sacrifice.[97]

The startling discrepancy between the two scenarios lies in the supposed energy growth rate. NERA assumes that electricity generation must grow at 4.5 per cent each year to the year 2000. Taylor proposes measures based on recent experience which would substantially slice this growth rate by forcing industry and other sectors of the economy to use electricity more efficiently.

The two scenarios illustrate quite different energy strategies. On the one hand, those that profit from selling energy technologies and fuel argue that a higher or even the present standard of living is based on increasing energy consumption. They emphasize the commercial energy technologies as being the only 'economic' means of satisfying this acceleration of energy consumption. They raise the spectre of an energy imperative dictating the need for a large-scale nuclear programme.

Others such as Taylor argue that economic well-being and growth in energy generation are not coupled, since the cheapest way to gain energy is to save on energy presently wasted. Further, they argue that other forms of generating energy, such as those based on solar energy or biomass (living or dead organic material) are available. Some are presently 'economic' while others are sufficiently advanced to be phased in as required by the retirement of

existing energy systems. Finally, they point to the enormous wastage of energy in a wide variety of products which could easily be replaced by less energy intensive substitutes. In this way the mode of living could be altered without decreasing the quality of life.

These arguments are often framed within the assumptions of economic determinism which lead to a seemingly inconclusive debate over the relative economics of nuclear and alternative energy technologies. There can be no doubt that a powerful case can be developed for the economic viability of alternative energy technologies.[98] But this is not really the ground on which the debate will be resolved. It has already been shown that the present economics of nuclear power rest on a bulky base of political decisions. So too the future economics of nuclear power and of the alternatives depend on present and future political decisions.

In 1952 the Paley Commission reported to President Truman that solar energy could play a greater role in 1975 than the then heady estimates for nuclear energy.[99] In 1973 the AEC concluded that solar energy could supply 15 to 30 per cent of total US energy requirements by the year 2000, and 30 to 50 per cent of all heating and cooling.[1] A major study by the US Congress's Office of Technology Assessment (OTA) concluded in 1978 that with modest federal encouragement it would be possible to build solar systems capable of supplying 100 per cent of all hot water and air conditioning needs for large US buildings by 1980. If the average price of electricity is assumed to rise by 40 per cent over the next twenty years (that is, to the present cost of electricity from new plants), these systems would be economic to install now.[2]

The relative economics of solar technologies depend sensitively on small marginal government policy initiatives. For example, the US Federal Task Force on Energy reported in July 1978 that if the present price of solar electric cells were to drop by 50 per cent the Defense Department could economically purchase $100 million worth of them each year. However, if the department were to replace 20 per cent of their petrol generators with photovoltaic arrays at a cost of $450 million phased over five years, this would drop the cost of the cells, then at $15.5 per peak watt, to a mere 75 cents.[3] The implication of all these reports is clear enough. As the OTA put it, 'the market in which solar competes is highly artificial'.[4]

If the nuclear industry and its supporters are to establish an

POLITICAL ECONOMICS 85

energy imperative for nuclear power they will not be able to rely on 'economic reality'. They will have to create that reality. To do this they must confront their opponents in a political battle for the future in which the need to recapture public opinion is crucial to winning.

THE BATTLE FOR THE FUTURE

The importance of public attitudes is increasingly well understood by the industry. This has never been clearer than in the aftermath of the accident at Three Mile Island, when the industry could be seen taking up position to counter the tide of public opinion with a frontal attack.

Only two months after the accident, on 18 June, Metropolitan Edison announced a plan to 're-educate' the public.[5] In the same month, interviews with industry officials, together with industry documents, revealed that utilities, reactor vendors, and nuclear trade associations were planning a $1 million public relations campaign.[6] It would mesh in with another launched a few months earlier by the OECD's Nuclear Energy Agency.[7]

During the final nine months of 1979 the industry's Committee for Energy Awareness, formed as a response to the accident at Three Mile Island, spent $1.6 million. In addition to the usual publicity paraphenalia of bill-boards, newspaper and magazine advertisements, and radio commercials, they produced a manual showing corporations how to organize citizens groups which aim to create 'a pro-energy environment'. Although these groups should be encouraged to develop their own identities, the manual made it clear that it would not be sink or swim. The 'corporate participant' would be ready to 'initiate, implement and sustain citizen action' as well as providing 'humour, linkages, resources and a sympathetic shoulder'.[8] Added to these efforts were those of the established nuclear lobby groups: the Edison Electric Institute with an overall 1979 budget of $14.6 million, and the Atomic Industrial Forum with $1.6 million reportedly budgeted for a 'public information program'.[9]

The accident had merely accelerated a campaign that had already begun. The month before, in February 1979, a National Energy Advocacy Conference had been held in Washington. Idaho Senator James Mclure had told the 800 participants: 'The supporters of nuclear energy must truly believe that nuclear energy is a moral

86   *GLOBAL FISSION*

necessity for mankind and that, without it, future generations will sink deeper into poverty and eventually dictatorship'.[10]

Drawing on those likely to be threatened by the industry's stagnation and those who are genuinely convinced that nuclear power is a 'necessity', pro-nuclear citizens groups have already been successfully created. Voice of Energy (VOE) chapters have been set up in Massachusetts, New Hampshire, and Vermont; Citizens for Energy (CITE) in San Jose, California; and the Committee for Jobs and Energy in Michigan. They mimic the successful activities of anti-nuclear groups, distributing buttons, tee-shirts, and stamps bearing slogans like: 'Nuclear Power is Safer than Sex', 'Nuclear Power the Recyclable Fuel', and 'A Little Nukie Never Hurt Anyone'.[11]

Courting the media, letter-writing campaigns, penetration into the schools, and films, pamphlets, and talks form an important part of their armoury. One particular target of the industry's battle to regain the community's hearts and minds is to bring over women. As early as 1975 a polling firm, Cambridge Reports Inc., warned the industry that it faced a serious credibility problem with women. Nuclear Energy Women (NEW) was set up in the same year. By October 1979 it had grown to the point where it was capable of organizing a 'nuclear energy education day' consisting of 'open door forums' and followed by more than 4 000 small 'energy coffees'. The message of the nuclear saleswomen, all from the industry, is that nuclear power is safe, and 'without sufficient energy to keep more jobs coming and GNP growing, women are not going to make it'.[12]

The small intimate energy coffees, where local women listen to an industry spokesperson who brings a toaster, coffee-maker, or crock pot 'gift for the hostess', are obviously considered important by the industry. Nearly 300 of them were held in the closing months of 1979 in New York City alone.[13]

The work of the industry's pro-nuclear groups is unspectacular but is important in the strategy to recreate the political climate necessary to reconstruct its economics. Although still dwarfed almost to insignificance by the corresponding network of anti-nuclear groups, on 26 August 1979 the pro-nuclear movement served notice of its potential. In what was the largest pro-nuclear rally so far in the USA, around 10 000 people assembled outside the Rocky Flats nuclear bomb trigger plant outside Denver, Colorado. The rally, put together by the three months old Citizens

POLITICAL ECONOMICS

for Energy Freedom, matched the size of the anti-nuclear demonstration that had been held there three months before. A CEF leader said that perhaps half of the rally was made up of Rocky Flats workers and their families, and CEF is reported to have been funded to the tune of 'thousands' by Rockwell International, the country's tenth largest defence contractor.[14]

The pro-nuclear demonstration was still small in relation to the numbers being attracted to the opposition rallies. One month later, on 23 September, some 250 000 people streamed through the gates of New York City's Battery Park.[15] Gathered together on the grass, the seemingly unending sea of people drove home the rally's demand for a phase-out and eventual shut-down of all nuclear plants, and by their presence gave eloquent testimony to the massive scale and formidable strength now attained by the US movement against nuclear power.

Throughout 1980 the battle over nuclear power continued to rage in the USA. The tactics included court cases, demonstrations, occupations, referenda, and publicity campaigns. The demonstrations were substantial although, perhaps not surprisingly, smaller than in the year following the accident. The results of referenda varied, in part indicating that this sort of contest is particularly vulnerable to the pouring in of large amounts of money in order to obtain a sometimes temporary level of public opinion. Over all, these still marginally favoured the nuclear industry.

In Maine, South Dakota, and Missouri, referenda that would have effectively precluded nuclear reactor construction were defeated. In Washington, the home of Hanford, one of the USA's major radioactive waste storage sites, a referendum was passed forbidding the importation of wastes into the state. In Oregon, a referendum was passed which prohibits nuclear plants being constructed without voter approval, a measure previously defeated at a referendum in 1976. But in Montana, a state which has already prohibited the building of nuclear plants without meeting difficult requirements, a referendum to prevent the disposal of uranium milling tailings was narrowly defeated.[16]

The industry continued to seek new ways to prop up its ailing economic fortunes, and the government too sought politically feasible ways of helping. The authorization in late 1979 of a $3 billion security bill allocating money to naval nuclear reactors, nuclear warheads, and the like, served as a reminder of the potential aid that could be given to the industry by a sympathetic administra-

88 GLOBAL FISSION

tion.[17] The election of Ronald Reagan as US president a year later, pledged to cut back on environmental regulation,[18] and the first Republican senate in twenty-six years, suggested that even though the House of Representatives remained in Democrat hands, a conservative swing was alive in the country. It seemed to make government aid to the nuclear industry even more likely over the next few years.[19]

Nevertheless, the opposition would be no insignificant force either in the forthcoming battle. Despite the manoeuvrings of the industry the defeat of some of the referenda in 1980 could be interpreted only as a maintenance of the status quo. Beneath the surface the blanket that now shrouded the industry's fortunes lay increasingly suffocatingly upon it. The plainest evidence of that was the steady stream of reactor cancellations which by the middle of 1980 had reduced the number of reactors on order by 45 per cent since the same time in 1979. The Department of Energy Information Office had over the same period reduced its 'high' estimate of the amount of nuclear power operating in the USA in the year 2000 by 30 per cent to 200 GWe. The NRC, projecting a maximum of 159 GWe, was assuming that the average licensing period would now be fifteen to sixteen years.[20]

The forecast long licensing periods, the decreased nuclear power projections, and the stark reality of cancelled orders were all further indications of the strength of the amorphous but pervasive opposition to nuclear power in the USA which continued to not only block the moves of the industry but also to develop the climate in which electricity utilities were turning away from nuclear power. It was the strength of this movement which had caused the industry to come out fighting at Rocky Flats and to accelerate its overall political campaign. But the limits to the effectiveness of the industry's campaign derive not from the limits to its resources, but from the strength of its opposition.

Each of the interlocked strategies available to the industry – lobbying for favourable legislation, pressure for energy growth, the search for subsidies, and manoeuvring for 'licensing stability' – are confronted by campaigns for further regulation of hazard, for energy conservation and alternative energy systems, for reduced subsidies, and for greater public participation in the licensing process.

The artificial division between the political and economic problems of the industry fades away to reveal the all-encompassing

POLITICAL ECONOMICS

social arena in which the battle over nuclear is being fought. With it dissolves any certainty over the result of the conflict. Whatever the present perceived strength of the opponents, the outcome, as with all battles, will be assured only when one side concedes or collapses. At stake is the form of the future. On the one side, the nuclear industry holds hopes for an ever more energy consuming world powered by their energy technology. On the other, the opposition strives for a world in which energy is used less wastefully and whose power is derived safely from the sun.

The outcome of this battle will not be decided by bare market forces. On the contrary, the market and the future of the nuclear industry will ultimately be shaped by the outcome of this contest of attitudes and political action now surging up around it. But as already noted, the contest is by no means confined to the USA. Wherever nuclear power stations or other parts of the nuclear fuel cycle are proposed manifestations of citizens opposition can be found. This struggle over nuclear power is an international phenomenon. It takes place in a political battlefield spanning the entire breadth of the industrialized world. The origin and nature of the major contestants in that battle, and the potential and limits to the force that they may exert, are the subject of the next two chapters.

5
ORIGINS OF THE OPPOSITION

Like the poor soul at the bottom of the heap in the tavern brawl the nuclear industry should not console itself that all those bone-crushing brutes sitting on top of him are all debating life-styles . . . In fact, I am filled with apprehension and despair at the thought that this sort of amorphous opposition has kept us off balance for the better part of ten years.

Carl Goldstein, Assistant Vice President,
US Atomic Industrial Forum[1]

The growth of the opposition to nuclear power can only be understood in relation to the properties of the technology itself. Irrespective of the country in which it is deployed, nuclear technology has several distinctive properties. Most striking of these is the scale of the nuclear enterprise. Huge investments – $1 billion for a single reactor – flow into the industry. Inevitably it is only the largest companies or governments which possess the resources to carry through the development and sale of nuclear technology. The very cost of the technology creates a dynamic for rapid expansion of nuclear programmes in order to cover investments already made.

In addition, nuclear technology lies at the frontier of scientific and technical development. The complexity of nuclear reactors, enrichment plants, and reprocessing facilities is of the highest level. Even now crucial sections of the nuclear fuel cycle, such as the reprocessing of wastes produced by modern reactors, are beset by technical problems. The reprocessing plant built by General Electric at Morris, Illinois, which not only did not function, but after six years construction was abandoned in 1975 when it proved incapable of being modified so that it would, testifies to the sophisticated problems encountered in nuclear technology.[2]

The sophistication of the technology meets the scale and rapid pace with which it is introduced and developed to greatly exacer-

ORIGINS OF THE OPPOSITION

bate problems inherent in the technology. Major hazards of hopefully small but unknown probability of occurrence accompany most stages of the nuclear fuel cycle. The fast pace of development, moving from small to large reactors through diverse designs and to rapid commercial deployment, has resulted in safety and other features translating from the drawing board to commercial use with only short pause for laboratory testing. The uncertainty thereby created and the modifications made necessary by experience further heighten the technical challenge.

The technical sophistication of the evolving nuclear industry has meant that large numbers of scientists, engineers, and technicians have been deeply involved in it since its inception. From the early days of the secrecy-shrouded Manhatten Project many scientists devoted themselves with enthusiasm and loyalty to the technical challenges presented by the nuclear industry. With the successes of the Manhatten Project behind them, and the steadily unfolding technological advances catalyzed by World War II all around them, there was a tendency to tackle the most immediate problems first, leaving others to be coped with later. Those set aside or still unknown would include problems with no easy technical answers. It would be in the 1970s that these would first return to haunt them.

Because of the technology's sophistication the accompanying problems remained hidden from the public for many years. Only those with both the benefits of a scientific background and access to the technical literature in which the unfolding saga of the nuclear industry was documented were able to comprehend the nature of the technology and its known and potential problems. Within the technical community which came into contact with this literature, a mixture of loyalty to the enterprise and a blinkering technological optimism shielded many of the potential problems from their focus of attention.

However, for those scientists who had become deeply concerned about the way technical expertise had been applied to devastate Hiroshima there was a fear less easy to suppress: would not the spread of nuclear reactors inevitably lead to the proliferation of nuclear weapons? To prevent this, less than a year after the bombing some scientists, including Leo Szilard and Harold Urey, were pressing for a limited moratorium on further development of nuclear energy. Most, however, considered this 'too high a price to pay for security against atomic warfare'.[3] A five-year moratorium

on large-scale production of fissionable materials was provided for in the 1946 Chicago Draft Convention for the Control of Atomic Energy, but it also was never taken up.[4]

For a few scientists concern extended beyond the proliferation of nuclear weapons. In particular, some could see that ionizing radiation presented a hazard only partly understood and requiring investigation before any major commitment to nuclear power. In 1946 a Dr Herman Lisco warned in the *Bulletin of the Atomic Scientists* that the problem of radiation 'has not been discussed with the frankness that it requires'.[5] It was a prophetic voice, but it was crying virtually in the wilderness.

The usual response to these problems was that they should be solved by international regulation. Where a 'technical fix' was plainly impossible proposals for a 'diplomatic fix' stood waiting in the wings.

The Baruch Report, a nuclear weapons control proposal put before the United Nations by US delegate Bernard Baruch in 1946, recognized that some problems raised by nuclear energy were not technical in nature: 'Science has taught us how to put the atom to work. But to make it work for good instead of for evil lies in the domain of dealing with principles for human duty. We are now facing a problem more of ethics than of physics'.[6] But the report's proposed solution was merely to place all atomic energy activities under a single managerial authority. Similarly, the proposal that the United Nations should be empowered to license countries 'to possess atomic weapons', made by the Endowment Committee on Atomic Energy, naively turned a blind eye to the dominant role of the USA in that organization.[7]

For many scientists such an approach can hold a particular attraction. First, it separates scientists' roles in research and development of technology from responsibility for its applications. Under this view of scientific work as politically neutral scientists are freed to do as they wish, safe from interruption by either public or personal concern. Second, it is consistent with the view that all problems are best solved by experts. Scientists are seen as those best fitted to solve technical problems, and diplomats and managerial experts are seen as best fitted to solve social problems. In the social model of a technocracy smoothly running society like a well-oiled machine it is, however, difficult to find any place for democracy or social accountability.

If this authoritarian approach is to be avoided the public must

ORIGINS OF THE OPPOSITION

somehow be involved in crucial decisions. Fourteen years after the bombing of Hiroshima, some scientists found themselves in a dilemma over this point. Argued Wallace De Laguna of the Health Physics Division, Oak Ridge National Laboratory, in 1959 when discussing 'What is Safe Waste Disposal?', the nature of the hazard must be explained if the public is to appreciate and ensure the enforcement of the 'peculiar precautions' required. But, he worried, the wastes may then be seen as a 'terrifying menace' which 'may lead us into unwise action'.[8]

Most of those involved with nuclear developments had yet to acknowledge that any serious problems accompany the nuclear fuel cycle. Any reference was usually to the problem of nuclear weapons proliferation and then couched in terms of 'beating swords into ploughshares', and uplifting phrases such as 'If we succeed in banishing this dark shadow then the fate, a happy fate, of mankind in the atomic age will be assured'.[9] Nevertheless, by the end of the 1950s a few scientists had emerged who were sufficiently concerned about the hazards associated with local nuclear activities to begin public agitation.

In September 1959 a letter in the *Bulletin of the Atomic Scientists* illustrated the start of what would be a successful campaign to stop the AEC dumping radioactive wastes in the sea 19 kilometres from Boston:[10]

> We are members of the Lower Cape Committee on Radioactive Waste Disposal which has been protesting proposed atomic waste dumps in Massachusetts waters. With the cessation of nuclear weapons tests – a cessation we hope will continue – we believe the disposal of radioactive wastes will become the 'number one' radiation problem of the nuclear age. Greater attention must be given to its hazards and their proper control.

The letter referred to twenty-eight dump sites and added 'strong opposition is expected from coastal areas'. More significant was the way the authors described themselves. We are, they said, writing in our capacity as 'citizens and scientists'.[11]

While the debate remained restricted to the technical community involved in nuclear developments it was constrained within the limits of optimism, loyalty to their employers, and the technical framework which infused their thinking. But as commercial nuclear reactors began to appear on the scene the issues came closer

94 GLOBAL FISSION

to others in the community. Yet it would be necessary for some scientists to both draw attention to the technical unknowns and translate their implications into language understandable by the public.

Many of the first to openly question the nuclear plans were life scientists – biologists and ecologists. This was partly because they were much less personally involved in the development of nuclear power, but also because they were aware that the state of human knowledge on the effects of radiation on biological systems was still rudimentary. They were therefore less inclined to the technical optimism of their colleagues in the physical sciences.

A few early actions against nuclear power occurred at the beginning of the 1960s. In 1962 David Pesonen, a forest ecologist, founded a group of citizens to try to stop the construction of a 325 MWe nuclear reactor at Bogeda Head, 25 kilometres north of San Francisco. In the autumn of 1963 they released 1 000 coloured balloons at the reactor site to show how wind would carry radioactive strontium and iodine.[12] Four years later a group of biologists at Cornell University attacked the thermal pollution that would be caused by an 830 MWe reactor which was to be constructed at Lake Cayuga in New York State. The result was an upsurge of community concern which five years later forced the plans for a nuclear plant to be abandoned.[13] In 1968 and 1969 a few outspoken scientists, including Dr Barry Commoner, a plant physiologist who had become involved in post-war science policy,[14] and Drs John Gofman and Arthur Tamplin, then employed by the US AEC, began to highlight the hazards of allowed levels of radiation.[15]

In 1969 a local Citizens Committee for Environmental Concern was also set up in Pennsylvania to fight a proposed prototype fast-breeder reactor to be built on the Susquehanna River at Meshoppen. Partly due to public pressure, it was eventually re-sited in Tennessee.[16] In the same year, at court hearings over a proposed reactor at Culvert Cliffs, only 48 kilometres from Washington D.C., a landmark decision was won to force subsequent hearings on reactor sitings to include environmental impact in their terms of reference.[17]

It was outside the USA, however, that the first sizeable citizens opposition to nuclear power surfaced. In Japan the local fishermen and farmers already had a clear indication that modern technology would not necessarily enhance their quality of life. They had seen

ORIGINS OF THE OPPOSITION

the devasting effects of two nuclear bombings. In 1950 they had also been subjected to seeing Minimata citizens suffering horribly from the effects of eating mercury-contaminated fish. Over 120 people had suffered irreparable brain damage. These events had given rise to a rapidly spreading network of citizens groups opposed to nuclear weapons and the contamination of the environment. As early as 1968, local fishermen took action against a proposed nuclear reactor in Onogawa prefecture. Twelve years later as the 1970s came to a close, the ground-breaking ceremony for the reactor had still not taken place.[18]

Nevertheless, in these early years, and continuing through the 1970s, there were always some members of the technical community who, by expressing their concern over nuclear power, played a crucial role in demystifying the issue for other citizens. In the USA, scientists like Dr E. J. Sternglass from Pittsburg, Dr Henry Kendall from the Massachusetts Institute of Technology, Nobel Laureate George Wald, and radiation specialist, Sister Rosalie Bertell, helped pinpoint crucial areas of concern about the nuclear fuel cycle. Later these scientists were joined by several high-ranking specialists from within the nuclear industry. These defections included Carl Hocevar, a reactor safety expert who resigned from the AEC in 1974 to work with the opposition.[19] Three scientists from General Electric resigned their management positions in February 1976 saying 'the cumulative effect of design deficiencies . . . makes a nuclear power plant accident, in our opinion, a certain event. The only question is when and where'.[20] In the same month, Robert Pollard, a project manager for the NRC, also resigned his post to work for the opposition saying 'I could no longer, in conscience, participate in a process which so effectively evades the single legislative mandate given to the NRC – protection of the public health and safety'.[21] In the same period in the USA alone, a statement of opposition to nuclear power was signed by some 2 500 members of the US technical community. It included the signatures of nine Nobel Prize winners.[22]

These scientists were joined in almost every country by other professionals such as lawyers. In the USA, West Germany, and some other countries, they played an important role in helping citizens slow down projects by taking court action. In the USA the cost of such actions eventually took their toll and, as the judge was often the AEC or the NRC, their limitations had to be recognized. Nevertheless, the opposition by some scientists, the defections

from the industry, the role of professionals, and later the taking up of the issues by state and federal politicians, in many countries played the vital role of demonstrating that both the scientific community and the establishment were divided over the issue. This imparted a stamp of credibility to the arguments of the local activists.

Although the professionals were important in strengthening the emerging citizens movement they formed only one part of that movement. It was the diverse actions of lay citizens which often encouraged professionals with concerns about nuclear power to express them, and which transformed those concerns into a political matter unable to escape public attention. As the barrier between the shrouded technical issues and the public consciousness began to dissolve the broad citizens movement began to spread, building momentum dramatically in the second half of the 1970s. Many citizens began to acquaint themselves with scientific areas which had previously seemed mysterious and inaccessible. Beyond the specific issues of nuclear power this movement was also fighting a broader battle for the right of citizens to control major influences on their lives. In that sense it was also a citizens struggle for access to science, and for control over the technological developments rapidly transforming their world.

None of this by itself, however, explains the speed and effectiveness with which popular opposition to nuclear power grew around the world in the 1970s. The awakening of concern, the preparedness to take often frightening actions, and the willingness to make sometimes major sacrifices to fight against nuclear projects, cannot be explained simply in terms of some diffusion process emanating from the technical and professional communities. This popular opposition can only be explained by understanding that the nuclear project was not being constructed on barren social soil. It was occurring in a context of other political and social developments already reverberating around the world.

THE SOCIAL CONTEXT

The technology of nuclear power and the roots of the citizen opposition stretch back to the bombing of Hiroshima. A wave of revulsion had begun, touching many people in many communities around the world. The epicentre was Japan. There, peace move-

ORIGINS OF THE OPPOSITION

ments sprang up and in 1954, in the wake of the US hydrogen bomb test at Bikini atoll in the Marshall Islands, they converged to form a unified Japanese Council Against Atomic and Hydrogen Bombs. Japanese opposition to such tests was widespread, and an estimated 35 million signatures were collected on petitions calling for bans on nuclear weapons.[23]

Much more slowly in other countries peace movements also started to develop. In the UK the Campaign for Nuclear Disarmament (CND) finally began in 1958. There were two catalysts.[24] The hydrogen bomb tests at Christmas Island in the Pacific reminded many people of the hazards of fallout, expecially to the unborn. Additionally, many opponents of nuclear weapons had hoped to force a motion through the British Labour Party declaring that the UK would abandon the bomb unilaterally as an example to the rest of the world. But in autumn 1957 that motion failed.

The first of CND's Aldermaston marches was held in 1958. Over the next few years and into the late 1960s, first thousands and then tens of thousands took part in the four-day marches from the Atomic Weapons Research Establishment at Aldermaston to Whitehall in London. The cause was also taken up by the left of the labour movement. Nevertheless, both the organizing committees and the participants spanned a broad political spectrum, and there was still a unifying theme. Wrote Peggy Duff, a central organizing figure throughout the campaigns:

> it was a community for which no vows were required. All you had to do was to step off the pavement and join it. While the bomb was its main occasion and theme, it was more than that. It was a mass protest against the sort of society which had created the bomb, which permitted it to exist, which threatened to use it.[25]

The movement grew rapidly because for many the nuclear bomb encapsulated the very worst of the direction in which society was moving. Later, nuclear power too would fulfil that role.

The movement against nuclear weapons was greatly stimulated by the rapid growth of CND not only in the UK but in many other countries. As early as 1959, 3 000 delegates gathered at the Paulskirche hall in Frankfurt, West Germany, in opposition to nuclear weapons. Styled the European Federation Against Nuclear Arms, it comprised the anti-bomb organizations from the UK, the

Netherlands, Sweden, and West Germany. By 1961 the federation was organizing conferences which included additional representatives from anti-bomb organizations in Denmark, Norway, France, Italy, Brazil, Yugoslavia, the USA, and Australia.[26] Most of those countries would later develop powerful movements against nuclear power.

The movement grew with great promise and then foundered. Its greatest victory, the enacting of the atmospheric Partial Test Ban Treaty in 1963, was probably also its downfall. An illusion of victory combined with an over emphasis on the problems of fallout left the opponents of the bomb with limited goals achieved and their final goal of a total ban on all nuclear weapons seemingly so remote as to be unattainable. The movement faltered and by 1966 CND had largely lost its ability to mobilize. It had also failed to find any other effective strategy.

Although the dramatic action had petered out, the nuclear weapons issue had been raised forcefully and would not soon be forgotten. Later it was to reappear in several countries, notably Australia, the Netherlands, and Canada, as a major reason for opposition to the export of 'peaceful' nuclear materials and technology. At the beginning of the 1980s it was to reappear in its own right. In the meantime a vast number of people and groups had discovered the ability of citizens to influence the actions of governments and community opinion by taking to the streets. Elsewhere in the USA the civil rights movement achieved major advances towards greater equality for blacks by similar tactics involving the medium of mass demonstrations and civil disobedience. Peace organizations had also been set up, and although the turmoil subsided these did not disappear. They were ready and waiting when, around 1965, it became clear that the US government and its allies were escalating the war against the people of Vietnam to the point that citizens must again take to the streets.[27]

The five years from 1965 to the end of the decade was a time of social turmoil and critical self-examination in many countries of the Western world. The period of post-war reconstruction was replaced in the 1960s by increasingly affluent consumerism. Here it was the Vietnam War that acted as a catalyst to enable many to realize that the rapidly escalating levels of consumption in the advanced industrialized countries relies on trade relations which pillage the Third World.[28] For those who understood this, and the nature of the political regimes that were supported in order to

ORIGINS OF THE OPPOSITION

maintain this situation, the rewards stemming from the growing affluence began to seem tarnished and hollow.

But for many the war meant more than that. Its immorality, the conscription of unwilling troops, its devastating effects on the Vietnamese people, and the lies, distortions, and cover-ups perpetrated by the Lyndon B. Johnston and Richard Nixon administrations to justify it, all served to shatter faith in the institutions of US society and its allies. In addition it helped break the sense of moral superiority which had previously enabled Americans to see themselves as the 'world's police'.

Discussing the factors that are crucial to a social system or 'order' being stable, Max Weber has observed that

> An order which is adhered to from motives of pure expediency is generally much less stable than one upheld on a purely customary basis through the fact that the corresponding behaviour has become habitual. The latter is much the most common type of subjective attitude. But even this type of order is in turn much less stable than an order which enjoys the prestige of being binding, or, as it may be expressed, of 'legitimacy'.[29]

For many the Vietnam War severely damaged the legitimacy of the government, the courts, the police, corporations, the military, the existing international economic order, and many other key institutions of Western industrial society. In doing so, it opened the way for a new period of increased social instability.

The Vietnam War was not alone in this. At the same time a current of dissatisfaction with the consumer society could also be discerned. From the early 1960s the division of cities into isolated suburban nuclear families was beginning to be recognized as leading to a growing feeling of alienation. A loss of a sense of community resonated with the recognition that in the midst of growing affluence severe poverty remained and that the modern pace of work and living was becoming more stressful.

Dissatisfaction with the values of the consumer society was evidenced first with the emergence of the counter culture. Finding the rewards offered deeply dissatisfying, a growing number of young men and women began to seek alternatives. The first wave, in the early 1960s, focused primarily on experimentation with socially proscribed drugs and less traditional patterns of sexual relationships. Later in the decade the emphasis on drugs decreased

and was replaced by a broader reaction against the results of industrial consumerism. Communes began to be set up in rural areas. Those who participated placed strong emphasis on restoring the disappearing values of communality, simplicity, and self-sufficiency.

Among the younger generation, opposition to conscription and the insights gleaned from the Vietnam War nourished a growing radicalism. In the universities students began to attack the curricula as being merely one cog in the military-industrial complex, turning out the technicians necessary for its operations. During the late 1960s and early 1970s mass student demonstration, occupations, and even battles with the police took place.

In France, in May 1968, with half a million people unemployed, students' and workers' discontent with the government and social structure boiled over, raising the prospect of revolutionary change. For a short time anything seemed possible. Proclaimed one leaflet from the Sorbonne:

> The revolution which is beginning will call in question not only capitalist society but industrial society. The consumer's society must perish of a violent death. The society of alienation must disappear from history. We are inventing a new and original world. Imagination is seizing power.[30]

It was not to be. The Gaullist forces of the right marshalled, and confronted, and then dissipated the challenge.[31] Nevertheless, an important lesson had been demonstrated.

The upsurge had begun with the isolated occupation of factories. Then, between mid May and mid June, in the largest labour revolt in French history, some 10 million workers paralyzed the country with a national strike. In doing so they showed they could successfully challenge the very heart of their social system by bringing to a halt the production of goods. Further, they achieved this in confrontation with one of the most powerful and centralized systems of government in the Western industrialized world, and demonstrated that even there the state is not all powerful. The modern industrialized state stood revealed as ultimately dependent on the cooperation of that large section of the community involved in operating the highly sophisticated technological base on which the consumer society has been erected.

At the same time, between 1965 and 1969, the form of scientific

ORIGINS OF THE OPPOSITION

practice and technology was beginning to be questioned as the Cultural Revolution swept through China. For a time, within China, the role of technical experts was subjected to intense examination. Judgements of 'efficiency' were made subservient to considerations of local self-sufficiency, reduction of waste, and the development of ways in which production processes could be made more accessible to the bulk of the population.

The ideological emphasis of the Cultural Revolution proved to be short-lived. Later it was submerged, at least temporarily, by the struggle within the Chinese Communist Party. Nevertheless the concepts raised diffused beyond China. The understanding that the social and technical aspects of 'progress' are inseparable and that the choice between the many possible directions of socio-technical development involves an array of social judgements was highly relevant to a new debate springing up elsewhere. As the 1960s drew to a close and the new decade opened the early rumbles of a storm of concern over the environmental effects of the path of industrialization being followed could be heard in much of the Western world.

Environmentalism, at least in a limited form, was nothing new. As early as 1908 President Theodore Roosevelt had called together the first conservation conference to be held in the White House to discuss policy for preserving aspects of the biophysical environment.[32] Over the next fifty years numerous governmental resource agencies were established in the USA. These were generally dedicated to the 'wise use' of the country's minerals, public lands, water, forests, wildlife, and fisheries.[33] But the comparative wisdom of different projects was judged mostly on the basis of potential economic gain. Some of these projects sacrificed rare and beautiful natural resources to which little economic value had been attached. Small citizens groups therefore also developed to lobby for the conservation of such resources in particular areas.

By the early 1960s concern was beginning to emanate from some scientific circles that this balance of controlling forces was no longer adequate. Suddenly they began to see that modern technology was creating new environmental problems of unprecedented scale and severity. 'For the first time in the history of the world, every human being is now subjected to contact with dangerous chemicals, from the moment of conception until death' worried Rachael Carson in *Silent Spring*, published in 1962.[34] It is necessary

to set up an 'intelligence agency' to 'forestall, counteract or rectify predictable future disruptions and imbalances of the human ecosystem' advised a National Academy of Sciences Report prepared at the instigation of President John F. Kennedy in the same year.[35] 'It is already too late to prevent a dramatic rise in the (global) death rate through starvation' concluded Paul Ehrlich in 1968[36] and 'We have been living under a vast and potentially fatal illusion: that we can enjoy the enormous benefits of modern technology without risk to the integrity of human life and the environment' warned Barry Commoner a year later.[37]

A new awareness was crystallizing that contemporary societies were producing, along with a spectrum of consumer goods, a compounding growth of often novel and cancer-inducing chemicals, pollution of water and air, and population increase, especially in the Third World, that was destroying vital energy and mineral reserves and threatening crucial elements of the biosphere. The concerns were initially expressed in simplified and politically naive terms. Even so they could not easily be dismissed except by the most stubborn exponents of technological optimism. The warnings resonated with other criticisms of the direction in which modern industrialized society was developing.

Among the most powerful was the critique being mounted by the burgeoning women's movement. It raised searching questions about the basis of the sexual division of roles and power in Western society. Over the 1970s the movement grew and produced major changes in social attitudes and national legislation. From the beginning of the 1970s it meshed with the emerging environmental consciousness. The emphasis within the women's movement that personal action is politically important reinforced a similar belief developing among environmentalists. At the same time women were gaining greater confidence in their ability to take part in shaping the direction of society. Often placing a greater value than men on the importance of health and the well-being of their children, women proved particularly sensitive to the environmental problems being uncovered.

Throughout the 1970s an environmental critique developed among large numbers of men and women in the community. It led to the creation of ministries of the environment, environmental regulation agencies, environmental impact legislation, and improved safety levels. It also led to the creation of expanding networks of citizens action groups. While these groups concerned themselves

ORIGINS OF THE OPPOSITION

with a wide range of problems their effectiveness was limited. However in 1972 nuclear programmes in many countries received considerable impetus from increases in the price of oil. A weakening of the USA's global military and political dominance enabled the oil-producing states in the Middle East to form a new alignment and dramatically raise oil prices. Suddenly it was clear that the supply of the few energy sources on which industrialized countries had become so deeply dependent could no longer be assured. Governments sought an easy technical solution and grasped hopefully at the nuclear industry. A rash of reactor orders and optimistic proposals for more spread across the world. The growing concern over the direction being taken by the industrialized world, and its social and environmental effects, became focused. The manifest hazards associated with nuclear power, and the urgency with which they threatened, provided a new unity of purpose.

Embodied in this one technology were all the threats posed by the direction in which society seemed to be moving: a gargantuan scale, a profound and terrifying potential to bring an already tottering world closer to the brink of nuclear war, extraordinary physical hazards, and entrenched centralized political and economic interests willing to deploy powerful forces to press their nuclear projects forward and protect them. Nevertheless, with many nuclear programmes conceived in the secrecy of government and industry discussions, a general appreciation of their implications took a long time.

In the scattered parts of the world where nuclear programmes began to take form local citizens started to come together and to seek ways to confront them. The emphasis of the concern over particular projects differed markedly from place to place, and so did the attempts made to stop the projects. It was not until 1975 in West Germany that the scattered citizens opposition finally decisively demonstrated that despite the potent coalition of forces which pressed the nuclear enterprise forward, citizens also had power to confront and begin to obstruct it.

THE BEGINNING OF THE BATTLE

In the middle of the Upper Rhine Valley, facing France across the river, lies the Kaiserstuhl, a famous grape-growing area and one of the warmest and most fertile areas of West Germany. The locals call the region in which it lies 'Dreyeckland', the place in which the

104 GLOBAL FISSION

south of West Germany, the north of Switzerland, and the east of France share common borders. Literally it means 'the triangle region' and it has its own local history, identity, and language (Alemannian) which stretch back to a time well before the national borders were drawn.

Before the oil crisis a great increase in industrialization had been planned for central Europe. Dreyeckland, with its close proximity to major cities of three countries, the Rhine River and other existing arterial transport routes, and large tracts of rural land ripe for development, seemed the obvious venue. This second Ruhr was to have extensive chemical works, oil refineries, and heavy industry. It was to be a new heartland for the consumer society and it was to be powered by nuclear reactors. But for the local people on whose land the developments were to take place the plans were in stark contrast to their primarily agrarian interests.

In the wake of the oil crisis, the plans for industrialization were partially set back, but the impetus towards the construction of nuclear reactors was strengthened. Already, as early as 1971, opposition to the plans had begun to surface. The government first announced that a large nuclear reactor was to be built in the Kaiserstuhl at Breisach.[38] However the local council was defeated at the local election after giving permission, and in 1973 the national government announced a new site. It would still be in the Kaiserstuhl, but now just outside the tiny rural village of Wyhl.

Here progress seemed likely to be much easier. The 3 000 inhabitants, encouraged by promises of a sports centre, swimming pool, and other public works, voted to sell the land to the reactor company. However, even in that vote 43 per cent opposed the sale. Afterwards many local farmers became progressively more concerned about the effects that radioactivity, and fog from the reactor's cooling system, would have on their crops and grapes. Together with city dwellers from nearby Freiburg, and school teachers concerned about reactor safety, they formed a loose coalition to oppose the plant. However before anything could be done a new cause for alarm emerged. Across the river in neighbouring France it was learnt that the French government intended to construct a lead factory in Markolsheim.

As would prove to be common, there were scientists prepared to help span the gap between the apprehension of the locals and the mystifying reassurances of government and industry. This time it was the Environmental Group at the University of Freiburg which

ORIGINS OF THE OPPOSITION

helped add flesh to local concern by demonstrating that the Markolsheim plant would produce seven times as much lead dust as claimed by the industry. Although the local townspeople voted for the plant on the grounds that it would provide additional jobs opposition grew rapidly, especially among the farmers.

On 20 September 1974 the opposition came to a head. From all over Dreyeckland citizens, led by the wine and vegetable growers and the Women's Institute, came to occupy the Markolsheim site and prevent construction.[39] Their concern was with lead but it had not arisen in isolation. Although not all may have seen it that way they were part of a broader movement and even their songs showed it: 'At Limberg over Sasbach, there grows a deep red wine. The taste is great as it should be, and we want to keep it that way . . . We have not forgotten DDT and Thalidomide'.[40] A 'friendship house' was constructed and the people settled in for a long siege. At times comprising as many as 30 000 people, men, women, and children shared the guarding of the site for five months. Victory was finally conceded to them, and the plans for the factory were cancelled. Now it was Wyhl's turn.

Construction workers moved into Wyhl five months later to begin clearing the heavily timbered reactor site, and the next day a few hundred locals arrived to block the work. It was a small beginning but it heralded a storm. Two days later the police arrived. The adults and children occupying the site refused to leave but they were eventually forced off by police water cannons and batons. Three days later 28 000 people from West Germany, France, and Switzerland poured onto the site. Sweeping aside the police, the stream of people tore down the fence, and moved in from all sides simultaneously. Another friendship house was thrown up. In it a 'peoples school' on the hazards of nuclear power and alternatives to it began. The occupation of Wyhl was under way.

For the state government the occupation posed a difficult problem. At the very beginning it had tried to brand the occupiers as 'left-wing extremists' but this had been made nonsense of by the detailed media reports showing local farmers and their wives and children being brutally removed. The rough treatment was widely condemned and made the wine-growers, clergy, and others in citizens action groups all the more determined. The government even found its use of the police limited. Media reports at the time noted that some 200 local police refused to take part in the action.

As the chairman of the local Policeman's Union told the press, the 'feeling of legal and moral conviction' which police must have for their work had 'not completely' been given in this case.[41] Even the additional 1 000 riot police placed at the disposal of the state government found their expectations of left-wing extremists strangely in contrast with reality.

The government discovered constraints on its ability to carry through its plans, and the occupiers found that they possessed a suprising degree of political strength. The source of this strength lay in the legitimacy attached to their action.

The government's justification for placing the reactor in Wyhl was that this power station would play an important role in increasing the commodity production of the West German economy. This would seem legitimate to many, as the belief that ever-increasing production was essential to social well-being had become a widely shared ideology. On the other hand, this society, and indeed all capitalist society, is fundamentally based on a respect for work and property. The claim by the local people that the government's proposal if implemented would not only threaten their health, but their property and livelihood, also appealed to an ideology deeply rooted in the society. The local people's opposition therefore was something that many could understand and identify with. The actions of the police in throwing people off their traditional land shocked people, and the response of the villagers in occupying the land seemed legitimate. In this case the legitimacy attached to the opponents of the Wyhl reactor, combined with the depth of their feeling and the perseverance with which they were prepared to fight, proved to be the stronger force. The government found it politically impossible to clear the site and the occupiers stayed for over a year.

The power of citizens to stand in the way of an industrial project that threatens their region, revealed at Wyhl, was not restricted to that area. Similar strength had already been demonstrated at nearby Markolsheim in France. Even as the occupation continued at Wyhl, at Kaiseraugst in neighbouring Switzerland 15 000 people rallied to support another occupation of a reactor site. As will be seen later, the result was a halt to the construction of the proposed reactor. As late as the end of 1980, construction of the Kaiseraugst reactor had still not begun.

In West Germany at Wyhl, the government and the Baden Power Company sought a way around the force confronting them.

ORIGINS OF THE OPPOSITION

One route might be the courts, even though prospects for an early solution seemed dim. In March 1975 the local court at Freiburg ordered a temporary halt to the construction of the Wyhl reactor. Seven months later the order was overruled by the high court in Mannheim and the Baden Power Company was given permission to begin construction of its administration building. These orders meant little, however, in the face of the occupation which continued to pose the same intractable barrier.

Further, opposition to the plant had now grown. By Easter some 65 000 signatures had been collected in opposition to the reactor. Outside the site support for the occupiers was demonstrated by 30 000 who gathered in solidarity. The legal framework may have been put together to enable construction to proceed but there were insufficient political foundations to use it.

It was not until 31 January 1976 that the occupiers left after signing an agreement with the nuclear company. This Offenburg Agreement was concluded against the bitter opposition of some of the occupiers. It proclaimed a pact in which the company agreed not to proceed with construction while the occupiers agreed to leave the site pending a second round of hearings before a panel of judges. A year later, in early 1977, the Administrative Court in Freiburg announced its verdict. It surprised many by declaring the reactor too dangerous to build. The judgment was based not on the broad safety issues but on the technical criticism that the containment vessel would be insufficiently strong to give confidence that it could safely contain a serious accident.[42]

The ruling left the reactor's opponents jubilant and the project seemingly inextricably stalled. Although further court hearings were subsequently initiated by the company the beginning of the 1980s came with the reactor site at Wyhl still a wooded nature reserve.

To halt a nuclear project is no minor achievement. At stake are hundreds or even thousands of millions of dollars, and behind it lies the concerted power of the nuclear industry, the government, and their agencies. The strength that the people of Dreyeckland had exerted derived not only from their commitment but also from their advantage in a contest of legitimacy. Behind the contest lay a conflict between two values simultaneously held by the community, and this conflict reflects the broader mechanisms which have thrown up the 'environmental movement'. The dominant ideology asks that society be judged in terms of the criterion of profitability

and economic growth. But this is by no means all that people value. Health, a sense of community, a feeling of control over one's life, and a satisfying job in which one can use one's abilities creatively and be respected for one's achievements are among the other things which people must believe they are receiving if they are to support the way society is operating. At least in part the environmental conflict in the West arises because despite the growth of economic indicators people feel that these aspects of their lives are being eroded. For many people nuclear power can symbolize much of this in one single project. The potential threat to health, life, and often traditional livelihood is deeply worrying. At the same time the proposed plant is pressed forward by powerful forces from afar. There are often few benefits for those who suffer the most immediate risk.

In Wyhl the early concern was over the effect that fog, caused by the proposed reactor's cooling system, would have on the vineyards. Later the analysis of the local opponents became much more sophisticated, taking into account the many technological concerns related to reactor hazards and problems with waste and radiation. Now, instead of opposing just nuclear power at Wyhl, they opposed nuclear reactors anywhere. They also began to argue for different energy sources and energy consumption patterns for the future.

Nuclear power is part of a more general trend in industrialized society, and opposition to nuclear power inevitably brings those who oppose it into opposition to other components of that trend. It is striking that the activities which constitute the fight against nuclear power also help break down some of the social barriers which have developed with modern industrialization. The very act of coming together to fight against the political forces behind the projected reactor helped weld the opponents at Wyhl into a more integrated community. The friendship house symbolized this. There people talked, worked together, ate, and studied. Townspeople formed new relationships with farmers, and young and old, women and men, and educated and uneducated, found new common bases of shared values and mutual respect. Long after the occupation was over the people's school that had begun in the friendship house still continued to hold regular classes and discussions in the surrounding villages. The ever-present threat of a reactivation of the project held people together and encouraged them to nourish the new links they had formed. The opposition to

ORIGINS OF THE OPPOSITION

the project had helped people find a new sense of community, and they had worked creatively together developing a new respect for each others' abilities and a confidence in their ability to control their own future.

Above all else, in Wyhl and the other sites in Dreyeckland, the local people had discovered that they had power which they could pit against those who sought to tell them what their future would be. The power that they had uncovered was not restricted to their locality or to their country. In other countries too, especially where the governments were particularly constrained in pressing forward their nuclear plans, other citizens were also beginning to flex their muscles. In several countries citizen power would soon be exerted at the national level. In doing so they would find themselves confronting not only the nuclear industry and its supporters within the broader community, but also everywhere the government and its agencies – the state. That the state should take a role in the nuclear conflict by pressing forward nuclear programmes as hard, and sometimes harder than the industry itself, is an important phenomenon. The reasons for this, and its significance for the nuclear conflict, require particular examination.

6
THE STATE AND THE INDUSTRY

The nuclear industry spans the world, but its corporate base is restricted to only a few companies in a handful of countries. Although a large number of small companies are involved in parts of the supply of the nuclear fuel cycle, the crucial area of reactor construction lies in the hands of only twelve corporations. They are located in the USA, France, West Germany, Sweden, the UK, and Japan. Outside these there is the state-owned reactor manufacturing complex in the USSR. It mainly supplies countries within the Soviet bloc, and its successes and problems will be discussed in chapter 10.

The market shares of the twelve companies which build reactors is shown in table 1. Of these corporations there are four in the USA – General Electric, Westinghouse, Babcock and Wilcox, and Combustion Engineering – which take the lion's share. Between 1950 and 1980, 58 per cent of all commercial reactors ordered were made by these four corporations.[1] In this, General Electric and Westinghouse were the prime actors, sharing 44 per cent of the total market between them.

The commanding position of General Electric and Westinghouse is not simply defined by their share of the reactor market. Westinghouse also owns 15 per cent of France's national nuclear company Framatome, and General Electric owns 30 per cent of the UK's national company, the Nuclear Power Corporation. Outside the USA (and the USSR), the commercial reactors produced in all countries except the UK, Sweden, and Canada, are produced under licence from either Westinghouse or General Electric.[2]

The domination of key sectors of the nuclear market by a few

THE STATE AND THE INDUSTRY

Table 1   Reactor supply companies

Country	Reactor supply companies	Ownership of company	Total number of reactors ordered to 1980
USA	General Electric	P	85
	Westinghouse	P	107
	Babcock and Wilcox	P	27
	Combustion Engineering	P	31
West Germany	Kraftwerk Union	P	30
Japan	Mitsubishi	P	10
	Toshiba	P	6
	Hitachi	P	3
Canada	Atomic Energy of Canada Ltd	100% Govt	30
Sweden	ASEA-ATOM	50% Govt	11
UK	Nuclear Power Corp.	35% Govt	42
France	Framatome	30% Govt	48
		Total:	430

P   Private

Source: Calculated from M. Lönroth and W. Walker, *The Viability of the Civil Nuclear Industry*, The Rockefeller Foundation/The Royal Institute of International Affairs, New York/London, 1979, p. 85, table 5; p. 15, figure 3.

very large companies is not surprising. The huge investments involved and the high levels of technical sophistication required make it an arena in which only the biggest, most powerful companies have any hope of survival. Indeed, even with these twelve corporations it is indicative of the financial strength required that several are owned at least partly by the state. Yet despite the size of these corporations the future of their nuclear activities is far from assured. For all of them it has now become realistic to ask how long they can maintain these operations in the face of the present drop in nuclear orders.

THE SLUMP AND THE INDUSTRY

For the reactor vendors the slump in orders has been dramatic. But as can be seen in table 2 it has not affected them all equally. In particular it is the US and West German companies, who up to 1980 had gained 65 per cent of all orders, for whom the collapse has been most traumatic. Their combined orders have plummeted

GLOBAL FISSION

Table 2   Reactor orders

	New reactor orders		Backlog of existing orders for which construction still incomplete		
	1971–75	1976–80	1980	1985	1990
General Electric	32	0	36	4	0
Westinghouse	51	6	48	17	0
Babcock and Wilcox	11	2	16	8	0
Combustion Engineering	15	4	23	11	4
Kraftwerk Union	17	2	14	6	0
Mitsubishi	3	3	5	1	0
Toshiba	2	1	2	0	0
Hitachi	1	1	1	0	0
Atomic Energy of Canada Ltd	12	6	16	6	0
ASEA-ATOM	6	1	3	0	0
Nuclear Power Corp.	0	6	15	8	0
Framatome	19	20	26	0	0
Total:	169	52	205	61	4
Average per year:	34	10			

Source: Calculated from M. Lönroth and W. Walker, *The Viability of the Civil Nuclear Industry*, The Rockefeller Foundation/The Royal Institute of International Affairs, New York/London, 1979, p. 85, table 5.

from 126 in the first half of the decade to a mere fourteen in the second.

Sharpest hit of all have been the US vendors who in 1973 supplied 82 per cent of all reactors sold. By 1975, however, their share of the world reactor market had dropped to 39 per cent, and in 1977 it reached zero when the USA sold no reactors at all.[3] The USA was receiving a diminishing share of a shrinking pie which still had to sustain the same number of nuclear companies.

But even for those nuclear companies hit hardest by the slump the result is not necessarily catastrophic. At least some of them possess a degree of forward momentum due to the orders they received in the 'honeymoon period' before the slump began. This existing momentum is illustrated in table 2. From the final three columns it can be seen that although orders have dropped as at 1980, the construction of some 205 reactors already ordered had not been completed. As noted earlier, the extremely long lead times required to carry through the planning and construction of a reactor add substantially to the costs of reactors and to the present

THE STATE AND THE INDUSTRY

plight of the nuclear industry. It is ironic that these same long lead times help provide the companies with a continuing income in the conditions of a market slump.

The nuclear vendors are still living on existing orders accumulated in the early 1970s. However, this situation cannot last indefinitely. Of the backlog of 205 orders only sixty-one will remain to be completed in 1985, and the number remaining drops to four in 1990. By then the industry will be in serious trouble if new orders have not flowed in. The question that must be answered, of course, is how many orders are needed over this period to keep the industry alive?

In a careful study commissioned by the Rockefeller Foundation and the Royal Institute of International Affairs, Mans Lönroth and William Walker seek the answer to this question. One of their conclusions, published in 1979, is that

> the early 1980s will be critical for a number of reactor suppliers, notably ASEA-Atom in Sweden, Babcock and Wilcox and General Electric in the USA, KWU in Germany, and possibly AECL in Canada. If they fail to gain significant new orders at home or abroad, their reactor manufacturing operations will be placed under severe strain . . . In the event of the de facto moratorium on nuclear ordering in the USA continuing, the other US reactor vendors would also begin to face similar difficulties in the mid-1980s.[4]

The nuclear industry is presently capable of manufacturing 50–60 GWe of nuclear capacity each year. In the second half of the 1970s, however, the orders achieved averaged out at 12.1 GWe per year.[5] During that five-year period the rate declined sharply, with cancellations and deferrals leading to a negative net ordering rate. The decline was most severe in 1979–80 when, according to the Australian AEC, the number of reactors on order dropped from 102 in June 1979 to sixty-five in June 1980. In the USA it slumped from forty-seven to twenty-six.[6]

Lönroth and Walker conclude that present official predictions of a revival of ordering to more than 35 GWe per year are highly unlikely to be met. It would require the companies to overcome 'the majority of economic, social and political obstacles'.[7] Unless this occurs they estimate that the annual average rate of orders in the 1980s is unlikely to exceed, in GWe per year, 2–7 in the USA, 0.5–1.0 in Canada, 3–4 in France, 1–2 in West Germany, 3–4 in

114 GLOBAL FISSION

Japan, 0–0.5 in Sweden, and 0.5–1.5 in the UK.[8] This would yield a combined average maximum order rate of 14–25 GWe per year for the non-communist world over the 1980s – considerably less than the 50–60 GWe which the industry is set up to produce.

Lönroth and Walker's maximum production rate is based on the assumption that the effectiveness of nuclear opposition will decrease over the 1980s, and that the effective moratoria on new orders in the USA and much of Europe will end relatively soon.[9] As has been shown, such a prospect is by no means certain. However, even if it occurs it is likely that the five reactor manufacturers most threatened – Babcock and Wilcox, General Electric, KWU, ASEA-Atom, and AECL – would be forced to withdraw at least temporarily from active manufacturing of nuclear reactors. This does not necessarily preclude them from returning if business picks up. But the longer the slump the harder it is for these companies to maintain the highly skilled teams and other backup needed to produce reactors. As early as 1979 Klaus Bartheld of KWU complained that his company was seeing a reduction in trained personnel. He warned that the situation was even more serious for smaller companies supplying nuclear components.[10]

The withdrawal of these five companies from reactor manufacturing would mean that 30 per cent of the Western world's reactor manufacturing capability would be lost. This, however, need not necessarily concern the remaining companies who would thereby gain an enhanced marketing position. Admittedly, for the remaining companies to survive as active reactor builders into the 1990s an ordering rate of 14–25 GWe would probably have to be achieved. But if it were, they would have greatly increased their power by concentrating the control of the market in considerably fewer hands.

Even with the above scenario for the continued health of the nuclear industry there is a problem. It assumes that withdrawals of the more vulnerable companies could occur at the same time as nuclear opposition decreases. However, these two are not independent. Even the withdrawal of one or two of the nuclear vendors from the market would greatly strengthen the spirits of nuclear opponents and weaken those of the proponents of nuclear power. Although it is conceivable that there could be a sufficient slump in opposition to offset this tendency, it certainly cannot be guaranteed. Should the level of opposition instead stay constant or rise further then it is possible that additional suppliers would be forced

THE STATE AND THE INDUSTRY

Table 3 State involvement in different stages of the nuclear fuel cycle

	Enrich-ment	Fuel fabrication	Reactor & supply	Repro-cessing	Waste disposal	Electrical utilities
USA	S	P	P	–	–	P
West Germany	SI*	P	P	SI	S	SI
Japan	SI	P	P	S	–	SI
Sweden	–	SI	SI	–	–	SI
Canada	–	P	S	–	S	S
UK	S	S	SI	S	S	S**
France	S	SI	SI	S	S	S**

P   Private
SI   State involvement in a substantial number of companies
SI* State involvement via utility ownership
S   State owned
–   Undecided or not relevant
S** Strong central authority of national government over decisions

Source: Adapted from M. Lönroth and W. Walker, *The Viability of the Civil Nuclear Industry*, The Rockefeller Foundation/The Royal Institute of International Affairs, New York/London, 1979, p.12, figure 2.

from the market. If this were to occur, warn Lönroth and Walker, 'say with the withdrawal of the remaining US suppliers, the risk of irreparable damage would be considerably greater, and an adequate response to any increases in demand in later years could no longer be assumed to be forthcoming from the manufacturing sector'.[11]

Even if the opposition were to decrease somewhat, survival of the nuclear industry cannot be guaranteed. The scenario in which competition for orders allows the more strongly competitive companies to eliminate the weak, thereby enhancing the chances of survival for the remaining companies, may not occur. This is because the assumption of a 'free' competitive situation is far from the truth. In four of the six home countries of the nuclear reactor manufacturers (Canada, Sweden, the UK, and France), the companies are owned either partly or wholly by the state (see table 1). As state actions are not governed solely by questions of profitability, it is possible that other considerations such as national independence or prestige may intervene to stop the withdrawal of these companies. In the long term this could prevent early withdrawals which later would weaken the industry even further.

The close ties between the state and the industry introduce a further element of uncertainty into the future of the nuclear indus-

try. A glance at table 3 is sufficient to confirm that in all countries the state is deeply integrated into the activities of the nuclear industry. The outcome of the battle over nuclear power will therefore depend partly on the state's motivation and ability to support the nuclear industry.

THE ROLE OF THE STATE

Since its inception the nuclear industry has been characterized by an exceptionally intimate relationship with the state. One explanation might be the profound relationship between commercial nuclear technology and its military antecedents. The bombing of Japan was the industry's birth cry and the military was its midwife. Nevertheless, it might have been that once the scientific possibility of nuclear energy had been so dramatically suggested companies producing electrical equipment would have jumped to harness it. But this was not the case. They instead showed extreme reluctance to divert their energies to the new technology.[12] In particular they felt unhappy about the financial risks and unsure about the returns. As J. Baner wrote in 1949:

> The relative over-all economy of atomic energy will thus probably be limited to the fuel item, and this for a large modern boiler plant comes only to about 1.5 mills (0.15 cents) per kilowatt-hour . . . If this view is correct, then at best the margin of total advantage for atomic energy will be small.[13]

The first investigation of nuclear power, by the Tolman group in the USA in 1944, found that it would be a long time before the development of nuclear reactors could be useful for civilian generation of electricity. Of this period one analyst, S. Kuhn, wrote:

> Such a timetable did not interest the Westinghouse executives . . . they were influenced largely by short-run considerations of immediate costs and returns normal to profit making firms. The AEC promised high costs and long-deferred profits. Despite fervent, informal suggestions from Rickhover that they get into nuclear power, the men at Westinghouse maintained their disinterest.[14]

First in the USA and then elsewhere, it was the military which

THE STATE AND THE INDUSTRY 117

created the early market for nuclear reactors. In particular, the possibility of constructing reactors which would take submarines on long undersea voyages without the necessity of surfacing entranced the navy. Westinghouse was the first to enter this potentially lucrative market. General Electric soon followed. By 1962 Westinghouse had built thirty-six reactors, of which fully thirty-four were for ships and submarines. Of General Electric's reactor production three were for submarines and only one was for civilian use.[15] As late as 1972 three quarters of the reactors produced in the USA were for powering ships and submarines.[16]

There was no clear commercial imperative to go ahead with the construction of commercial power reactors which would have to be both much more sophisticated and bigger. Instead of fulfilling a military need for which there were no competing technologies available, these would have to compete against large coal-burning stations at a time of coal surplus and rapidly improving coal-burning technology. As early as 1955 it was clear that the first nuclear reactor plants could not be economic. Ralph Cordiner, president of General Electric, years later expressed the doubts then current: it 'is a very difficult and expensive technology. It was – and is – possible to lose millions on premature ventures or endless engineering studies'.[17]

In the USA private ownership of nuclear plants was encouraged by the 1954 Atomic Energy Act. But the relevant corporations were frankly unenthusiastic.[18] The same was true in West Germany when its unofficial nuclear five-year plan was floated three years later.[19] In the USA, the birth-place of the nuclear industry as it is now known, the government sought to overcome industry reluctance with a smorgasbord of overt and more subtle incentives. They included all the research and development from the Manhattan Project made available free of charge, an offer of further research and development assistance in 1955, design assistance in 1963, and the waiving of fuel charges for five years for utilities buying reactors. They also included bonuses for uranium exploration, a guaranteed minimum price, subsidized enrichment, and the operations and research of the AEC. By the end of 1962, $1.25 billion had been poured into nuclear technology by the AEC, more than double the cost to private industry.[20]

When even these lavish subsidies failed to draw sufficient enthusiasm from the corporations the government tried a different tack. As A. Roberts put it in his penetrating analysis of the period,

118   GLOBAL FISSION

'when the carrots failed, the government resorted to the stick'.[21] At
the time AEC chairman Lewis Strauss put it hardly less bluntly:

> It is the Commission's policy to give the industry the oppor-
> tunity to undertake the construction of power reactors. How-
> ever, if industry does not, within a reasonable period of time,
> undertake to build types of reactors which are considered
> promising, the Commission will take steps to build the
> reactors at its own initiative.[22]

It was a threat, and behind it lay an historic enemy of the US
electricity utilities, the spectre of public power. Some 80 per cent of
US electricity production is supplied by privately owned utilities.
The limited monopoly granted to them had not gone unchallenged.
It had been subjected to attack by numerous politicians as being a
violation of free-enterprise principles.[23] The alternative put forward, since true competition was impossible, was to make electricity production publicly accountable by vesting ownership of it
in the state. At least one successful precedent, the government
owned Tennessee Valley Authority, already existed and the utilities
were fearful of any moves to further erode their position. Confronted with an offer that it could not refuse the power industry
began reluctantly to move towards the construction and use of
commercial nuclear power reactors. Once Westinghouse and
General Electric began to enter the reactor construction field other
firms traditionally concerned with boiler construction and the like
joined in fearful of being left behind.

Together with the subsidies the threat of public power was sufficient to force the industry to move into nuclear power at 'a faster
pace and on a larger scale than the immediate economic outlook
warranted'.[24] What was the motive behind the state's actions?
There may be no simple answer to this question. But one
immediate suggestion is that despite the economic uncertainty the
state perceived the development of commercial nuclear power as
advancing its overall interests.

NUCLEAR POWER AND THE INTERESTS OF THE STATE

There were of course other motives. There was the current of
technical optimism flowing from the successes of the Manhattan
Project and other war-time research which suggested to many that

THE STATE AND THE INDUSTRY

no technical problem could withstand a concerted attack from science, especially given sufficient support from the state. There was also the enthusiastic lobbying of the highly influential scientists who had taken part in the Manhattan Project. Later there was the institutional momentum of the AEC. However, behind this lay another important ideological factor much reinforced by the war. It was the belief that in the future, political and economic power would rest with nations in proportion to their grasp of what technology and science had to offer and their speed in developing it. From that perspective, whatever the technical problems, to face the problems presented by nuclear power and hurdle them seemed undeniably in the interests of both the national corporations and the state. There was no time to hold back wondering what might lie on the other side.

That perspective was reflected in the statement by Thomas Murray, an AEC commissioner, on hearing that the USSR had successfully tested its first hydrogen bomb. It was, he said,

> less dangerous for the free world's hopes for the end of godless tyranny than would have been the case if the Soviet had announced that day a practical, industrial nuclear power plant — and was that day offering nuclear power technology in exchange for uranium . . . Once we become fully conscious of the possibility that power-hungry countries will gravitate towards the USSR if it wins the power race, regardless of the aversion they have for tyranny, it will be quite clear that this power race is no Everest-climbing kudos-providing contest.[25]

Underlying this immoderate comment lay a conviction that US national prestige and power are linked to victory in the 'power race'. As J. Hogerton, when writing on the era, put it: 'the thrust for an accelerated nuclear programme was not sparked by any specific technological development, nor did it develop solely from the buildup of economic pressure within the US power industry. Considerations of national prestige and foreign policy were of central importance'.[26] These considerations were not only prevalent in the USA. As two scientists prominent in the West German nuclear programme in 1969 made clear, 'It is no longer military power which proceeds parallel with political power. Rather we believe that political power is acquired today to a large extent from the pursuit of civilian technological projects. The capacity to act politically derives from this pursuit'.[27]

The terms in which national interest demanded nuclear power to be pressed forward at a rate faster than commercially warranted no doubt varied from country to country. In the USA, the heartland of the nuclear industry, a crucial component was the desire to exert national and thus technological leadership. Added to this was the promise of substantial returns although, once stripped of its mantle of technological optimism, that promise was obscured by layers of unsolved problems. But in addition there must have been some understanding of nuclear technology's potential for supporting the US national interest in a different way. That was through its role in reinforcing the international terms of trade.

THE TERMS OF TRADE

Advanced industrialized countries depend for their high levels of consumption on a highly asymmetric trade flow. This is particularly evident in their trade with the 'underdeveloped' countries which make up the Third World. Among the countries of the capitalist world in the 1960s the major proportion of exports, 75 per cent, from the advanced industrialized countries such as the USA, Australia, and Western Europe, were to other advanced industrialized countries. By contrast, only 21 per cent of exports from Third World countries were to each other. The bulk of these exports, 73 per cent, went from the Third World to the advanced industrialized countries of the capitalist world.[28]

With its low level of industrialization the Third World has historically found itself in such an economically disadvantageous position that even to pay for a meagre inflow of manufactured goods it has to produce and export vastly more of its natural resources than it consumes. It produces 36 per cent of the world's copper, 20 per cent of all zinc, and 30 per cent of all superphosphate, but consumes only 3 per cent, 7 per cent, and 6 per cent respectively of these resources.[29] Its trade in other resources follows this pattern, although less dramatically, with few exceptions. This is the basis of the material wealth of the advanced industrialized countries. Nowhere is this more obvious than in the trading relations of the richest country in the world with the highest per capita levels of consumption – the USA.

It has at times been considered a little tasteless to mention the basis on which industrialized society's wealth is built, but not all

THE STATE AND THE INDUSTRY

have shared this reticence. Former US president Richard Nixon is quoted as saying:

> There are only 6 per cent of the world living in the US, and we use 50 per cent of all the energy. That isn't bad; that's good. That means we are the richest, strongest people in the world. That is why we need so much energy and may it always be that way.[30]

Although not all would share this relish at the fact, the levels of consumption are indeed so distorted.

Despite the claims of some immoderate proponents of nuclear power, the riches in countries like the USA did not accrue simply because of the consumption of energy. Rather, the extraordinarily high consumption of energy is a symptom of the same asymmetric terms of trade which enable the USA to consume an equally high level of almost all other resources. This situation did not arise by accident. It was produced by a complex web of actions, strategies, and alliances of the corporations, national governments, and international agencies which make up the modern global economy. Nor is the situation static.

In the 1970s there was a tendency for transnational corporations to reorganize their operations. Taking advantage of a lack of environmental protection laws, often prohibited unionism, and an availability of subsistence labour, many re-sited the labour-intensive parts of their manufacturing operations in Third World countries. There, small but rich elites profit from the outward flow of resources and cheap mass produced commodities. In exchange they receive a much smaller volume of resources in the form of sophisticated manufactured goods which supplies them with a standard of living equal or superior to that of their counterparts in the advanced industrialized countries. It need hardly be said that these benefits are not available to the majority of Third World inhabitants, the condition of many of whom is seriously deprived and deteriorating. In the USA and the other advanced industrialized countries, corporations receive and process the incoming flows of raw materials and mass-produced items and sell the resulting commodities at often substantial profits. It need also hardly be mentioned that these goods and profits are also not distributed evenly.

For the USA nuclear power technology was capable of playing

an important role in relation to these asymmetric terms of trade. Because the technology is highly sophisticated it is capable of being produced by only a few of the most industrially advanced countries in the world. Because it is enormously expensive to assemble the necessary backup facilities, to carry out the necessary research and development, and to attain the scale of production necessary to bring the price of key components to anything like commercial levels, the technology is capable of being produced by only a very small number of very large corporations.

For most countries nuclear technology is impossible to duplicate. Because of this, the first receipt of a nuclear reactor is not unlike a first dose of some addictive drug. It may produce desirable effects for the user, but the habit demands continuing servicing. In the case of nuclear power the USA pre-eminently, but also a handful of other countries, seem likely to be able to reserve for themselves the role of supplier.

For those countries unable to produce the major nuclear fuel cycle facilities, fuel has to be purchased, reactors maintained and serviced, parts purchased, waste handled and stored, the reactor finally decommissioned, and often loan payments made. All this represents a constant outflow of money to the reactor suppliers. Since the USA controls much of the market the major proportion of these money flows are destined for that country. In return, the dependent country is forced to sell other goods to pay for the nuclear facilities. In the case of Third World countries the bulk of what they have to sell is their natural resources.

For Third World countries this is only part of the process by which their nuclear technology could reinforce their dependency. As J. Sabato, former manager of technology for the Argentinian AEC points out, another 'negative aspect' is that

> Nuclear power is perhaps the most striking example of how the introduction of a new technology leads a developing country automatically to become a market for virtually the whole spectrum of the developed countries' technology. Once the door is open to one technologically advanced country the others follow, resulting in more technological dependence and cultural alienation.[31]

The market for these goods is of course generally the elite, which explains much of why a Third World country enters into the purchase of nuclear technology. A more direct beneficiary is the

THE STATE AND THE INDUSTRY 123

technically trained elite which will manage the technology. From this develops a 'nuclear lobby' which, ironically, has at times argued that the purchase of the technology is necessary to help lift their country from its position of dependence.[32]

In these ways nuclear technology reinforces the terms of trade between suppliers and purchasers and strengthens the position of existing elites. In that sense it has political relations 'build into it'. While this was probably not understood in detail when the US government moved so decisively to press forward the development of nuclear power, its potential was probably understood in outline. As Sabato notes: 'The development paradigm in the early 1950s equated development with progress and modernisation . . . The developing nations believed that if they did not invest in techno-logically sophisticated programs the gaps would widen'.[33] No doubt this view was not restricted to the planners in the Third World.

These aspects of nuclear technology are of course most clearly relevant to relations between the USA and countries of the Third World. However to a lesser degree they also apply to the relations between the USA and many less powerful countries of the advanced industrialized world. Another aspect of the technology which could enhance the USA's power in relation to these coun-tries stems from the crucial importance of electricity in these countries' production systems. For a country whose electrical supply has become substantially dependent on US nuclear techno-logy there is a consequent decrease in its autonomy from US national policy. It stems from a power that every drug supplier knows well: the great influence which can be exerted through an implied or overt threat of limitation or withdrawal of supply.

During the 1970s the potential of such threats of withdrawal was demonstrated several times. One of the most notable was associ-ated with the provision of enrichment services. In 1972, when hopes for nuclear power were at their peak, the US AEC supplied over 90 per cent of the world's enriched uranium to fuel commer-cial reactors. Moves were then afoot within the USA to create a private US enrichment industry. Once again, however, this was not left to face the international market unassisted. Instead, the virtual monopoly held by the USA over enrichment capacity was used to ensure a market for the intended private interests.[34] For six months to May 1973 the AEC suspended all new enrichment contracting. Then it announced that all customers interested in contracting to

purchase future enriched uranium would have until June 1974 to submit their proposals. After that the AEC would close its order book.

The new contracts were to be under very different arrangements than before. Previously they had been 'flexible'. Now they would be long-term fixed commitment contracts. Instead of requiring only 180 days notice before delivery the AEC now required purchasers to buy a fixed supply of fuel over ten years. If the purchaser should default then he would be liable to a penalty of 50 to 75 per cent of the total enrichment service costs. Significantly, allowance was made for transfer of the contract without penalty to a private US enrichment company. The would-be purchasers had nowhere else to turn, and the response was galvanic. Over the following year more than 145 proposals were made to the AEC, more than doubling its holdings of contracts. Its capacity to supply was exceeded by 25 per cent, so when it finally accepted the proposals in August 1974 it made the forty-five foreign customers' contracts conditional on the USA beginning to recycle plutonium in its own reactors.

The threat of withdrawal had been successful in creating the kernel of an assured market for the still to be created private US enrichment industry. Later, plummeting orders for nuclear reactors reduced interest in private enrichment. Nevertheless, the USA had served an object lesson to other nations on the effectiveness of withdrawing supply. The lesson was not lost on other countries which set about trying to build their own reprocessing and enrichment plants to increase their independence. Later, in April 1977, when President Jimmy Carter attempted to head this off for the announced reason that it would lead to proliferation of nuclear weapons, his move was met with less than enthusiasm.[35] Said Egon Bohn, West Germany's then Social Democratic Party leader, 'A new form of colonialism is no answer'.[36]

Both the initial threat and the subsequent reaction illustrate the attraction that nuclear technology has as a tool of US power. The capital intensiveness and sophistication of the technology imparts to it a particular social role. It reinforces the power of the few nations and organizations capable of making it. This understanding was not restricted to US agencies. The mere possibility that another country might obtain a decisive advantage over them was sufficient to create pressure for the governments of other countries also to press ahead with nuclear programmes at a pace that might other-

THE STATE AND THE INDUSTRY   125

wise not have been justified. The result, combined with other factors mentioned earlier, is a nuclear industry tightly intertwined with the state. It represents a mutually supportive relationship between two megalithic structures, each of which stands to gain.

The equating of technical mastery with political strength, the relationship between the commercial and military nuclear establishment, and the role of nuclear technology in maintaining favourable terms of trade and political power, all mesh with the nuclear industry's need to have its shaky economic foundations supported by the state. The establishment of this relationship is now history. However, as the industry slides deeper into its present trough it is natural that its intimately bonded partner, the state, will seek to reverse that trend.

Some of the ways the state may act to support the nuclear industry have already been discussed. That is, to reshape the economics of nuclear power by granting more subsidies, or by passing regulations to decrease the impact of citizen concern on the industry's cost structure. However the possibility of such action is constrained by the strength of the opposition to nuclear power. In the attempt to enhance the prospects for nuclear power the state may thus find itself manoeuvring to constrain the effectiveness of the opposition itself. Nowhere has this been more clearly illustrated than in the battle that ensued in West Germany after the initial success of the occupation of Wyhl.

THE STATE AND THE OPPOSITION

The battle over nuclear power had been initiated in West Germany by the occupation of Wyhl. The legitimacy of the local people had exceeded that of the government, and in the end the occupiers had won. Bewildered, the government turned to the international research organization the Batelle Institute for an explanation. One million dollars later, the institute reported. The answer was simple. Not surprisingly, it found that country people who believe their lives and livelihood threatened by a nuclear project are even more likely to actively resist than city dwellers whose jobs are less dependent on the preservation of a healthy environment.[37] What the report proposed to do about this 'problem' was not revealed. Nevertheless the government responded by launching an extensive 'information campaign'. In this particular case it had more the character of a propaganda broadside. Free booklets, eight series of

126 GLOBAL FISSION

newspaper advertisements totalling 76 million copies, television advertisements, and information packets flooded the community. Some $2 million were poured into the campaign in 1975–76 and a similar amount earmarked for the next year.[38] But neither this nor a 'public dialogue' launched by Federal Minister for Research and Technology Werner Matthöfer, acted to dampen concern. Instead it was heightened. In order to protect its nuclear plans the state soon began to search for other more direct tactics.

In West Germany the maintenance of public order is principally the responsibility of state governments. However the federal government also controls the 'border police'. When the opposition finally found itself directly confronted with the force of the state it was not merely local but also federal police who faced them. The first major confrontation was at Brokdorf.

On 25 October 1976 permission was given to begin the first stage of construction of the proposed Brokdorf reactor on the Elbe River, only 80 kilometres from Hamburg. But five days later, when the local coalition of citizens action groups, the BUU (*Bürgerinitiative Umweltschutz Unterelbe*) held their demonstration in opposition to the decision, they found themselves in a new situation. At the plant site, facing the 8 000 demonstrators, stood a high barbed-wire fence, and around that was a water-filled moat. It was guarded by several hundred police armed with shields, clubs, chemical repellents, and dogs.

Initially even these fortifications proved inadequate. A section of the fence was torn down, and some 2 000 demonstrators occupied the site. The result was very different from that in Wyhl. Later on the same day the police attacked in force. Several demonstrators were injured and all were driven off. When the demonstrators returned on 13 November, the 20 000 people found themselves facing a virtual fortress. Broad ditches, strengthened fences, and large amounts of barbed wire formed an apparently impregnable barrier between them and the site. Roadblocks confronted them as far as 11 kilometres out from the site. Overhead a helicopter hovered, dropping tear gas on demonstrators who deviated from the body of the crowd, and below a solid mass of well-equipped police waited.[39] The strategy of the state seemed clear enough. It had already been articulated by Matthöfer in the middle of his 'dialogue': 'We have to build this plant, and we will build it. The rights of a few must be sacrificed for the benefit of the nation as a whole. Otherwise we will fall into anarchy'.[40] It was not the first

THE STATE AND THE INDUSTRY

time that Law and Order had been raised as an effective slogan in West Germany.

This time the forces of the state proved more than adequate. Those who tried to penetrate the fence were beaten back. As the crowd retreated in the evening, pushed along by the police and nursing their injuries, arguments were already beginning to flare over what strategic response to the naked power of the state was possible. The issue temporarily split the opposition. The next time two demonstrations were held against Brokdorf. One was organized by the national BUU and was held at Itzehoe, about 20 kilometres from the site. The other, organized by the Hamburg BUU at Wilster, some 5 kilometres from the site, contained groups prepared to attempt to force entry. But that preparedness proved academic.[41] For the two demonstrations, each of 20 000 to 30 000 people, the massive police presence made it impossible to implement any such desire.

At subsequent demonstrations the state carried out its strategy with even greater determination. March 1977 saw 5 000 demonstrators beaten back at Grohnde, and 100 arrested.[42] At Ohu, at Easter, 2 000 protesters found themselves confronted by a miniature army of some 7 500 police equipped with eight helicopters, 1 400 vehicles, including armoured cars, and twenty water cannons.[43] A few months later, in September, as 50 000 demonstrators marched on the site of the Kalkar prototype fast-breeder reactor, even larger police offensives took place.

By now the state's strategy seemed to have extended to attempting to prevent protesters from reaching their planned point of assembly. According to one commentator,

> The nationally coordinated stop and search operation began the night of the 24th and continued into the next day. Its most striking characteristic was its massive scale: 146 000 individuals and 74 485 vehicles were stopped and thoroughly searched. The French, Belgian and Dutch borders were meticulously controlled. With its ever widening definition of 'weaponry' the government claimed a booty of no less than 8 000 objects . . . Over 10 000 would-be demonstrators – by some estimates 20 000 – of whom 141 were arrested, never reached Kalkar.[44]

Although the police claimed that dangerous weapons, including pistols, were seized, most of what was confiscated was classified as

'passive weaponry' and included scarves, gas masks, and motorcycle helmets. Among the more impressive symbols of state power was the helicopter used to halt a train carrying demonstrators who were forced to submit to a search and then left to complete the journey on foot. The organizers had intended the demonstration to be a show of numbers and, given the moat, concrete fence, ring of police, armoured cars, helicopters, and water cannons making up the $1.2 million operation, anything else would have been impossible.

Even this extraordinary display of power had not prevented the assembly of a large and impressive demonstration of opposition to nuclear power. In other countries, as was shown earlier, even the growth of such 'passive' action can be part of a political process which may paralyze or dismantle a nuclear programme. Therefore the state may seek not merely to prevent the physical disruption of nuclear construction, but also to weaken the opposition itself. Since the information campaign had proved a failure in this, the West German government tried another tactic.

At Wyhl the government had attempted to label the opposition as 'communist subversives' but without success. The legitimacy attached to local adults and children prepared to withstand police water cannons to protect their livelihood made such a claim laughable. It was a different matter, however, with the demonstrators who had converged on Brokdorf from all over West Germany. Many were young and looked it. They were not identifiable and their genuine concern could more easily be obscured. For many Germans it was easier to accept the claim that these were 'terrorists'. That belief was reinforced because some of the demonstrators arrived equipped with helmets, gloves, wire cutters, and occasionally even gas masks. The government acted with dispatch to capitalize on this and destroy what legitimacy the demonstrators possessed and, by implication, that of the entire opposition movement.

West Germany had been subjected to a series of kidnappings and murders by the small Red Army Faction. Only three weeks before the first Brokdorf demonstration, Hans Schleyer, head of West Germany's Employers Association, had been kidnapped by the group. The government used this as a device for associating all opposition with 'terrorism'. In this way the government generally was able to render legitimate the increased application of social

THE STATE AND THE INDUSTRY

control measures. The effectiveness of the tactic was reflected in an editorial in the *Frankfurter Allgemeine* newspaper on 14 May 1977:

> it takes only 100 organised terrorists to drive the State into a corner. And if the communist groups which up to now have drawn attention to themselves with occupations of town halls, nuclear power plants and building sites, but have refrained from murder, one day attempt to link up with terrorist groups? . . . This is sufficient reason for the State to prepare itself; not only by increasing the number of police and upgrading equipment, but also by changing the law where up to now it has been too easy for the terrorists to protect themselves and too difficult for the State.[45]

For some opponents of nuclear power, the mere label of 'terrorist' might well be sufficient to intimidate them from any further expression of their opposition. In West Germany the intimidating effects of labelling all opposition in this fashion does not end there. There are state powers which may be applied to 'subversives'. One of the most insidious of these is the 1972 *Berufsverbot* law. Meaning literally 'forbidden to work', under this law any state employee may be placed under surveillance and investigated. If it appears to the Offices for the Protection of the Constitution that they are in some sense acting contrary to the constitution they may be denied employment unless they can prove otherwise. Public servants, school teachers, train drivers, university academics, and postal clerks may all find themselves under scrutiny for alleged indiscretions.

Dr Andreas Dress, a leading campaigner against this legislation's applicability to academics, has on file some 4 000 cases of interrogations carried out under this law.[46] In 3 000 of these cases the subjects were either reinstated or forced to find employment in the private sector. However, as Dress points out, for academics there is often no private sector to which to go. So powerful has been international concern over the existence and enforcement of this law that the German government has spent some $37 million (DM 70 million) on a campaign to counter this opposition.[47]

Activities likely to precipitate investigation include attending a demonstration, writing an article in a supposedly subversive magazine, or associating with someone considered to be subversive. Since July 1976 some 500 people are believed to have been

130 GLOBAL FISSION

dismissed, but estimates for the number who have been investigated go as high as one million.[48] Whatever the actual numbers the effect is to create an extremely intimidating atmosphere, especially among the technically trained members of the nuclear opposition who are more likely to be in public sector jobs. The experience of a Dr Traube, a physicist who was surveilled, wire-tapped, dismissed from his job, and finally found innocent, does not reduce this anxiety.[49] Nor does the experience of Hans Krause, a mechanical engineer who expressed doubts about the fast-breeder reactor programme. Because of his membership of the works committee at his place of employment he has so far managed to survive attempts to dismiss him.[50]

THE NUCLEAR STATE

The power of the state is turned more overtly against dissent, and in particular nuclear opponents, in West Germany than in most other Western countries. However, as Robert Jungk has documented, the trend for state control to increase with the spread of nuclear power is quite general.[51] There are two reasons why this should be so. The first is that the nuclear enterprise is the project of very large corporations intimately connected with the state. This gives a strong political motivation to protect it from any activity which may obstruct its progress. The second is that nuclear technology is extremely vulnerable. It requires extraordinary measures to protect the community from its unusually severe hazards. Once in place, the consequences of sabotage or theft of fissionable material are conceded to be unacceptable. For this reason the technology legitimates the arguments of those within the state who may wish, for any number of reasons, to see further encroachments on civil liberties and stronger measures of social control. US energy analyst D. Hayes encapsulates the problem in the following 'simple paradox': 'Commercial nuclear power is viable only under social conditions of absolute stability and predictability. Yet the mere existence of the fissionable materials undermines the security that the technology requires'.[52] The attempted solution is to restrict anything which might threaten that stability. Already a trend towards this can be seen in several countries.

It is not easy to identify the social effects of nuclear power. They do not happen suddenly, and they are not easily untangled from other causes for a tightening of social control. What occurs is what

THE STATE AND THE INDUSTRY

London barrister P. Sieghart calls a 'creep effect', in which 'the minimum security measures found to be necessary in the light of experience . . . will expand and escalate so slowly that the subtle erosion of civil liberties will not become visible until it is too late to do anything about it'.[53] One of the earliest expressions of concern about this possibility was a 1975 report by Russell Ayres on the 'civil liberties fallout' of policing plutonium.[54] Other reports, including the UK's Flowers Commission,[55] the Australian Ranger Uranium Inquiry,[56] and the US Ford–Mitre group, [57] all stressed the extensive security measures that would have to accompany a major nuclear programme, especially where reprocessing was involved, and also the consequent danger to civil liberties. In April 1977 the International Commission of Jurists, a group of leading lawyers concerned with human rights, expressed 'grave concern about the dangers to human rights created by nuclear weapons and materials, and nuclear power programmes'.[58]

The way in which the vulnerability of nuclear technology may make infringements on civil liberties seem legitimate was illustrated in *Nuclear Prospects*, a careful study by R. Grove-White and M. Flood published in the UK in 1976. They examined the procedures that would be necessary to ensure the security of the nuclear programme projected by several government agencies for the UK in the year 2000.[59] According to these some 130 GWe of nuclear power would then be operating, split evenly between breeder and conventional reactors. This would involve 1 700 cross-country shipments of over 80 tonnes of plutonium each year. The study concludes that necessary security could include the extension of security vetting to 20 000 electrical supply workers, an armed nuclear police force with special powers and at least one third the size of the present London metropolitan police, the surveillance and infiltration of conservation and amenity groups, police powers for house to house search without warrant, and under certain circumstances the right to use torture to extract critically urgent information.[60]

This conclusion might at first sight appear to be extreme and alarmist. But it is not without suggestive precedents. As the authors point out, in 1974, after a spate of IRA bombings had destroyed pubs in Birmingham, the British government passed the Prevention of Terrorism Act which entitled police to arrest without warrant. Under that act some 1 330 arrests were made but only sixty-five of these persons were ever charged. The lesson would

seem to be twofold. First it is clear that when such legislation is available it is used, and not always against the guilty. Second, if this severe piece of legislation seemed justified and reasonable to protect the community against bombs that had caused a handful of deaths, would not more stringent measures seem reasonable and urgently desirable if a group had illegally obtained sufficient plutonium to make a nuclear bomb, had delivered credible plans for its construction, and was threatening to explode it with the possibility of causing tens of thousands of deaths?

Several studies have demonstrated that the 'creep' towards this situation has already begun. In the UK there is already the Atomic Energy Agency's armed Special Constabulary which operates on a constitutionally unique basis with its actions shrouded from the public and parliament for security reasons. Over 1976–79 its strength was increased by 50 per cent to 600, although there was no growth in civil nuclear capacity.[61] In the USA, the Federal Bureau of Investigation (FBI) already takes an interest in the anti-nuclear movement. FBI Deputy Assistant Director James Adams has confirmed that the FBI investigates groups such as the Clamshell Alliance or the Abalone Alliance who have occupied nuclear facilities. Additionally he notes, 'I would be incorrect if I took the position that we don't have anything in our files dealing with the peaceful uses of nuclear energy pro or con'.[62] In fact the FBI provides the NRC and the Department of Energy with advance warnings of individuals and groups which it believes may pose a threat to nuclear facilities, and FBI files include the results of over 2 000 investigations per year carried out for these US agencies.[63]

Many more details of instances of increasing surveillance of individuals, including nuclear workers and the infiltration of anti-nuclear groups, which have been carried out in the USA in the name of nuclear security, are included in a detailed report by D. Warnock published at the end of 1978.[64] However these should come as no surprise. In 1974 the AEC's until then secret Rosenbaum Report was released to the press. It concluded that

> The first and one of the most important lines of defense against groups which might attempt to illegally acquire special nuclear materials to make a weapon is timely in-depth intelligence ... As part of this there must be '... an ongoing analysis of the attitudes of the people in the plant and the community around the plant'.[65]

THE STATE AND THE INDUSTRY

Similarly the Mitre Report, a year later, advised that any group which organizes large demonstrations should be suspect since it could 'attract extremists to its cause'.[66]

For some people, the fact that what they say may be recorded on an anonymous FBI file may be enough to intimidate them from taking active part in a group opposing nuclear power. For others this would not be a barrier. However, this present action may be only the thin edge of a more substantial wedge. In some other countries nuclear technology can already be seen to be legitimating more potent attacks on the rights of the individual.

An extreme case is given by South Africa, which in 1978 set about extending the powers of its already far-reaching Atomic Energy Act. The new powers would not only restrict publication of information on South Africa's uranium reserves, output, contracts, and processing, but would also prohibit 'speculation by the press' on such subjects as South Africa's atomic energy and uranium enrichment programmes.[67]

A less extreme case is offered by Australia where in 1978 the government faced a possible confrontation with unions applying bans on the handling of uranium. One strategy already available was the use of the Atomic Energy Act enacted years earlier for military projects. This act provided tough penalties for interfering with uranium projects. So harsh were they that the Ranger Inquiry reports had specifically recommended that this act should not apply to commercial uranium mining.[68] However the government found the act eminently satisfactory for its purposes. It authorized mining under it in May 1978.

Under the act and its associated Approved Defence Projects Protection Act, any person who boycotts, threatens to boycott, hinders or obstructs uranium mining, or who advocates in speech or writing the hindrance of uranium mining may be fined $A10 000 and imprisoned for twelve months.[69] Ancillary legislation enacted included seven-year prison sentences for anyone committing 'any action which interferes with or impairs the efficiency of' the projects, twenty-year prison sentences for any employee who communicates any 'restricted information' – a very broadly defined category – and the ability to declare an effective state of emergency if any harm might stem from a 'situation resulting from a nuclear activity'.[70] Editorialized the prestigious Melbourne *Age* newspaper:

The original legislation was regarded by those concerned with

civil rights as repressive even when it applied only to defence projects and national security. Now that the restrictive and punitive provisions may be extended to cover highly controversial commercial enterprises, it becomes potentially much more repressive and repugnant.[71]

These examples simply evidence a trend that accompanies the use of nuclear technology. Further evidence of this trend was provided on 29 October 1979 when the International Atomic Energy Agency announced in Vienna that after two years of negotiation fifty-eight countries had agreed on the text of the first international Convention on the Physical Protection of Nuclear Material. A crucial element of this agreement was to create new categories of international crime − a crime which can be tried in any country regardless of where it is committed. Under such a category, a US citizen, for instance, accused of a nuclear-related offence could be tried for it in Iran. Understandably, the placing of crimes in such a category is very rare, and until then was limited to crimes such as genocide.

The agreement creates eight new categories of international crime, including not only theft or fraudulent obtaining of nuclear material, but also threats to steal nuclear material 'in order to compel a person to act'.[72] The vagueness of this and other clauses is worrying given the seriousness of such legislation. Perhaps more serious still are Articles 5 and 6 which provide for cooperation between the countries to exchange information relevant to such offences which they undertake to keep secret. Says Sieghart, 'the moderate analyst will wonder what effect the gradual installation of those measures − and of future and even more stringent ones, if those too are found one day to be necessary − will have on the open society in which he lives'.[73]

The trend then that accompanies the use of nuclear technology is towards the increased power of the state, and the decreased rights and power of the individual. The point is not that such measures are wrong. On the contrary, the vulnerability of nuclear technology often makes them supremely rational. In that sense nuclear technology contains within it certain political relations. Once deployed, its very nature legitimates a constriction of civil liberties and a growth in the power of the state that otherwise might seem sufficiently unreasonable for it to prove politically impossible to implement.

THE STATE AND THE INDUSTRY

In addition, as has been seen in the West German example, the opposition to nuclear power can itself be a catalyst for increased social control by the state. This is partly because billions of dollars have already been poured into the nuclear enterprise by both the state and the nuclear industry, and partly due to the reason that those billions were invested. This was, at least to some extent, because nuclear technology can play an important role in sustaining the asymmetric trade flows that are the basis for the wealth of the advanced industrialized countries. More generally, however, the reason is that nuclear technology is supremely compatible with the needs of the major forces shaping the global economy.

NUCLEAR POWER IN THE GLOBAL ECONOMY

Since World War II the development of the global economy has been characterized by the increasing size and power of the state, and the increasing size of, and economic concentration in, a few gigantic transnational corporations. Both these trends are well documented elsewhere.[74] As B. Miliband writes in his seminal book *The State in Capitalist Society*:

> More than ever before men now live in the shadow of the state ... It is for the state's attention, or for its control that men compete; and it is against the state that beat the waves of social conflict. This is why as social beings they are also political beings, whether they know it or not.[75]

Professor Howard Perlmutter of the Wharton School at the University of Pennsylvania predicts that by the year 1985, 200 to 300 global corporations will control 80 per cent of all productive assets of the non-communist world.[76] Whether or not the figure is exactly right the trend is broadly acknowledged. Judd Polk, senior economist of the US Chamber of Manufacturers, also estimates that by the turn of the century a few hundred corporations will own productive assets in excess of $4 trillion, or about 54 per cent of everything worth owning for the creation of wealth.[77]

Not only is this concentration of power increasing, but for the transnational corporations the rate of increase is itself increasing. Write R. Barnet and R. Muller: 'during 1955–9 the largest 200 industrial corporations increased their share of total industrial assets each year by an average of 1 per cent. Ten years later this

average rate of concentration had doubled: they were taking over an additional 2 per cent of assets each year'.[78] Prominent among these largest corporations are the oil corporations which have now expanded their activities horizontally into all the major energy resources including uranium.[79] During 1979 the seven largest of these had annual sales of $249 billion, equal to 34 per cent of the combined annual sales of the fifty largest corporations in the world.[80] Prominent also are the two key reactor vendors, General Electric and Westinghouse, which in 1979 ranked ninth and twenty-ninth by sales in the list of the world's largest corporations.[81]

Nuclear technology is highly compatible with this general trend of development in the global economy. Its enormous capital requirements and its sophistication enhance the power of the few large corporations which can produce it, and forge closer links between these corporations and the state. The same features enhance the political power of the countries that can sell it, while within these countries the technology's vulnerability extends the legitimate area within which the state may exercise its power.

It should be noted that these features are not unique to nuclear power. Other energy technologies which also would be compatible with this trend are on the drawing board. Most notable of these is the proposed Sunsat project. This is based on a 1968 proposal by Dr Peter Glaser that large satellites be assembled in space to catch the incoming rays of the sun. The collectors would be enormous, perhaps 10 kilometres long and 5 kilometres wide, and covered with silicon photo-electric cells. These would convert the light to electricity which would then be converted to microwave radiation beamed down to a collector on the Earth and converted back to electricity. The US government has already spent $20 million on a feasibility study, and the official projected cost of sixty such satellites is $830 billion.[82] Among the major lobbyists for the project are the largest aerospace corporations in the world, including General Electric and Westinghouse.[83] They are the shadow which lies behind the sun.

The entry of such corporations into solar energy is part of a prudent series of moves to control all energy options. According to M. Stewart of the Council for the US Small Business Administration, two-thirds of the top fifty companies in solar technology are now subsidiaries of the world's largest 500 corporations.[84] A study of the US Department of Energy's solar research funding by

THE STATE AND THE INDUSTRY   137

Critical Mass concluded that 70 per cent of the funds went to big corporations such as Rockwell International, McDonnell-Douglas, and, of course, General Electric and Westinghouse.[85]

While these developments show that nuclear power is not the only technology compatible with the general trend of development in the economy, it carries the greatest momentum. This derives from the huge investments by both the state and the industry which have been poured into the industry's infrastructure over the last thirty years. Neither the state nor the industry will lightly see this investment abandoned.

Because nuclear power has both substantial momentum and a form compatible with the objects of the more powerful corporations, these seek to deflect, muzzle, or devitalize the opposition to nuclear power. They cannot however do this without the help of the state. It should not be assumed that the state will always assist. As will be seen later, there may be competing demands from other corporations or from the community, and the state may become an arena for conflict over whether it will assist or how. But because of the intimate links between the nuclear industry and the state, and the compatibility of the nuclear enterprise with the state's interests, the state will often assist. Then the state becomes a protagonist in the nuclear conflict. Here the vulnerability of the technology helps enlarge the severity and diversity of the means of social control which the state may legitimately deploy. But even when the state becomes a protagonist the success of its intervention is far from certain. This is because the power of the state is anything but unbounded.

LIMITS TO STATE POWER

At first sight the phenomenal growth in the structures of government, authorities, educational and welfare infrastructure, and administrative and policing organs which make up the modern state combines with the increasing concentration of economic power of the larger corporations to present a spectre of almost limitless power. This view is only sustainable if the state and corporations are seen as a single monolith. But seeing them thus is to oversimplify.

First, not every corporation is in the business of every other and therefore to an extent the desires of one limit the actions of others. Nevertheless, the larger corporations will have an area of common

interest, such as the stability of the economic system and the protection of their property, which they desire the state to guard. It might be tempting to follow a number of contemporary political theorists and base a picture of the state on Engel's conclusion that it acts as the 'ideal collective capitalist' simply complementing the needs of the corporations. Since these are now moving into a period of cohesive centralization and concentration of corporate power, it would then follow that the state's function is to complement this trend, diminishing the action of market forces and reinforcing a centralized authoritarian social structure. However, as other analysts have pointed out, the function of the state has not been to eliminate the action of market forces. On the contrary, the state has expanded rapidly in the attempt to supplement the market's deficiencies.

In order to preserve the capitalist market it has been necessary to erect a towering edifice of banking controls, international monetary agencies, subsidies, restrictive trade practices legislation, and a vast total outlay of payments to millions of people displaced from work during the periodic economic crises. The list of such types of state intervention in the modern market is virtually endless. Their central thrust is to attempt to prevent the economic system from plunging into depression, raging into inflation, or shattering under the political turmoil that could result from unrestrained growth in mass unemployment.

The ideology on which the whole system is founded, however, is that the free market provides the optimum allocation of goods and services for the society as a whole, acting as an 'invisible hand' to produce the best possible social distribution. It is this ideology which provides the justification for the particular spectrum not only of goods and services but also the 'bads' and gross inequalities characteristic of Western society. As some political scientists have pointed out, the state growth necessary to prevent economic crisis is increasingly in conflict with the ideology of the free market which legitimates the system.[86] This provides a crucial boundary to the growth of the state's power. The more the state intrudes into the market and into people's everyday lives, the more it weakens its own legitimacy. It moves closer to a legitimation crisis in which people begin to ask whether the particular sort of unfree market that has developed could not better be managed in a different way to provide a better or perhaps safer life, and whether they are prepared to cooperate.

THE STATE AND THE INDUSTRY   139

Required of the state in addition to the management of the economy are the functions which have developed over a long period as a result of a struggle by workers for a greater share of the products of their labour. In many advanced industrialized countries it has become accepted that the state must supply some level of unemployment benefits, old age pensions, medical assistance, education, and a host of other benefits. The provision of these have become a condition regulating the legitimacy of the state to exercise its power in other areas. The reduction of those benefits has the potential to bring the legitimacy of the state into serious question and, eventually, crisis. In this sense, the large subsidies required by the nuclear industry place the state in a dilemma. On the one hand, subsidies are required for the continual survival of the nuclear enterprise which accords with the state's interests. On the other, they drain money from welfare expenditure, a vital factor in maintaining the state's legitimacy. The opposition to nuclear power has not yet made maximum use of this dilemma, but to the extent that it is exposed the state's ability to prop up the nuclear industry may be substantially constrained.

The concept of legitimacy is central to an analysis of the battle over nuclear power. But it enters in many ways. In particular, some characteristics of the nuclear enterprise – centralization, capital intensiveness, and vulnerability – reinforce and legitimate the trend towards the concentration of power characteristic of the entire economy. Opposed to this, the perceived 'bads' accompanying nuclear programmes – the hazards to life and property, the threat to livelihood, and the assault on freedoms by the measures used to protect the programmes – are seen to be assaults by the state on the very things it is supposed to protect. This leads to a political reaction challenging the legitimacy of the state, the trends towards centralism, and the mechanisms which produce it.

Even the West German state, with its preparedness to use the massive police presence seen at Brokdorf, cannot exercise unlimited power. This was seen at Wyhl, where the legitimacy of the local community's actions overrode that of the state. For a different reason, even at Brokdorf, the state soon found its freedom to manoeuvre to keep the project going severely limited. This was demonstrated nine days after the Brokdorf demonstration, on 10 February 1977, when a Schleswig-Holstein court refused the permit for the construction of the Brokdorf reactor on the grounds that the waste problem was still unsolved.[87] The court reflected a

political process within the state which led to the Schleswig-Holstein government calling for a moratorium on nuclear power.[88] Reinforced by that process in the region where the reactor was to be sited, the court decision still prevented construction as late as the beginning of 1981.

Events in West Germany will be discussed in more detail in chapter 9. Two conclusions may however already be drawn. The first is that in a head-on confrontation between the opposition and the state over nuclear power the course of the resulting battle is bounded by the legitimacy attached to the actions and arguments of each side. The second is that the confrontation need not necessarily be head on. In some countries, including West Germany, opposition to the nuclear enterprise may penetrate into the decision-making structures of the state, with technical advisors, political parties, regional governments, and even judges and the police becoming opposed to part or all of the nuclear programme. Here the state becomes fractured by the issue, and its ability to support the industry and press forward with its nuclear plans is weakened. In some countries this process has already dramatically constrained nuclear programmes.

7
REFERRED TO THE PEOPLE

In Austria and adjoining Switzerland, and in the countries which make up Scandinavia, the nuclear debate has deeply divided not only the community but the national parliaments. Symptomatic of the power with which the issue has surged through several of these countries has been the action of their governments in constraining their own ability to act by referring their nuclear plans to a referendum.

In the nuclear debate the holding of a referendum has become an important strategy, at times used by either side. At the national level it can be a major political act accompanied by considerable risk as its result cannot be guaranteed. Although reasons for initiating a referendum will vary, one motive is quite common. By calling out an entire community, or sometimes an entire national population to vote on the issue, it is hoped to produce a result which is not only favourable but binding. But the holding of a referendum represents just one move in an unfolding saga of strategies and counter strategies in the contest over nuclear power. And although it may reveal something of the balance of contesting forces, its implications for the future are far less certain.

In the seventeen months between November 1978 and March 1980, Austria, Switzerland, and Sweden went to the polls to vote on their nuclear future. The backgrounds to each of these votes are quite different, but one general conclusion can be drawn. In each of these countries, as in several of their neighbouring countries, nuclear power has become an issue of profound national significance.

Although there are major differences between the countries, the

142   GLOBAL FISSION

events that led up to their nuclear debates fall naturally into similar
periods. The first of these, as in most Western industrialized countries, was nuclear power's 'honeymoon period' in the 1960s and
early 1970s.

THE HONEYMOON PERIOD 1960–73

The period 1964–69 marked Switzerland's honeymoon with
nuclear power. The country's remaining unused hydroelectric
potential was dwindling, and with little information available on
the hazards of nuclear power many of the arguments for 'non-
polluting' nuclear power were well received.[1] Construction of three
nuclear reactors was begun, and by 1972 all were in operation,
pumping out a total of 1 GWe, a full 20 per cent of Switzerland's
electricity consumption. This rendered the country the largest
consumer of nuclear power per head of population in the world.
Plans for a further seven reactors had already been announced.

Sweden followed a little more slowly but with even more
ambitious plans. During the 1950s, although there was strong
popular opposition to embarking on any nuclear weapons programme, all political parties supported the 'peaceful use of atomic
power'.[2] A first small-scale reactor using heavy water as a
moderator began operating in Agesta on the outskirts of Stockholm
in 1964.[3] A similar larger scale plant built in Marviken proved to
be a technical and economic disaster and was abandoned in 1970.[4]
It was later rebuilt as an oil-fired station, but despite the internal
scandal that this caused, little penetrated to the public, and the
central power company CDL (*Centrala Driftsledningen*) embarked on
an ambitious nuclear programme based on imported light-water
reactors.

In 1970–71 the national parliament unanimously approved the
construction of the first eleven nuclear reactors. The following year
CDL put forward plans for a further thirteen reactors to be built
between 1975 and 1990. These plans were endorsed by the ruling
Social Democrats in 1973. Realization of the programme seemed
likely to make Sweden the world's largest consumer of nuclear
power per head of population.[5]

In Austria, 1971 marked the beginning of the construction of its
first nuclear power plant. Situated on the Danube about 40 kilometres upstream from Vienna at Zwentendorf, the small 700 MWe

REFERRED TO THE PEOPLE 143

plant was intended to provide some 10 per cent of Austria's electricity generation.[6] This would only be the beginning, and the government had plans to build two or three plants before 1985.

In many countries it was a time of ambitious plans for nuclear programmes and little if any citizen opposition. The exceptions were the few countries where commercial nuclear plants had already come into operation. In Switzerland, as early as 1960, opposition to nuclear plants began to develop locally, substantially delaying their implementation and preparing the ground for the time when a wave of opposition would spread across it and then to several other countries.

THE SPREAD OF OPPOSITION 1973–75

Formed as a federation of twenty-two cantons, the Swiss political system reserves important local powers for the individual cantons. Included among these is the power for cantonal governments to issue licences for nuclear projects.[7] Adequate provision exists for local people to challenge reactor projects, and it was not until mid 1973 that three of the seven planned reactors – Kaiseraugst, Leibstadt, and Gösgen – were untangled from the legal hearing and construction began. However eighteen months later, in the Swiss part of Dreyeckland, a new obstacle was thrown up to confront the nuclear plans.

On 1 April 1975, 500 members of the Non-Violent Action Kaiseraugst, following the recent example of Wyhl in West Germany, illegally entered and occupied the site of the Kaiseraugst reactor.[8] Support for the action was widespread. Five days later 15 000 people came together in the first of several demonstrations aimed at giving the occupiers courage to continue. Although threatened with police action (which did not eventuate), and legal action against them by the company building the reactor, the occupiers held their ground. It was not until mid June that the occupiers left after having negotiated a substantial concession.

There was to be a one-month halt to all work on the site to enable further negotiations to take place, and a binding commitment was made to discuss the matter seriously. Later the halt to building was further prolonged and in November it was announced that no building permit would be granted in the following year. That was later extended, and as late as 1980, permission to begin

144   *GLOBAL FISSION*

construction of the plant had still not been granted.[9]

The effects of the opposition rippled well beyond the confines of the Kaiseraugst site. On 26 April 1975, when 20 000 people marched against nuclear power and in support of the occupiers in Berne, the capital, it was clear that a new phase had developed in the Swiss movement. Now the crowd included diverse groups opposing reactors planned for other parts of Switzerland. The opposition had coalesced into a national movement.[10]

The growth of the Swiss movement undoubtedly helped nourish the seeds of opposition in other countries of Western Europe. In Austria the nuclear plans had been much slower to appear and so too had the opposition. The construction and licensing of Zwentendorf had been shrouded from public attention by the provisions of the Radiation-protection Law and opposition was restricted to a handful of scientists and other citizens. It was not until 1974, when it was publicly announced that the government intended to build a second nuclear reactor near Linz in Upper Austria, that opposition began to grow rapidly.

The new proposal, for the St Pantaleon plant, involved the creation of a huge industrialization programme that would destroy large areas of beautiful recreational land surrounding the Danube River. Citizens action groups sprang up to campaign against the project. By the spring of 1975 they had collected 80 000 signatures opposing the plan.[11]

In Austria and Switzerland the opposition sprang up spontaneously from the people. But this was not the only way opposition could develop. In Sweden the general public suddenly awoke in spring 1973 to find a heated debate on nuclear power already occurring in the national parliament. The debate was over a private bill introduced by Birgitta Hambraeus of the Centre Party. Formed in 1921 as an agrarian coalition, this party had in recent years increasingly found itself courting the urban vote in the steadily industrializing country. The leader of the party, Thorbjörn Fälldin, had reportedly been greatly influenced by Hannes Alfvén, a Nobel Prize-winning scientist who had become steadily more critical of nuclear power. Fälldin's personal convictions, together with those of some colleagues, were able to assume a role in the political arena when it became clear that an alignment with environmentalists could be of value to the future of the party.

The powerfully presented arguments were sufficient to create government recognition of the need to proceed cautiously. In May

REFERRED TO THE PEOPLE

1973 the parliament resolved that 'no decisions in favour of further expansion of nuclear power should be taken until new basic material, including information on the state of research and development of radioactive waste disposal, has been put before the parliament'.[12] Not long afterwards the Organization of Petroleum Exporting Countries (OPEC) oil price rises intensified the debate.

By 1974 it was clear the debate's effects stretched well beyond the walls of parliament. In January of that year an opinion poll was released by SIFO, an official Swedish opinion survey institute. The government and the power companies who had commissioned the poll were shocked to find that 59 per cent of the people interviewed wanted to 'hold back' nuclear power.[13]

The debate in Sweden naturally affected discussions about nuclear power in other countries of Scandinavia. The debate proved especially contagious in Finland, a country with a sizeable Swedish-speaking population.[14] In 1973 the Swedish-language newspapers, read by about 8 per cent of the population of Finland, began to cover the growing concern over nuclear waste disposal and reactor safety. News of Finland's nuclear programme reached its media at about the same time as the Swedish reports. It became known that the national Energy Policy Commission was about to release an interim report with recommendations on nuclear power. When it finally appeared before parliament in 1974 the rumours that an extraordinarily ambitious nuclear programme was planned for Finland proved justified.

Under the plan one third of all Finnish electricity production was to be provided by nuclear power by 1985. This 5 GWe of nuclear capacity was to expand rapidly. By the year 2000 it was envisaged that 6 GWe would be located at Olkiluoto near Rauma and Kopparnäs near Helsinki, 3 GWe at Lovisa near Helsinki, and 24 GWe at another eight sites.[15] All in all the plan proposed forty nuclear reactors at eleven locations.[16]

The first substantial opposition was at Sibbo to the east, and Ingå to the west of Helsinki. The focus of concern was the announcement in 1973 of a plan for a nuclear reactor at Kopparnäs in the Helsinki region to provide hot water for the city's heating. Within six months membership of the Sibbo group had risen to over 1 000, and the group at Ingå had begun to collect a petition. By the time the petition was presented to the prime minister in April 1976 it carried 7 000 signatures, including more than 70 per cent of the population of Ingå.[17]

146 GLOBAL FISSION

By then the opposition had consolidated. The previous December a broadly based action group with members from all the major political parties had been formed to further develop the campaign against the plant. Despite the pro-nuclear views of the Helsinki Council, the local authorities in Kopparnäs refused permission for the plant's construction.[18] The national government could override that decision but as late as the beginning of the 1980s opposition was so strong that such an action still seemed unlikely.

In Switzerland, Austria, and Scandinavia, opposition to the early ambitious nuclear plans had developed rapidly. Whether initially sparked by local projects, debate in the parliament, or from news from a neighbouring country, the concern over the hazards of the proposed nuclear programmes had diffused through the general public in the early to mid 1970s and helped fuel an increasingly active citizens opposition. The activity of this opposition still differed markedly from region to region. However, it resonated with the less visible but growing undercurrent of general concern to create a political challenge not merely regional but national.

Slow to recognize its potential but now turning to meet that challenge were the nuclear companies and the national governments. However, even the governments were beginning to perceive constraints to strategies they might apply. One ominous indicator of the political limits was that the debate was not only polarizing the community. Now, not only in Sweden but in the other countries, it was creating divisions within the parliaments.

DIVISIONS WITHIN PARLIAMENT

Typical of the divisions were those within the Finnish parliament. Late 1974 and early 1975 had seen some important Finnish national citizens organizations such as the government-based Environment Protection Council, the Environment Protection Union, and twenty-three Swedish-speaking people's organizations call for a moratorium on any new nuclear projects beyond the four, at Lovisa and Olkiluoto, already under construction.[19] This call was soon taken up by some of the political parties.

Initially no political party, apart from the small Christian League, had opposed nuclear power. Of the four large parties – the Social Democrat Party, the left-wing Finnish People's Democratic League (which contains the Communist Party), the right-wing National Coalition Party, and the Centre Party – which together control

REFERRED TO THE PEOPLE

about 80 per cent of the vote, only the Centre Party had not endorsed nuclear power. However, as national pressure began to develop some of these parties responded. Of the major parties, the Centre Party and the National Coalition Party took up the call for a moratorium, at least until the waste and other associated problems had been solved. They were joined by the smaller Liberal People's Party and the Christian League.[20] Notably absent, however, were the major parties of the left who, compared with the other two major parties, control the majority of the votes.

Finland, of course, is a very special case. Both the fact that it borders the USSR and its history link it intimately with that country. Its first reactor at Lovisa is a hybrid of a Soviet pressurized water reactor, and a US Westinghouse cooling system – an 'Eastinghouse' reactor. Since that reactor was bought the bulk of Finland's nuclear purchases have been from the USSR and its other neighbour, Sweden. As might be expected, the Finnish left has particularly strong links to the policies of the USSR, and the country's Communist Party is an ardent advocate of nuclear power. As early as 1966, and again in 1968, the Communist Party stressed at its party congress that 'the construction of the first nuclear power station under state supervision and in cooperation with the Soviet Union has to be accelerated'.[21]

However the attitude of the left-wing parties in Finland is not unique. In Austria and neighbouring West Germany, and Sweden also, the largest parties were social democrat and the major force pressing forward the ambitious nuclear programmes at government level. Some of the reasons why many parties of the left, at least initially, embraced nuclear power with enthusiasm will be discussed in chapter 10. Here it is simply noted that to challenge the nuclear programmes in the national parliaments the opposition to nuclear power often had to look to the centre or conservative parties for support. Only in Switzerland did the social democrats seem at all divided on the issue.

In Switzerland the occupation of Kaiseraugst had precipitated a major parliamentary debate. Two parties, the Radical Party and the Christian Democrats, had come out in favour of limited development of nuclear power. The situation was less clear-cut with the Social Democratic Party which now found the debate surging up through its own ranks. Along with the remaining parties it was unable to muster a majority, and left it to its members to make up their own minds. The controversy over nuclear power was now

148   *GLOBAL FISSION*

not only dividing the community. It was also cutting across traditional party barriers.[22] In Switzerland, a country governed by a multi-party coalition, there now seemed to be the potential to undermine the nuclear programme at the parliamentary level.

It is often necessary to change the balance of power in parliament in order to obstruct a major plan of the ruling party. Even if insufficient power exists regionally to prevent the implementation of the plan, it is often possible to create a situation where the government fears that polarization over the issue is reaching enough strength to seriously threaten its future. Occasionally a small party can play a crucial role in this.

In Austria it was the small liberal Freedom Party which first took up a position of opposition to nuclear power at the national level, prior to the federal elections scheduled for October 1975.[23] With public concern rapidly increasing over the St Pantaleon proposal, and with electricity demand having grown much less quickly than predicted, the Socialist Party government found the logic swinging out of its control. Worried about the elections, it sought to defuse the situation, and took a step that would later prove its undoing. It announced that the important national decision over 'whether Austria should go nuclear' would be made with the participation of an informed public.

In Sweden the ruling Social Democrats' ability to govern had been seriously weakened at the 1973 elections. They now required the cooperation of at least one of the non-socialist parties to exercise a majority of votes in the parliament. The Centre Party had already launched its attack on the nuclear programme and the issue looked increasingly divisive. To make matters worse the left flank was fracturing. The Swedish Communist Party had reversed its previous policy, becoming the first communist party in the world to come out against nuclear power.[24] In an attempt to calm the issue and heal the potentially very damaging division, the government turned to a well-worn strategy.

The Social Democrats had established a system of Royal Investigative Commissions which could be used to examine politically controversial matters. It was part of the process providing the stability that had allowed them to rule continuously since the war. The first commission, into energy, was established in 1973, and in 1974 three more commissions, into the siting of nuclear plants, the problems of radioactive waste disposal, and the possible directions of energy research, were also set up.[25]

REFERRED TO THE PEOPLE

When a government sets up an inquiry its purpose is often more than to merely uncover and summarize 'facts'. It often hopes that the inquiry will produce a statement that will be highly credible to the general public and will impart to government policy a sufficient level of legitimacy to enable it to be implemented.

The legitimacy derives from the fact that the inquiry is an authoritative, apparently independent body which has listened to all the arguments. The government's hopes stem not only from its confidence in its own case but also from its ability to choose the inquiry's commissioners. Even so, as several governments have discovered to their chagrin, even 'respectable' commissioners may not necessarily produce a report which favours the view of industry and the establishment. There is, therefore, an element of political risk. The government usually embarks on such a venture because the issue has become so potentially divisive that the risk of not invoking some such measure is even greater. In the case of the Swedish government, the appointment of the commissions was not enough to guarantee political security in the face of the burgeoning nuclear issue. It therefore embarked on an additional strategy, one which was to be attempted by several other governments in similar situations.

THE INFORMATION CAMPAIGNS 1974–77

During 1974 the Swedish government decided to institute 'a major project of public education and consultation' on its nuclear programme.[26] The mechanism would be the 'study circles', a system of small study groups dating back to the beginning of the century and managed by adult education associations. They are linked to the political parties and other major community organizations such as trade unions and religious groups. For the groups it organized the Social Democrats issued information packets on the nuclear debate. Other political parties soon followed suit. By the end of 1974 an estimated 100 000 Swedes taking part in some 7 000 study groups were engaged in the giant study programme.[27]

Undoubtedly the government's expectation was that increased information would lead to decreased opposition to its nuclear programme.[28] However the result was the opposite. By the end of the programme the membership of the Social Democrats had become deeply divided over the issue. In contrast, opposition in the Centre Party had consolidated to form an overwhelming majority

of the membership.[29] As the political parties struggled with the results of the programme, political positions on the energy issue began to emerge. All the parties agreed that strong energy conservation programmes should be introduced and that electricity growth rates should be reduced by at least half.

The Conservative Party endorsed the government's original nuclear programme 'in principle' but expressed some concern over unsolved problems. The Liberal Party opposed resumption of the original programme until some of the problems had been solved. The Centre Party and the Communist Party came out flatly opposed to further nuclear plant construction and raised the possibility that those under construction should be cancelled.

The Social Democrats now sought to contain the debate by making some concessions. In February 1975 they announced a new policy. Energy consumption growth would be kept to less than 2 per cent per year, and after 1990 it would be halted altogether. Only two more reactors would be added to the eleven already approved by parliament pending a review of the whole programme in 1978.

The policy became law in May, but the informed public was not soothed. The new law entailed exactly the same nuclear growth rate as before, and was only qualified by the woolly promise of a 'review'. A SIFO survey taken just before the enactment showed that an increased majority of voters, 63 per cent, were now against the building of further reactors. Only 31 per cent were in favour. Another poll, taken on 28 May after the parliamentary vote, showed the Centre Party had picked up support from the Social Democrats.[30] It was an ominous augury for the coming elections.

Adding to the worries of the nuclear proponents, and the government, was the growth of the organized citizens opposition. The loose collection of environmental groups which had been the mainstay of the opposition (including *Miljöcentrum*, a research and lobbying group;[31] FOE; and MIGRI, a union of environmental groups[32]) had begun to orient themselves towards more direct expressions of opposition within the community.[33] In spring 1976, a new federation (*Miljöforbundet*) was formed to pursue these objectives and it grew rapidly. The debate that had been sparked in the parliament, was now igniting the grassroots.

The new mood of activism crystallized in early June 1976 at a demonstration of 10 000 people in the hills of Billingen near the town of Skövde in west Sweden. It was the largest anti-nuclear demonstration ever held in Sweden. Beneath 500 square kilometres

REFERRED TO THE PEOPLE

of the tranquil landscape lay vast deposits of uranium, comprising around 80 per cent of Europe's known reserves.[34] The low-grade ore is difficult to extract and it was only recently that the state-owned company LKAB had come up with a method. It had extracted 200 tonnes and was now applying to extract 30 000 tonnes annually. In the face of the mounting opposition, however, it would soon withdraw its application.[35]

For the Social Democrats the outcome of the September national elections was distressing. Their forty-four years of power ended as they were voted out of office. A coalition of the Centre, Liberal, and Conservative parties came together to form the new government with Fälldin as prime minister. Significantly, key members of the Social Democrats attributed their loss of power to the effectiveness of the anti-nuclear vote, an effectiveness they had unwittingly strengthened through their failed information campaign strategy.[36] The government had been keen that people join the information campaign. They had poured public funds into it and pushed forward the theme 'Learn more, and you will have more influence. Join an energy study circle'.[37] People had, in their tens of thousands, but the result had scarcely been to the government's taste. What had gone wrong?

Lying in the background of this information campaign and similar ones mounted by other governments, such as the one in West Germany carried out after the action of Wyhl, was undoubtedly a simple equation. The nuclear experts favour nuclear power. Make people more technically expert and they too will favour it. The fatal flaw may well be that the technically trained proponents of nuclear power and the concerned public view the issue from opposite sides.

From the technologists' side, the complexity and difficulty of the problems to be solved is seen as an attraction. From the public's point of view it reduces their confidence that the technology is safe. The technologists find the sophisticated approaches to their solutions soothing and satisfying; the public find the extent of what has not been solved alarming. The technologists take the boundaries of the problem – the need for increasing energy consumption – as fixed and then seek the 'sweetest' technical solution; the public do not like the solution and instead question the 'need'. The technologists believe that their perspective is the only valid way to approach the problems, while the public fear that many of the key problems are social in character and not amenable to technical solution.

152   GLOBAL FISSION

In short, the issue is viewed by each side through different paradigms and on the basis of different values. As the public becomes more informed the technical language becomes less opaque and the nature and extent of the conflict is further clarified.[38] With increased understanding comes a tendency to ask not whether nuclear power is the right or wrong solution, but whether it is the solution to the right problem.

This difference in paradigms was evident in Switzerland in the autumn of 1975 when the construction of Kaiseraugst was suspended and the occupiers withdrew from the site. Then followed a 'dialogue' between the government's 'experts' and the occupiers'. But, as Deputy Energy Director Claude Zangger put it, 'as far as the crux of many issues was concerned they could only be described as the dialogue of the deaf'.[39] People were talking past each other at least partly because they were concerned about different things.

In Austria the same lesson could be drawn when, in the wake of his announcement in July 1975 that the nuclear decision would be made with public participation, Chancellor Kreisky announced that the Ministry of Commerce and Industry had been directed to conduct a nuclear energy information campaign. It was launched in February 1976.

Intended to justify the nuclear power programme and quieten public fears, the information campaign quickly proved to be doing just the opposite. The official pamphlet 'Nuclear Energy – A Problem of Our Time' evoked widespread criticism and debate. The first of the ten proposed 'discussions of facts', in October 1976, showed that the public was not satisfied to merely sit and listen to an expert panel discuss a narrow selection of the issues. Subsequent meetings were stormy, and revealed the extent of the problems associated with reactor safety and waste disposal. One meeting, in January 1977, was taken over by objectors and transformed into a discussion on how to stop the Zwentendorf plant.[40]

The last meeting, for March 1977, was cancelled by the dismayed government. On the same day, demonstrations against nuclear power took place in many Austrian cities and towns. In what was for Austria a substantial demonstration, 3 000 people marched in Vienna protesting against the nuclear proposals.[41] Undoubtedly the tide of public opinion was rising against nuclear power. Said Dr Peter Weish, an ecologist and prominent opponent of nuclear power:

REFERRED TO THE PEOPLE   153

A year ago you were regarded as a crank in Austria if you talked about stopping the nuclear programme. Now all that has changed. You can argue in favour of stopping Zwentendorf and people will listen. I am optimistic that Austria will decide to scrap its nuclear programme.[42]

Weish's comment would soon prove remarkably prescient.

The experience of the Austrian government reinforced the conclusion that can be drawn from the result of the information campaigns in Sweden, Switzerland, and West Germany. As an industry or government strategy for containing opposition the information campaign is fraught with difficulties. This gives some support for the earlier conclusion that it highlights hidden but profound differences in viewpoints and values rather than overcoming technical misunderstanding. Even more support is given to this by the observation that the information campaign can be a powerfully effective weapon for the opposition. Nowhere was this more clearly visible than in the debate over nuclear power in Denmark.

In other countries the opposition had developed as a reaction to an announcement of an ambitious government nuclear programme. But in Denmark opponents began to organize before any plans had been announced, and when the announcement finally came the opposition was ready and waiting. There had of course been some talk of nuclear power. Indeed, Denmark's atomic scientists and electricity utilities had been talking vaguely about harnessing nuclear power for some twenty years. However, in 1971 ELSAM, the electricity conglomerate for the country's western region, backed a preliminary search for suitable sites.[43] The signs were there that nuclear power was becoming a reality. The grounds for this conclusion were strengthened in May 1971 when the electricity utilities began a publicity campaign, complete with glossy brochures and seminars for teachers and journalists, to promote the 'peaceful use' of nuclear energy.

In the summer of 1973 a handful of citizens met to form a working group to confront these developments. Over the next six months they prepared themselves for the announcement they knew would come. On 30 January 1974 the government Environment Ministry published a list of ten proposed sites for nuclear power plants. The next day, at a press conference, the opponents to nuclear power replied by formally establishing The Organization

for Information on Atomic Energy (*Organisationen til Oplysning om Atomkraft*) or the OOA.

Because they had time to plan, the members of the OOA were able to fashion their strategy with particular care and sensitivity. Central to this was an information campaign based on establishing a dialogue with those who did not yet oppose nuclear power and those who presently supported it. They were confident that the result of such a dialogue would be a spread of opposition.

The OOA demands reflected this strategy. Instead of calling for a government commitment against nuclear power they asked merely for a three-year stay of any decision to allow the public to discuss the questions involved. Further, they asked that any decision on the establishment of a nuclear programme be made by the parliament, not by administrative decree.[44]

The government was quick to accede to the latter demand since it sensed the looming political force of the issue. In addition, it felt unable to proceed precipitously. Accordingly, it fell in with much of what the OOA demanded by instituting its own information campaign. This ran between June 1974 and May 1976. An official Energy Information Group was set up which produced six volumes containing a comprehensive survey of the arguments for and against nuclear power. In parallel, the OOA worked hard to spread its own message.

In particular, the OOA began to go beyond the question of nuclear hazards. In 1900 some 100 000 windmills had dotted the Danish landscape, pumping, grinding, and generating electricity. The advent of cheap fossil fuels had rendered them obsolete in the 1930s, but now windmills were again beginning to look attractive.[45] Danish scientists such as Sørenson at the Niels Bohr Institute showed that, with appropriate short duration storage systems, windpower could produce electricity competitively with nuclear reactors. In late 1974 a small college community at Tvind had put the theory into practice by building the world's largest windmill.[46] The giant three-bladed windmill carried a 2 MWe generator capable of producing 75 per cent of the town's energy needs. It was to begin producing electricity at Easter 1978.[47]

By tapping this sort of expertise in alternative energy systems, the OOA with another group, the Organization for Continuous Energy, produced an alternative energy plan for Denmark. They argued that if implemented their proposal would create more employment and sufficient energy at similar cost to the govern-

REFERRED TO THE PEOPLE

ment's Energy Plan but without any need for its five projected nuclear reactors.[48] Together, the flows of competing information helped stoke the already smouldering controversy. So also did the government's actions.

In April 1976 the government became impatient to implement its nuclear programme and passed a law to permit the establishment of nuclear power plants. But now the result of the information campaigns surged up to confront it. A further Coming Into Force Law specifying the actual reactors to be built would have to be enacted before any construction could begin. It was against this that the opposition now mobilized. In six weeks the OOA gathered 170 000 signatures to a petition calling for delay. They distributed over 200 000 badges and sold over 1 million letter stickers.[49] A four-page newspaper was printed, and its three editions attained a circulation of 900 000 copies sold. The momentum was visible, growing and, at least temporarily, irresistible.

In May the government postponed discussion of the bill and on 10 August withdrew it altogether. The special session of parliament planned for the debate on the issue was cancelled and no new date set for any further decisions. An opinion poll taken in the same month showed only 16 per cent of the population in favour of nuclear power and 56 per cent against.[50] The prevalent belief was that the government would not attempt to move further on the issue, at least until some time in 1978. For the time being the opposition had won.

Whether initiated by government or by citizens groups, in each of the countries considered here the information campaigns had created among the public a deep distrust of nuclear power. Informed by the debates that had ensued, many had become actively opposed to nuclear power, joining local opposition groups and taking part in various activities intended to further spread the message. The issue had become divisive, threatening the future power of political parties and the plans, if not the existence, of the nuclear industry. The pro-nuclear groups sought to develop new strategies to head off the growing threat. Now, however, the options seemed limited by the strength the opposition had achieved. Head-on confrontation seemed politically impossible, yet the information campaigns had proved a disaster. The issue was proving too difficult to resolve by parliamentary edict. A method would be needed that wielded greater legitimacy than a parliamentary majority could. Steadily the parliaments were being drawn to

156 GLOBAL FISSION

what seemed the one remaining strategy: to refer the issue to a decision by the people.

THE ROAD TO REFERENDA 1977-79

The first to succumb to the pressure was the Socialist Party government of Austria. May 1977 saw the opposition merge into a national coalition of over fifty citizens groups. They demanded that the next stage of the government's rapidly disintegrating information campaign be public, and that politicians be allowed to state their positions. The government refused, and the groups announced they would no longer participate.

Seven months later the flames were further fanned when opponents of nuclear power uncovered government plans to secretly import fuel from the start-up of the now completed Zwentendorf reactor. In early 1978 the fuel rods were carried to the plant by a military helicopter and barricaded in by the police. Bringing the rods had proved physically possible, but using them looked increasingly like political suicide. The report of the official information campaign with its pro-nuclear message did nothing to dampen the opposition. In fact it inflamed it further.

The anti-nuclear movement, while still comprising a minority of voters, was now sufficiently large to threaten the political future of any party that could be blamed for starting up Zwentendorf. The People's Party reconsidered its position and announced itself against the fuelling of the reactor because of the unsolved safety problems.

Even within the ruling Socialist Party division over the issue was now visible. In the western-most province of Austria, Vorarlberg, the local people had just fought desperately against a Swiss plan to build the Rüthi reactor in close proximity to the Austrian border. Socialist members of parliament from Vorarlberg indicated they would not be able to support their party's pro-nuclear policy. Neither the Socialist Party nor any other was now prepared to place itself alone on the tracks in the face of the onrushing opposition. In June 1978, with the support of all parties, Chancellor Kreisky, leader of the Socialist Party and the government, announced that the issue would be put to a referendum on 5 November 1978.

In Switzerland a referendum on nuclear power was also imminent, but launched by the opposite side and for opposite reasons. Whereas the Austrian government sought to uphold

REFERRED TO THE PEOPLE

nuclear power and end the debate, the Swiss opposition, flushed with the success of its occupation of Kaiseraugst, began moves for a referendum to constrain the nuclear programme and keep the nuclear issue alive. Occupations of reactor sites at Kaiseraugst, Gösgen, Aarau, Lucens, and other parts of the country would continue,[51] but additionally the opposition now began to collect signatures nationally. By 1976 they had collected 120 000, sufficient under Swiss law to require such an initiative to go before the people at a referendum.[52]

The text to be voted on proposed a series of conditions to be placed on the granting of operating licences of new nuclear reactors. In addition it required that the operation of the four existing plants must be approved by parliament if they were to continue.[53] Of greatest concern to the nuclear industry was a clause which provided that in order to be approved a project must gain the consent of a majority of the people living in all cities, communities, and cantons within 30 kilometres of the proposed reactor site. Proponents of nuclear power claimed, with considerable justice, that this clause would in practice mean an outright ban on any further nuclear developments. The referendum was to go to the vote on 19 February 1979.

In Sweden the road to its referendum was much more circuitous. With the election of the Fälldin government in September 1976, high hopes had been held briefly by opponents of nuclear power that this success would halt the operation of any further nuclear reactors. All eyes were turned on Sweden's sixth nuclear reactor, Barsebäck-2, which was completed but not yet primed with fuel.[54]

Opponents were to be disappointed. They had forgotten the fragility of Fälldin's coalition government and the compromises he would have to make in order to govern. Within one week of taking office he was forced by his coalition partners to allow the fuelling of Barsebäck-2.[55] The decision was not just a disappointment to Swedes.

Located near the large city of Malmö on the south-west coast of Sweden, the two Barsebäck reactors are only 19 kilometres across the border from Copenhagen, the capital of Denmark. Thus centred on the most populated region of Scandinavia, the reactors have been a source of outrage to Danes as well as Swedes. Danish concern was demonstrated as early as 1974 when 6 000 people marched on the plant from Denmark.[56] During 1977 the opposition in Denmark expanded rapidly, with a network of over 170

citizens action groups covering the country.[57] In September of that year Barsebäck was again the target of action as 8 000 Danes crossed the straits separating the country from Sweden to take part in a joint 19 kilometre march of 20 000 people on the silo-shaped reactor.[58]

Disappointment over the fuelling of the reactor was, however, partially compensated. In January the Fälldin government had passed its nuclear wastes Stipulation Law. This required Barsebäck-2 to close down unless its owners could demonstrate before 1 October an agreement covering a safe method of reprocessing its waste.[59] More generally, the Stipulation Law provided that no utility could obtain permission to load fuel into a new nuclear reactor until it had either produced a contract which adequately provided for the fuel to be reprocessed and had shown where the resulting wastes were to be stored with complete safety, or until it had shown where the spent fuel rods would be stored with complete safety.[60]

The Centre Party may have thought this a clever political compromise, but it was more an expression of the party's political powerlessness than an effective obstacle to the nuclear industry. It did nothing to slow or prevent the continued building of the four reactors under construction which when completed would bring Sweden's total to ten. On the other hand, by specifying the disposal of the wastes as the obstacle to commissioning the reactors, the government had effectively reduced the vast manifold of social and technical problems associated with nuclear power to a single technical deficiency. In response, the nuclear industry sought to find a credible technical solution.

In December 1977 the Swedish State Power Board submitted an application under the Stipulation Law for permission to load fuel into the completed Ringhals-3 reactor at Varberg. It provided a study commissioned by the nuclear industry, the KBS Report, which purported to demonstrate a safe method for disposing of the liquid high level wastes produced after fuel reprocessing. In addition, a contract with the French company COGEMA (*Compagnie Général de Matières Nucléaires*) was provided. This company undertook to reprocess the wastes, separate out the uranium and plutonium, and cast the residue into blocks of glass. The long-term safety of this technique is still the subject of scientific controversy. The reprocessing was to be carried out by an as yet unfinished plant, and the contract contained an escape clause which

REFERRED TO THE PEOPLE

allowed COGEMA to return the rods if it could not reprocess them.[61]

The trap which the Centre Party had set for itself had been sprung. Was this or was this not an absolutely safe answer to the waste problem? Who should provide the answer? And, even if the answer was yes, should the reactor be fuelled? The coalition split over its attempt to answer these questions. Members of the Centre Party opposed the fuel loading. The government tried to manoeuvre and announced that the applicants must demonstrate to the satisfaction of the Nuclear Power Inspectorate that there was suitable rock available to entomb the solidified waste. They thus passed responsibility for a decision to an agency of the technical bureaucracy. This was immediately criticized as being a 'soft yes' to loading. As criticism mounted, Fälldin demanded that the coalition partners agree to hold a referendum on the issue. They refused, and on 5 October 1978 Fälldin handed his resignation to the Speaker, and the coalition government fell from power.[62]

The sudden fall of the Fälldin government left a political vacuum which was filled by a minority Liberal government. Its new energy minister, Carl Tham, favoured nuclear energy. However the depth of opposition made it politically hazardous to move any distance forward on the nuclear programme, and the new government fell back on the conditions that had already been set by the Fälldin government. The industry indicated that it intended to demonstrate a 'safe' repository by the end of 1979, while geologists involved in the drilling tests said this could take as long as ten years.[63] But, at least temporarily, it was an uncomfortable stalemate. Four months later the accident at Three Mile Island made headlines and with its shock waves the political situation suddenly crystallized.

Less than a week after the accident at Three Mile Island, acting with seemingly indecent haste, the Nuclear Power Inspectorate announced its decision. Rhinghals-3, and also the Forsmark-1 reactor at Uppsala should be allowed to be loaded with fuel.[64] The decision overcame the last formal barrier presented by the Stipulation Law. But it was too late. The ripples from the accident were now visibly reinforcing the opposition. As 20 000 people marched in Denmark, and 10 000 to 15 000 demonstrated in Stockholm against the fuelling of Barsebäck, a petition of 40 000 signatures against nuclear power was presented to the Swedish government.[65]

The opposition was strengthening, and both sensing the political

damage that could result at the forthcoming elections to any party which endorsed nuclear power, and genuinely shocked by the accident, the Social Democrats did an abrupt about turn. They now declared that they wanted a referendum on nuclear power. Party leader and former prime minister, Olof Palme, told a press conference that he was 'deeply shaken and worried' and that he had always felt nuclear power should be stopped if proved dangerous.[66] A few hours later the Liberal prime minister, Ola Ullsten, announced that the government had changed its mind also. The two nuclear reactors awaiting fuel would have to wait for the results of a national referendum to be held in March 1980.

In Austria, Sweden, and Switzerland the nuclear debate had achieved sufficient momentum and so polarized the community that one or more of the contestants had seen it as politically desirable to force the issue to a national plebiscite. Their reasons varied. For the Swiss opponents of nuclear power the referendum promised more than its end result. It would provide a national forum through which the debate could be continued, thus building further opposition. For the political parties, on the other hand, it could provide a convenient mechanism for shifting a politically hazardous issue away from election day. Each side also hoped to achieve a favourable result on the day of the referendum. In relation to this the pro-nuclear governments and the nuclear industry had some reason to be confident, for despite its appearance, a referendum is usually not an evenly balanced contest. The object is to obtain a surge of public opinion on a particular question on a single day. In striving for this, large corporations with the financial resources to saturate the media, and governments with their ability to take action to change the context in which the debate takes place, are particularly well equipped. The discrepancy between the capabilities of each side to obtain a strategic advantage in this particular form of contest was clearly illustrated in the final weeks leading up to each referendum.

FINAL MANOEUVRES

As Austria moved towards its referendum day in December 1978, Chancellor Kreisky announced that he was confident of obtaining a clear majority for nuclear power. But he seemed reluctant to leave it at that. He went on to declare that a vote against Zwentendorf would amount to a vote against the Socialist Party. Having

REFERRED TO THE PEOPLE 161

separated the issue from the elections he still wanted the advantage of party loyalty. In fact this manoeuvre merely confused the issue and divided both the opponents and supporters of the Socialist Party. Still aiming to use the advantage of party organization two weeks before the referendum, Kreisky further increased the confusion by threatening that he might resign if the vote went against the plant.[67]

In Switzerland the opposition to nuclear power had already confused the issue by framing an initiative which was not clear in its effect, although many believed that it would effectively prevent further nuclear construction. Fearing the initiative's potential, the government acted to try to defeat it by capturing the middle ground. It announced that there would be a second referendum which would be held three months after the citizens' initiative, in May 1979. Then people would vote on a 'partial revision' of the Atomic Energy Act. If carried, this second referendum would impose more stringent conditions on plant construction and require utilities to prove the need for a proposed reactor.[68]

This manoeuvre certainly added to the already prevailing confusion. Nevertheless there was general agreement that the implications of the citizens' initiative would be the more far-reaching. The atmosphere was tense, and the political parties were divided in their attitude to the initiative. The Social Democrats and independent Landestring Party supported it while the Liberal Democrats and Christian Democrats recommended a 'no' vote. Even the Swiss Federation of Trade Unions was split, not only on the issue but also over whether it should make any recommendation to its membership. In the end, none was made.[69]

In Sweden, the build up to its referendum, in the closing months of 1979, was also characterized by the manoeuvring of the political parties. With six reactors operating, four more awaiting start-up, an eleventh half-finished, and a twelfth designed but not yet under construction, much more was at stake than in the other two countries. Here the battle would be over the precise wording of the questions to be asked.

The debate in Sweden had reached a point where unqualified support for nuclear power was politically untenable. During the Centre Party's term of office, non-nuclear energy paths had gained the credibility of government sponsorship. Funds for energy conservation had been doubled, research and development money had been redirected from nuclear to alternative energy sources, and

additional funds had been allocated to synthetic fuels, public transport, and wind, tidal, and other alternative forms of energy production.[70]

The report of the government-funded Secretariat for Future Studies, which had made proposals which went much further along the non-nuclear road than that of the Energy Commission, had also entered the debate with considerable presence. It had described how the country's energy supply could be developed in such a way as to yield an energy economy based entirely on renewable energy sources by the year 2015. It assumed a doubling of goods and services but a steady phasing out of energy waste along with the phasing in of renewable energy systems.[71]

Energy conservation had been officially embraced by all the major parties, and when it came to framing questions for the referendum, even those who most strongly favoured nuclear power aimed for more discreet wording. In the end it was decided that three questions would be put to the people. This in itself was unusual. Perhaps even more unusual was the fact that the first two options were essentially identical.

All three options proposed the phasing out of nuclear power. What they did disagree about was when and how this was to be accomplished. Options 1 and 2 proposed that 'at most' the twelve reactors in operation, complete or under construction, should be used. There should be no further expansion of nuclear power beyond this, and the implication was that these reactors would be phased out over their economic lifetimes. Option 3, on the other hand, proposed that no nuclear reactors should be allowed to operate beyond the six presently in operation, and that these should be phased out within ten years. In addition, no uranium mining would be permitted and safety measures must be intensified.[72] The previous September, after the decision had been made to hold the referendum, the national elections had returned a non-socialist majority, and a coalition of the Centre, Conservative, and Liberal parties, had again formed the government with Fälldin as prime minister. But the alignments on the referendum's options were very different. Option 1, usually referred to as the most pro-nuclear, was supported by the Conservative Party. The only substantial difference between option 1 and option 2, however, was that the latter specified that nuclear power stations must be owned by the state. This mattered little, since the state was already heavily involved in electricity production in Sweden. It was supported by the Social

REFERRED TO THE PEOPLE   163

Democrats and the unions. Option 3, for a more rapid phase out, was supported by the Centre and Communist parties.

In Sweden, Austria, and Switzerland, the referendum questions had thus been muddied by the manoeuvring of the political parties. In Austria people were not merely voting about nuclear power, but about the resignation of Chancellor Kreisky. In Switzerland the choice had been spread between two options separated by an interval of several months, both of whose implications were not entirely clear. And in Sweden three options were presented, all of which limited the nuclear programme but in different ways.

The resulting confusion may have been deliberate. But whether or not this was so, it highlighted an important aspect of the pre-polling day contest. If the implications of the various choices to be voted on were unclear, they would require interpretation. Voters get much of this through the newspapers and electronic media. In addition, the media play an important role in setting the 'agenda' of the contest by presenting particular issues, groups, and statements as newsworthy. Here, the money and influence of the nuclear industry and its supporters within major areas of media may play a particularly influential role. As well, the timing and terms of the contest, including the wording of the referendum questions, may frequently be decisively influenced by the state and the industry. In this sense, when the opposition agrees to contest a referendum, it may well be agreeing to fight in a set-piece battle on the home ground of the other side. Although this does not guarantee a win for the nuclear proponents the terms are so favourable that should they lose they may well prefer to contest another referendum rather than recontest the issue in another way.

In Austria the opposition consisted of a wide network of diverse citizens groups. It included Mothers Against Nuclear Power, Teachers Against Nuclear Power, and students, Catholics, artists, scientists, trade unionists, and many others formed into a variety of different anti-nuclear action groups. Significantly, it included a group of dissenting Socialists against Zwentendorf which, despite threats of party discipline, continued to press forward their slogan, 'Yes to the government of Kreisky, no to Zwentendorf'.[73]

On the other side were the state-owned electricity utilities and their supporters. The utilities had none of the power to mobilize possessed by the opposition, but made up for this by access to large amounts of money. They spent some $2 million on spreading their message. Millions more were poured into their campaign by the

164   GLOBAL FISSION

Industrialists Association, the umbrella organization of the trade unions, and the Socialist Party.[74]

In Switzerland the National Coordination of citizens groups promoted their campaign with demonstrations and brochures. Their case centred on the twin issues of radioactive waste, and the energy options. In particular they stressed that the energy from new nuclear reactors could be replaced by various energy conservation measures, including the trimming of Switzerland's sizeable electricity exports and the use of solar energy systems.[75] On the other hand the Association for Swiss Electrical Industries emphasized its belief that it was only through the import of electricity and the production from existing nuclear plants that Switzerland had been able to meet its winter demand the previous year.[76]

Although the debate was intense it was generally considered unlikely that the initiative would succeed. Not a single referendum proposal had been accepted since the war and, as the government pointed out, there had already been one referendum on nuclear power. In 1977 nuclear power had received an overwhelming 'yes'.[77] The nuclear industry, however, was not prepared to rely merely on these historical trends. Over the months preceding the referendum it pumped enormous amounts of money into its campaign. Opponents estimated that the industry's publicity expenditure amounted to some $50 million or 28 million Swiss Francs.[78]

In Sweden too the still relatively amorphous Folk Campaign Against Nuclear Power pitted itself against the more formal organization of the other side. It used its ability to mobilize people, bringing an estimated 120 000 out against nuclear power at demonstrations in 150 Swedish cities and towns a week before the referendum. On the other side stood not only the campaign committees for options 1 and 2, but also the business community and substantial parts of the trade union movement. Of the two large labour federations, the largest, the Swedish Trade Union Confederation, had strong links with the Social Democrats and came out in favour of option two. The other was forced by internal opposition to tread a more neutral path. The press tended to favour nuclear power. One estimate is that 75 per cent of newspapers, in terms of circulation, and 90 per cent, in terms of different newspapers, were favourable to nuclear power. This inherent bias was substantially reinforced by the business sector which set up an information service and poured many millions of Kronar into a large scale pro-nuclear advertising campaign.[79]

REFERRED TO THE PEOPLE

In each country the forces which confronted each other were different in character and capabilities. The anti-nuclear side had in general much larger access to active helpers, while the pro-nuclear forces had greater access to the established organization of society, newspapers, and money. The varying capabilities of these forces to manoeuvre in the last few weeks of a referendum might make a considerable difference to the result. Nevertheless, when the polling day arrived in each country, it was clear that the result was anything but pre-ordained.

RESULTS

On 5 November 1978, 74 per cent of the Austrian population turned out to vote on the future of the Zwentendorf reactor. When the results were announced it was clear that in the fluid fortunes of the nuclear industry a new watershed had been reached. Despite the prestige and threats of Kreisky, the power of the Socialist Party, the support of the organized trade union movement, and the big spending of the utilities, the pro-nuclear forces failed to achieve their majority. By a narrow margin of 50.5 per cent to 49.5 per cent, the people had cast their votes against the use of the completed but inert reactor. For the first time in the world a nation had voted to halt a nuclear programme.

A month later, on 15 December, the Austrian parliament unanimously passed a law prohibiting nuclear power in the country. The planning company for the reactor was dissolved, the uranium enrichment contract with the USSR cancelled,[80] and investigations were begun into the possibility of transforming Zwentendorf into a fossil-fired station.[81] For the time being at least the future of Austria seemed set on a nuclear-free road.

A few months later, during February 1979 in Switzerland, the vote went equally narrowly the other way. Nearly half the population voted on the initiative. Of their votes, 965 271 were cast against it, and 919 923 for it. Had a mere 1 per cent more voted for the initiative it would have been carried.[82] Three months later the second referendum on the 'partial revision' of the nuclear legislation initiated by the government was carried by an overwhelming 70 per cent of the votes. This time cantons which had supported the stricter measures in the first citizens' initiative, lined up in support of the government proposal.

In Sweden, on 23 March 1980, a lower than usual 74 per cent of

166   GLOBAL FISSION

the population voted on the three available referendum options.[83] Numerically the result was clear enough: option 1 ('no more than twelve reactors') received 18.7 per cent; option 2 ('no more than twelve reactors, which must be owned by the state') received 39.3 per cent; and option 3 ('no more than six reactors') received 38.6 per cent of the vote.[84] But the meaning of this result was open to interpretation.

One thing was fairly clear. Combining the vote for the nearly identical options 1 and 2, 58 per cent had voted for at most twelve reactors in Sweden, with the implication that these should all be phased out within about twenty-five years. Although this represented a majority of the vote, a solid block of almost 40 per cent had been prepared to vote for the four completed reactors to remain inert, and for the presently operating six to be phased out within ten years. Although the opponents of nuclear power were not overjoyed, neither was the nuclear industry.[85]

The immediate implication seemed to be that the four completed reactors would be allowed to operate, bringing Sweden's operating total to ten. Beyond that, a maximum of two more seemed to have been set. Whether these two would be completed and allowed to operate, and the speed at which the ten would be phased out, now seemed to be the area in which the battle over nuclear power in Sweden would concentrate.

AFTERMATH

At the beginning of this chapter it was noted that those who initiate a referendum may hope to achieve several objectives. One of these, of course, is to win the vote. But even if the vote is won, there is something more which may be accomplished. That is to persuade those who lost that the referendum's result is binding, not only for the present, but also for the future.

The strength of the claim that the result of a referendum should be permanent is that the whole nation has had an opportunity to consider and vote on the issue. Its weakness is that the opportunity occurs at one instant in the circumstances of the conflicting information with which the people are presented.

If the side that loses chooses to accept the claimed permanency of the referendum's verdict the referendum will mark the end of the conflict. But if it does not the referendum will represent simply a

REFERRED TO THE PEOPLE 167

single test of strength, on a particular day, in an on-going battle. In the wake of the referenda over nuclear power in Austria, Sweden, and Switzerland there was little evidence that any side believed that the battle was over.

Writing from Sweden several weeks before the referendum had been held, nuclear commentator Wendy Barnaby noted that 'it is clear that the referendum will not be the final curtain of the nuclear drama. It will simply set the stage for another long act'.[86] Afterwards there was no evidence that the political forces that had destabilized the situation and forced the referendum, had magically faded.

Fälldin remained in office noting that although he was committed to allowing the operation of the four waiting reactors, which duly were approved,[87] his party was also best equipped to supervise the phasing out of all the reactors.[88] The opposition responded to the referendum result with the slogan 'the fight goes on', and by September a demonstration of 10 000 people against Barsebäck showed that they meant it.[89] In addition the opening shots were already being sounded in the fight to prevent the construction of the twelfth plant.[90] On the other side, the nuclear industry did not accept the limit of twelve reactors set by the referendum. Said Curt Nicolin, chairman of ASEA-Atom, 'No political decision is valid for a long time'.[91] There could be no doubt the fight would go on.

In Austria, the nuclear industry showed equal reticence in accepting the result of the referendum as final. This had been anticipated by the opposition. As Peter Weish had commented in 1979: 'Far from being complacent about the narrow victory, anti-nuclear groups are well aware of the fact that the nuclear industry will sooner or later try to reverse the vote'.[92]

By the end of 1980 the nuclear industry campaign for a new referendum was well under way. The country's three political parties had agreed that the Austrian Nuclear Deterrent Act passed the previous November could only be changed by a second referendum and a two-thirds vote of the parliament. By July 1980 the nuclear proponents had gathered together the 10 000 signatures necessary to place before the parliament a 'people's request' for a referendum on abolishing the act. By then, the opponents of nuclear power had put together a similar number of signatures requesting a referendum on the conversion of Zwentendorf to a fossil-fuel powered station. The request was a preparatory step

towards achieving a petition from 200 000 people. This would force the parliament to vote on whether they would hold a referendum. There was little doubt that both sides would achieve their 200 000 signature targets.[93]

There was no guarantee that the parliament would agree to either referendum, nor which way the referendum would go if it were held, nor what the parliament would do even if a decision to overturn the act were carried. What was clear was that in Austria too the fight would go on.

In Switzerland also, although no new steps were yet under way for a new anti-nuclear initiative at the federal level, action was now visible at the cantonal level. In June 1977, 76 per cent of the population of Basel had voted for a moratorium on reactor building.[94] Initiatives in other cantons had been overshadowed by the federal referendum. Now the movement was striving to further increase its momentum by obligating cantons to oppose nuclear reactors by all legal and political means at their disposal. The results were soon visible. In November 1979 the canton of Geneva asked the federal government to cancel its plans to build a reactor near Verbois, in the neighbourhood of Geneva.[95] In the same week a power transformer at one of Switzerland's three operating reactors was damaged by a bomb.[96] The cantonal councils of Basel-City and Basel-Country also asked that site permission for Kaiseraugst be formally revoked.[97] In March 1980 this call was reinforced when the people of Basel-Country voted 4 to 1 against the building of the reactor.[98] About the same time the villagers of Haegendorf voted 3 to 1 against allowing test drilling for waste disposal in their areas,[99] and in the St Gallen area the chairman of the company that was to have built the Rüthi reactor announced the plan had been cancelled.[1]

The nation was still clearly deeply divided by the nuclear issue, and the constraints on the development of the nuclear programme seemed increasingly strong. It was estimated that even the provisions of the revised law would stretch out the time to bring a nuclear plant into operation to about thirteen years. As an industry spokesman said, 'If we move ahead it will be slowly'.[2] It seemed an understatement.

With nuclear plans already dramatically retarded it was clear that the conflict over nuclear power in Austria, Switzerland, and Sweden had not been substantially deflected by the referenda results. Indeed the referenda seemed merely to be symptoms of the

REFERRED TO THE PEOPLE

extraordinary growth of opposition to nuclear power that had reduced the plans of these countries' nuclear industries almost to a standstill. This was not peculiar to these countries. In the other countries of Scandinavia also, wherever nuclear plans had been formulated, they seemed now to be undergoing a similar attack.

Only Iceland, which sits virtually on top of a live volcano, had escaped untouched. With geothermal energy supplying all of the heating of its capital, Reykjavik, and a substantial surplus of hydro-electric power, it has no use for nuclear power.[3] On the other hand, in Norway, despite its large hydroelectric resources, nuclear power was being considered as a possible source for power. But in early 1980 the release of a government white paper suggested that the lesson from neighbouring countries had been learned. It recommended that Norway should have no thought of nuclear power, at least before the turn of the century.[4]

In contrast to Norway, Finland's nuclear programme continued to progress, but in the face of a growing opposition. A proposal for a fifth nuclear reactor was met by the distribution of 200 000 copies of a four-page newspaper opposing it.[5] According to two Gallup polls, taken a year apart, the number of Finns who believed nuclear power safe had dropped from 43 to 34 per cent.[6] Results were seen in mid 1980 when 6 000 people marched against nuclear power, the largest demonstration in Finland since the war.[7] Further progress of the nuclear programme seemed certain to be accompanied by escalating conflict. The situation was very different in Greenland, with no plans for nuclear power but with substantial uranium reserves. In February 1980 the government's majority leader, Lars Johanson, announced his party's decision not to allow uranium mining to take place.[8] Although Greenland is a territory of Denmark, the decision met little resistance.[9] It was hardly surprising since Denmark's nuclear plans had recently suffered a major setback.

In Denmark, as in Sweden, it was the accident at Three Mile Island which had raised the tempo of the debate decisively. A few weeks before, the Minister for Commerce had announced that a decision on the first reactor could be expected by the end of the year.[10] Then came the accident, and by mid May 300 000 signatures had been collected demanding that the Danish government request Sweden to close down its nearby Barsebäck reactor.[11] The signatures represented more than 6 per cent of the population of Denmark.[12] Two weeks later, over 50 000 Danes demonstrated in

170   *GLOBAL FISSION*

fifteen villages and towns for a permanent closure of the reactor.[13]

The OOA had never called for a referendum, believing that it could do nothing to resolve the hazards of nuclear power and that it might prematurely alter the political balance before the public had fully grasped the extent of those problems.[14] Nevertheless, as the opposition grew the two government coalition partners recognized that to proceed with their nuclear programme in this climate could prove politically disastrous. Four months after the accident the two government coalition partners, the Liberals and the Social Democrats, announced that all decisions on the possible construction of nuclear reactors had been postponed. They promised that a referendum would be held before any vote would be taken by the parliament on the introduction of nuclear power in Denmark.[15]

The key countries examined in this chapter all form a pattern. Initially, ambitious nuclear programmes gave rise to an unexpected and increasingly powerful citizens response. The ensuing conflict, in marked contrast to that in some other countries, was essentially non-violent in character. Instead opponents and proponents of nuclear power clashed through the media of non-violent occupations and demonstrations, courts, information campaigns, elections, and referenda. Although the detail differs between countries, in those countries examined in this chapter, in which the conflict has been most intense, a general pattern is clear. Initially ambitious nuclear programmes crystallized a citizens response either locally or nationally. Each country moved through an information campaign and many more citizens penetrated the curtain of technicalities which shrouded the proposed developments. They did not like what they saw.

As the community and the political parties divided over the issue, governments sought to contain the conflict. The divisions and pressure became so intense that governments were forced to initiate referenda with questions which could limit the government's own nuclear programmes. This tactic may have taken pressure off forthcoming elections or, in the case of Switzerland, helped to defeat a more substantial referendum question, but in the longer term it did nothing to end the conflict. On the contrary, whether resorted to by nuclear proponents or opposition, the net effect was to add to a process which had reduced the ambitious nuclear plans of government and industry to a virtual stalemate through the development of opposition and action within broad sections of the community.

REFERRED TO THE PEOPLE

In these countries the power exercised by the opposition has stemmed not only from the strength of broad community opposition to the government's nuclear programmes, but also from the divisions forced into the unity of the parliaments and, more broadly, the state. It might therefore be expected that in those countries in which the central government is most united, strong, and least subject to challenge, and where the state is most cohesive and integrated with nuclear activities, the national nuclear programmes are in their healthiest condition. The situation of nuclear power in two such countries, and the nature of the opposition that rises to meet it, are the subject of the next chapter.

8
CENTRAL CONTROL, REGIONAL REVOLT

In the home countries of the West's nuclear reactor vendors, it is the UK and France in which the nuclear industry and the state are most integrated. Additionally, in these two countries there are few formal constraints restricting the power of the central governments to carry through their nuclear plans. It is therefore not surprising that both countries still have particularly ambitious nuclear programmes, and that France has been uniquely successful in the 1970s in putting its programme into practice.

As in all Western countries the implementation of nuclear programmes in the UK and France has been accompanied by growing citizens opposition. But during the 1970s, in both of these countries, that opposition was insufficient to markedly retard their nuclear programmes. In examining why this is so for them, but not for the other Western countries with large nuclear programmes, light can be shed on the nature of both the opposition and the battle in which it is now engaged.

As already noted in chapter 6, the centralized and sophisticated nature of nuclear technology gives it certain political properties. Most importantly, it makes it useable as an instrument in the growth and centralization of power in both the state and the larger corporations, reinforcing the overall trend of development in the post-war capitalist economy. This provides considerable incentive for the use of nuclear power additional to any technical advantages that it might be considered to possess over other energy sources. This may not always have been understood in its full generality by state or corporate planners. But it need only enter into their thinking piecemeal to greatly help explain why they have pressed nuclear

CENTRAL CONTROL, REGIONAL REVOLT   173

power forward with such vigour. In short, the development of nuclear power is a social project ('the nuclear enterprise'), and instead of being motivated simply by technical or economic considerations, it is motivated by the sum of all the implications that it carries for the shape of, and distribution of power in, the future.

In this chapter this argument is taken further by also examining the opposition to nuclear power in this way. Is it not possible that the opposition has grown up not merely in response to the technical problems – the 'hazards' associated with nuclear technology – but also in response to the other social and political implications of the nuclear enterprise? There is evidence that although the individual groups opposing a particular project may not always realize it, the basis of their opposition is not only the hazards implicit in the project, but also concern over the political relations built into it. In particular, there is evidence that the opposition to nuclear power is a response not just to the technological problems of the nuclear enterprise, but to the centralization of power which it implies. This can be seen most clearly by examining those countries in which the centralization of power is already strong and combined with a nuclear programme.

In a country with a strongly centralized system of government, and without any strong federal structure which could block the exercise of central authority, a pattern tends to develop. The opposition to nuclear power is usually strongest in regions where opposition to the state's exercise of central power is also strongest. It is particularly strong in those areas which through history, culture, and sometimes language, have a sense of regional community identity which can at times outweigh any allegiance to the central national government.

It is misleading to view the state and the corporations, or even the state alone, as simply a monolith. The state has an institutional structure which has grown up as a response to a history of conflicts of interest within the society. This is not to say that the state acts neutrally. On the contrary, its principal role is to administer the society so as to facilitate the stability of the existing social order and the smooth working of its economy. In this way the state supports the interests of the prevailing power group which is, in the countries presently being considered, the dominant economic class. But the state is not a single, completely united entity. Its departments comprise large numbers of workers who are themselves increasingly unionized. And it contains a variety of structures reflecting

174   GLOBAL FISSION

different competing interests, some of which are regional in character and can limit the exercise of central power. Prominent among these are regional and local governments.

Regional governments have a variety of characteristics, some of which reinforce the exercise of central state power and some of which challenge it. Miliband puts it this way:

> In the countries of advanced capitalism, on the other hand, sub-central government is rather more than an administrative device. In addition to being agents of the state, these units of government have also traditionally performed another function. They have not only been the channel of communication and administration from the centre to the periphery, but also the voice of the periphery, or of particular interests at the periphery; they have been a means of overcoming local particularities, but also platforms for their expression, instruments of central control and obstacles to it.[1]

Although it has obvious relevance to a federal system of government, or one in which there is at least some relatively strong system of local government, the concept of a periphery which may limit the exercise of central power by the state need not be restricted to formal administrative sub-units nor even to informal geographically defined regions. Instead the society can be seen as spread, not necessarily evenly, across an arena. At the centre, perhaps densely clustered, are those for whom the actions of the state carry greatest legitimacy, including those within the formal central organization of the state itself. Around the periphery are those for whom the state's actions are in general more open to challenge. Prominent among these, as already mentioned, are likely to be groups which possess a sharply different culture and historical identity from that of the dominant groups in the society: rural people whose way of life is suffering from the overall trend of development in the society, and displaced native peoples such as the Australian Aboriginals and the native peoples of North America.

Because of the technology's centralized form, decisions about nuclear power are usually made and implemented from the centre. But it is in the periphery that opposition to those decisions is likely to erupt most fiercely. This polarization is most pronounced in countries where there are fewest formal barriers to the exercise of the centre's political power. It is not reduced by the tendency of central governments to place their most polluting projects in places

CENTRAL CONTROL, REGIONAL REVOLT   175

where they will alienate fewest of their supporters. The export of pollution from the centre places nuclear plants where people on the periphery are clustered, giving rise to a heightened eruption of opposition.

France and, until recently, the UK have little formal federal structure. The exercise of central authority is relatively unrestricted and all sections of the nuclear fuel cycle are tightly and formally incorporated as operations of state agencies. It is perhaps not surprising that these countries' nuclear programmes have progressed with uncharacteristic ease. Where they have encountered resistance, it has flared up with great ferocity and effectiveness on the periphery. This can be seen by examining the course of the battle over nuclear power in these countries.

CENTRAL DEVELOPMENTS 1950–79

In the British Isles only Ireland has achieved full independence. Northern Ireland, Scotland, Wales, the Orkney Islands, and the Isle of Man, despite considerable regionalist sentiment, have at best a few limited powers for home rule. Outside these they must rely for influence over central policy on their minority representation in the national parliament in England. France is even more directly consolidated into a single nation subject in its entirety to the central authority of the president and the two houses of parliament. Especially in France, the authority of the centre is consolidated through a strong spirit of nationalism.

Nevertheless, France also has its areas of strong regionalist sentiment emanating from historical factors which pre-date the present French borders. In the north-east corner bordering Germany is Alsace, encompassing the French part of Dreyeckland. In the north, bordering Belgium, is Flanders. On the west coast is Brittany, with its own local language and, just above it, Normandy. In the south, bordering Spain, are the French Basque and Catalan regions, and around them the larger region of Occitanie. These areas have played a key role in the development of opposition to the country's central nuclear plans.

In the UK and France, the 1960s and early 1970s were a time for ambitious nuclear plans. In the UK, there was also early action. In 1956 the first 'commercial' nuclear reactor began operation in the UK. It was a plutonium producer for the nuclear weapons programme, to which had been added a small generator. Seven more

176 GLOBAL FISSION

were completed by the end of the 1950s, making the UK the then world leader in civilian nuclear power.[2] The pace did not slow. Eighteen more were completed in the 1960s and eleven planned for completion by 1975.[3] Only in the late 1970s, when a substantial surplus generating capacity had been achieved, did the programme come to a halt. By then the UK had almost 12 GWe of nuclear power, most of it situated away from major population centres in East Anglia, the Severn Estuary, north England, Scotland, and Wales.

Opposition was slow to develop in the UK. Even the unhealthy release of radioactive iodine which occurred when a plutonium producing reactor caught fire at Windscale in 1957, caused little long-term outcry.[4] It was not until 1975 that opposition to nuclear power became at all noticeable. A network of FOE environmental groups had been set up over the past five years, and they and some existing, mainly conservation, organizations began to take up a position of concern over nuclear hazards. But the burgeoning opposition was faced with a parliament in which both the ruling Labour Party and the opposition Conservative Party were enthusiastically in favour of nuclear power. The nuclear opponents, especially the well-respected consultants to the London FOE office, spent a great deal of effort attempting to persuade the Labour government to change its position, but were unsuccessful. Nevertheless, their efforts were crucial in documenting and making more widely known the fragility of the economic case for nuclear power and the extent of its hazards.[5]

In terms of the argument, the opponents were very successful. In September 1976, those arguments were given a solid stamp of legitimacy when a royal commission on nuclear power and the environment, chaired by Sir Brian Flowers, produced a report which expressed grave concern over the hazards associated with a large nuclear reactor programme.[6] Despite the increasing credibility of their arguments the political situation remained unshaken, with all parties supporting nuclear power. The nuclear programme continued with implacable momentum, only faltering when it was realized that more generating capacity was being built than the UK seemed able to use.

The only success achieved in the centre of the political arena was tacit agreement by the government that major projects should be subjected to a process of open public inquiry. For a central government such a process can have two important advantages. It can

CENTRAL CONTROL, REGIONAL REVOLT

maximize the legitimacy of its final decision by demonstrating that all reasonable channels were exhausted before the 'difficult decision' was made. It can also defuse the opposition by diverting its energy into the substantial effort of preparing testimony, raising money for legal representation, and the like. On the other hand, as was demonstrated by the Australian Ranger Inquiry and the Canadian Porter Inquiry (both to be discussed later), this can sometimes produce uncomfortable results.

In March 1977 British Nuclear Fuels (Limited) (BNFL) submitted a formal application to build a reprocessing plant which would be capable of extracting some of the plutonium and uranium from modern reactor spent fuel rods. This conjured up the spectre of the 'plutonium economy' with the physical and social hazards of regular cross-country shipments of plutonium, and eventually the construction of fast-breeder reactors. In June, the Windscale Inquiry under Mr Justice Parker began a long set of hearings into the proposal. One hundred days of sittings, 146 witnesses, much argument by the Queen's Counsels of FOE and BNFL, and 1 500 documents later, the inquiry closed.[7] All eyes were turned on 'Inspector Parker' for the penetrating analysis which the media and the comments from both sides at the hearings had firmly encouraged everyone to expect.

In the event, the framework through which Justice Parker viewed the world allowed little room for understanding of the objectors' case. The result was that the report, released in January 1978, failed, as *New Scientist* editorialized, to 'reflect the detail and character of their case. One need not be on their side to note that the report fails to do this and, more seriously, misrepresents views and obfuscates or distorts the context in which those views were tendered to the inquiry'.[8] The result was anger. Noted Jeremy Bugler in the *New Statesman*, 'until now the anti-nuclear forces in Britain have shown themselves willing to protest "within the system" . . . it will not be surprising if the anti-nuclear movement here now changes its approach'.[9] He was right. On 29 April 1978, 10 000 people packed Trafalgar Square to demonstrate against the Windscale expansion plans. It was the first large anti-nuclear power demonstration ever to be held in England.[10]

Opposition was building, but it would have to be built much more if it were to impinge on the nuclear momentum emanating from the pro-nuclear sections of the community and the state. That momentum was greatly increased a year later when, on 3 May

178   GLOBAL FISSION

1979, the Labour Party fell from office and the movement found itself facing an apparently implacably pro-nuclear Conservative government.

In France the unanimous support of all the parliamentary parties was only one contributor to the solid central authority with which nuclear plans entered the political arena. As in many other countries when technical matters are at issue its elected government acts more as a rubber stamp for policy already decided by its technical establishment. In 1946 an elaborate process for framing national plans had been instituted under Charles de Gaulle. It imparted to the decisions of the administration's technical branches a high level of authority. As N. Lucas notes in a study of this process, the result has been an 'acquisition of the authority of the state by public enterprise' which is 'not inhibited by the political assemblies which are weak'. This authority is strengthened by the consensus between senior administrators and directors of public agencies such as the electricity authority EDF, (Électricité de France), caused by their common schooling, technocratic philosophy and, usually, Gaullist politics. The result is 'homogeneity between political power, administrative power and economic power which permits policies in France to be driven through'.[11]

For nuclear power, the responsible planning agency is the Commission on Nuclear Electricity Production, PEON, (Commission sur la Production d'Électricité d'origine Nucléaire). Although a few small reactors of French design had been built, it was not until May 1969 that substantial plans for an additional four to five large US designed commercial reactors were announced by PEON.[12] In 1973, when the OPEC oil crisis struck, PEON responded by setting a target of 200 GWe of nuclear power for the year 2 000. Early the following year this became the national target when the president announced his government's intention to shift from oil to uranium as the country's basic fuel for electricity generation. To achieve this, seven 1 GWe nuclear reactors would be built each year. It was too ambitious a programme to be realized, but nevertheless EDF would achieve sufficient momentum towards this goal to become the envy of every other nuclear authority in the world.

These bold plans, emanating with powerful central authority, did not impinge on a politically unaware populace. One year before the first PEON plan was announced in May 1968, an upsurge of revolt over widespread unemployment and discontent with the

CENTRAL CONTROL, REGIONAL REVOLT

social system had swept through the major cities paralyzing the economy. In that month, fired by ideals, hope, and a spirit of rebellion, unionists had gone on strike nationally, and citizens had taken to the streets. When the crowds had dispersed and the barricades had been dismantled, a deeply divided society remained, strongly polarized into left and right ideologies, factions, and parties.

The government too had learnt from those tempestuous days, and the militant demonstrations which afterwards became commonplace. It reinforced the available means of confronting and repelling citizens revolt. In particular, it strengthened the CRS, the national riot control police (*Compagnie Républicaine de Sécurité*), and equipped them to form a powerful riot control unit.

In the major cities, opposition to nuclear power was visible by 1972, when a coalition of environmental groups (including FOE) initiated a petition for a moratorium on nuclear developments.[13] By the time their campaign had finished they had collected over 100 000 signatures.[14] The campaign could be seen to be flowering not only in the major cities when, on 6 May 1973, small demonstrations occurred in Strasbourg, Toulouse, Montpellier, and Cherbourg – all areas with a regionalist tradition – as well as in Paris and Lyon. These provided an important springboard for the 'ecological' campaign of René Dumond.

Already ecologists in groups such as FOE were developing a vision of an alternative to the future for France implied in state plans. Overtly recognizing the political relations built into nuclear technology, this vision was based around the decentralization of both energy sources and political power. They envisaged the restructuring of French society into self-managed communities. These would feature planned growth, renewable energy systems, integrated agriculture, neighbourhood democracy, and the control of factories by workers. Their ideas echoed the concerns of many over the direction of development of French society, and Dumond was a particularly able exponent of the new philosophy.

On 2 April 1974 the death of President Georges Pompidou precipitated a national election. The opportunity of presenting an ecological candidate was grasped and over the months leading to the election, millions of people in the cities were introduced to Dumond's environmental philosophy and the problems of nuclear power. Dumond received only 1.3 per cent of the vote, but the ecologists were far from disappointed. The anti-nuclear campaign

had been greatly stimulated and, for the first time in the community, both the left and right had been forced to question the problems and limitations of perpetual economic growth.

The growth of opposition was impressive but at the centre it confronted an implacable wall. In the parliament in May 1975, when the government held its promised debate on the nuclear programme, not a single member challenged the government's intention to base France's energy future on nuclear power.[15] The major opposition parties in the parliament were the Socialist Party and the Communist Party. In both, the leadership shared with the Gaullists the nationalism, the belief in the progressive nature of all modern technology, and an implicit faith in the advice of the technical bureaucracy which held them firm to a position of almost unquestioning support for nuclear power. In particular the hefty Communist Party, with its leadership's still evident allegiance to the USSR, saw nuclear power as an inevitable adjunct to its vision of a socialist France.

Only one political party, the tiny Unified Socialist Party (PSU), had seriously questioned the nuclear proposals. It had developed a radical critique of society which already accorded well with the ecologists' views. It had declared itself in favour of a moratorium on nuclear construction in 1974 but its influence was insignificant in the face of the ponderous bulk of the Socialist and the Communist parties.

At the level of central government the growth of the ecologists therefore had little impact. However, in the 1976 municipal elections the ecologists showed they were capable of inserting a wedge at the local government level when their candidate for the district of Paris received 7 per cent of the vote. A year later, when the ecologists ran again, their 'green list' achieved a national average of 10 per cent of the vote. In some places the vote reached 20 per cent. In Creys-Malville the ecologist candidate received a vote of 60 per cent and was elected.[16]

These successes, however, did not inhibit the relentless progress of the state's nuclear programme. Although practicalities had necessitated the reduction of the original extremely ambitious rate of seven reactors per year to a more manageable five or six, EDF's activities continued. Early 1976 saw preparations for reactors in Cruas, Cattenom, Nogent-sur-Seine, St Maurice l'Exil, and a major complex at Flamanville. An enrichment plant would be built at Val-de-Sâone.

CENTRAL CONTROL, REGIONAL REVOLT   181

The government's only clear acknowledgement of public concern was the development of a familiar tactic. As early as 1974 it mounted an information campaign. The result, as usual, nourished the growth of the opposition. By 1977 the public relations service of the French AEC had a budget of $1.9 million (8 million Francs) and a library of fifty-eight films which it circulated to over 3 000 schools. It had produced 40 000 manuals, complete with slides, and 320 000 brochures on nuclear power.[17]

In the forthcoming battle with the emerging opposition, the government, like the state in many other countries, would have many more tactics available than that of the information campaign. In the crucial battle for public opinion it would be able to make use of many publicity channels, from ministerial statements to glossy brochures and advertisements, to press forward its message. Even so, the outcome of the battle would not necessarily be decided by which side possessed the best access to the media. It would also depend on the power of each side's respective messages.

The opposition arguments can be very effective. But the state also has effective messages available. It may appeal to nationalism, citing the need to maintain the nation's prestige and balance of payments. It may draw on its role as manager of the national economy by arguing that disruption of the nuclear programme jeopardizes both the entire national plan and the country's economic future. It may appeal to the broadly held faith in progress and its ready access to experts, and define its plans as technically pre-determined and thus inevitable, and those of the opposition to be naive and their supporters latter-day Luddites.

All these are powerful messages which have at times proved very effective, and their use constitutes part of the armoury of 'soft' tactics available to the state to counter the opposition. But other more coercive tactics can also be employed. These include directly confronting the opposition with force. In France, on 31 July 1977, at Creys-Malville, the government demonstrated that in certain circumstances such action is possible. On that day, citizens groups from all over France converged on the tiny village of Creys-Malville to protest the building of a nuclear reactor there. By launching such concerted action they hoped to achieve much greater political pressure than could be developed by isolated actions in diverse regions. When they arrived they found their centralized citizens action met by the full central power of the state.

MALVILLE   Creys-Malville, situated on the banks of the

Rhône, lies in the middle of a triangle formed by Lyon, Grenoble, and Geneva. There are regionalist sentiments in the area, but they are mixed with the heavily industrialized and capitalist atmosphere of Lyon, where some hold pretentions to make the city the leading economic centre for France. The site of the reactor may be rural in character, but within 100 kilometres of it live six million people. It seems a curious choice for the site of a prototype breeder reactor.

The breeder is an experimental technology whose purpose is to turn the relatively inert Uranium-238, which forms 99.3 per cent of the uranium dug from the ground, into plutonium which can be used as a fuel. The hazards are enormous; much greater than with conventional reactors. It is theoretically possible for a loss of cooling accident to cause a nuclear explosion or, as it is sometimes euphemistically described by the industry, a 'core disassembly incident'. Further, the coolant itself is liquid sodium, which catches fire spontaneously in air and explodes on contact with water.

An early demonstration against these hazards at the plant's site was organized by FOE and local branches of the PSU and the Socialist Party, as well as some local trade union branches in July 1976. It drew 25 000 people. After several days of occupation the CRS arrived and the occupiers were dispersed. It was only a mild taste of things to come. However, even this shocked the local inhabitants. A week later farmers paraded their tractors in protest, accompanied by some 6 000 people, many of whom were locals.[18] By 1977 over 100 Malville committees had been formed across France to try to stop the project.[19]

On 31 July 1977 the opposition to Creys-Malville from across France crystallized into an attempted occupation of the site by 60 000 people. Included were some 20 000 drawn from Germany, Switzerland, Spain, and other parts of Europe. Confronting them was a substantial detachment of the CRS. Although most demonstrators intended to use non-violent tactics, it took only the provocations of a few small groups for the CRS to launch the first of several brutal baton charges. It was in one of these that 31-year-old school teacher Vital Michalon was killed.

The CRS used rifle launched tear-gas and percussion grenades. Not content with the effects of the gas, they fired these directly at the protestors. One of them exploded on the chest of Michalon who died shortly after. Before the day was over, more than 100 protestors had been injured, several losing arms and legs from the effects of the grenades. Wrote one commentator:

CENTRAL CONTROL, REGIONAL REVOLT

It was as if the worst nightmares of the ecologists had come true. The scenario of unequal conflict between a centralised, technocratic state and citizens, powerless to influence the decisions that affect their lives, was vividly dramatised. After the demonstration one inhabitant of a village near Malville commented, 'When the protestors said that a nuclear society was a police society, we didn't believe them. Now we know that they are right'.[20]

For the demonstrators the lesson was devastating. In their most powerful national expression of concerned opposition to date, they had been totally overpowered, outmanoeuvred, and put to rout.

The effect was divisive and disheartening. Even before the demonstration some had warned that violence was likely and had opposed it on that basis. But the CRS's violent action also shook many who had supported the demonstration, and bitter arguments followed over what tactics were now reasonable to even consider using. It was little consolation to the nuclear opponents that they had taken a cavalier attitude to coordination, tactics, and strategy before the demonstration.[21] The rout was profoundly disheartening not only for opponents in France but in several other countries. The whole value of a centralized confrontation with the state had been thrown into question. In France it took several years before the movement began to recover as a national entity.

The next year, in a centralized confrontation of a different kind, the ecologists were also unsuccessful. At the national elections held at the end of 1978 the ecologists ran a new 'green' list. But despite much campaigning they received less than 3 per cent of the vote. Even at Creys-Malville they got only 4 per cent. The press were quick to interpret it as the 'death of the ecology movement' but they were wrong. It was more the strategic error of taking on a contest they could not win.

It was one thing to lure voters from their traditional party politics in a municipal election on a matter that could severely change the nature of the locality. It was quite another to ask voters to shed their party loyalty on the question of who should govern France. It is absurd to expect the ecologists to manufacture the national image and policy of a fully fledged national political party. To do so they would have to gain consensus on a vast array of issues which were quite different in character from those that held them together, not the least of which was a resistance to national decision-making.

184   *GLOBAL FISSION*

The failure of the ecologists as a political party at the national level said nothing about the degree to which the nuclear, or indeed the ecological, issues had permeated the community. It merely showed that people were not prepared to thrust all other issues aside for these. In the autumn of 1978 an opinion poll showed that 77 per cent of the community still considered nuclear power dangerous.[22] But so far that concern could find no effective expression at the centre of French politics.

In France and in the UK, national opposition stood over-powered by the national power of the state. It would be easy to believe that this meant there was no threat to the progress of the central government's nuclear plans. But this was not so. The point is that where resistance was effective, it was on the periphery.

ON THE PERIPHERY 1970–80

In France, as in many other countries, the growth of opposition was strongest in the communities where nuclear plants were to be built. The seeds were sown in the late 1960s at Fessenheim in Alsace, where a committee formed around Jean-Jacques Rettig, a local school teacher, to try to prevent the construction of two large reactors. A little later, in 1970, a group of local farmers and towns-people formed another group opposed to the reactors, and in April some 1 500 marched in opposition to it. It was no surprise that opposition should begin here. The region adjoins the by now familiar area of Dreyeckland.

Near Lyon another group, Bugey-Cobayes, was set up by locals to oppose the Bugey-1 reactor. By July 1971 they were strong enough to organize a festival of 5 000 in protest against the plant. Regional opposition spread over the next few years. In Bordeaux at Braud-Saint-Louis local farmers and fishermen began organizing against the Blayais reactor complex. They gathered together some 25 000 letters and documents opposing it.[23] At Creys-Malville locals forced an, admittedly token, inquiry.[24]

The government's response was the information campaign. They sent a map of the thirty-five reactor sites to the regional councils. This stirred opposition further. But it also gave EDF a major strategic advantage. It was now able to monitor opposition to the plans and place its reactors in those regions where it was weakest. This enabled the programme to proceed but did nothing

to reduce the growth of opposition. In April 1975 a 'week of anti-nuclear action' showed that it had spread widely. As 25 000 people marched in Paris, 3 000 occupied the site of a four-reactor complex at Gravelines in Flanders, and 6 000 gathered at Paluel. Others demonstrated at Royan, Mans, Narbonne, and Lyon.[25] Regional opposition was now widespread and still relatively low key. However, a more powerful battle was already stirring at Flamanville.

FLAMANVILLE   On the tip of a wave-lashed promontory known as Cap de la Hague, on the spectacular wind-blown coast of Normandy, lies Flamanville, the site of a proposed complex of four nuclear reactors. Carved into verdant pasture land, its planned output of 5.2 GWe would render it one of the largest nuclear complexes in the world.[26] For many of the local farmers and fishermen, however, it represents a major step towards transforming their homeland into a giant nuclear park.

The first step was the construction of the la Hague reprocessing plant on the promontory over 1961–65. Since then it had continually discharged radiation into the sea and air, and local action groups had documented what they believe is a serious and increasing accumulation of radioactive isotopes in the environment.[27] Many accidents had added further weight to their concern.[28] The plant's red-striped chimney is clearly visible from the Flamanville site.

By the time EDF was ready to begin preliminary site works on the Flamanville complex there was already a history of demonstrations against it, and two opposition organisations, CRILAN, the Regional Committee for Information and Anti-Nuclear Struggle (*Le Comité Régional d'Information et de Lutte Anti-nucléaire*), and CCPAH, the Committee against Atomic Pollution at la Hague (*Le Comité Contre la Pollution Atomique dans la Hague*), had been set up. When construction workers arrived in February 1977 several hundred demonstrators were waiting. Before the day was over the fence had been torn down, trucks overturned, and the construction workers forced away. A week later the locals began an indefinite occupation of the site and the workers were withdrawn. Six weeks later the CRS arrived.

Using a phalanx of bulldozers escorted by several bus-loads of police armed with tear gas, the 200 occupiers were swept aside and arrested. Nine days later, when 6 000 demonstrators gathered, they

found the site rendered impregnable by the batons and rifle-launched gas grenades of the CRS. Direct confrontation seemed useless. The local people therefore shifted their strategy to a programme of local education and harassment aimed at delaying the operations and further building opposition.

One strategy was the use of legal channels. The local farmers set up a cooperative and bought land on the site. Now EDF would have to battle an expropriation order through the courts before it could begin actual construction. Another strategy became possible after the CRS were replaced by the company's civilian guards. Small guerilla raids were launched to sabotage bulldozers, and damage also began to mysteriously afflict the company's equipment in other parts of France. However, in the long term, none of this was enough. It also became necessary to build links with the broader community. One important additional source of support were the workers at the la Hague plant.

Among the la Hague workers, concern over the hazards of the plant had existed for some time. Built to treat the wastes from the French programme, the la Hague plant had recently begun to process substantial quantities of wastes from elsewhere. If the peninsula was on the way to becoming the power station for France, it was also becoming the nuclear dustbin of the world. The implications for the workers were serious, with accidents and routine radiation exposure rapidly increasing with the swelling volume of wastes.

Some of the workers at la Hague had begun to discuss these problems in their union federation, the CFDT, the French Democratic Confederation of Labour (*la Confédération Française Démocratique du Travail*). As early as March 1974 the CFDT had released a series of brochures canvassing the nuclear issue, and these later formed the basis of an influential book.[29] In April 1975 it went further and declared itself opposed to the government's nuclear programme, criticizing 'this snowball of secret decision'.[30] Later still it declared itself in favour of a three-year moratorium.[31] At la Hague, however, the issue was becoming pressing.

Between May and October 1976 the workers at the plant went on strike. Backed by the CFDT, they demanded an overhaul of the entire installation, and opposed its expansion under a private company, COGEMA. It was a vital opportunity for local farmers and workers to form greatly strengthened links. It also highlighted

CENTRAL CONTROL, REGIONAL REVOLT 187

their joint problems across France. On 6 June 1977 a demonstration of 10 000 at the plant showed that the people of Flamanville were no longer completely isolated.

Within the plant, the subsequent debate broadened the concern of many workers from the specific hazards that they faced to the more general problems posed by Flamanville. However there was still a serious obstacle to further development of the concern. Apart from the CFDT, the other major union confederation is the CGT, the General Confederation of Workers (*la Confédération Générale du Travail*). It covers those unions linked to the Communist Party, and it still strongly supported nuclear power. But a process of erosion had begun, and by November 1978 the grip of the national leadership over its local branches on this issue had begun to loosen. At la Hague a coalition of political forces described in the local paper *La Manche* as 'inconceivable even a few weeks before' signed a common statement of opposition to the 'demented programme' to extend the plant.[32] Included among the signatories were the local branches of the CGT, CFDT, and the Communist and Socialist parties. Slowly, at the periphery, a wedge was being driven in which might eventually crack the impassive solidity at the centre.

Two months later, in January 1979, eighteen organizations including the local sections of the CFDT and the CGT called a demonstration against the first load of Japanese spent fuel rods arriving aboard the *Pacific Fisher* for reprocessing at la Hague. More than 5 000 demonstrators were ready when the containers were unloaded. The confrontation with the CRS lasted several hours, although their tear gas and water cannon finally prevailed.[33] Now not only were important links developing but, at least on the periphery, tempo was returning to the post-Malville citizens movement.

The tempo increased with the accident at Three Mile Island in May. In one of the more notable reactions, the locals in the town of Chooz near the border with Belgium locked the mayor in the town hall for several hours in protest against a planned reactor extension.[34] Elsewhere there were also demonstrations.[35] Of these, the most impressive was the battle between local people and the CRS that had erupted in Brittany around Cap Sizun.

The south side of Cap Sizun, a peninsula jutting into the sea off Brittany, has been chosen as the site of four reactors with a total capacity of 5.2 GWe. These may be of value to Paris, but for the

local fishing community they represent a threat to everything they value. Brittany is one of the poorest regions of France, and a substantial proportion of Bretons have supported, at least nominally, an active independence movement.[36] In January 1980 the combination of these two factors came to a head.

French law requires EDF to consult the local people on such plans. But when EDF moved to do so, both the residents and the local council of Plogoff, the nearby village, refused to cooperate. The mayor shut down the town hall, so EDF brought in two vans to act as 'city hall annexes'. They also brought in the CRS. For six weeks, from 31 January to 17 March, the local people fought a running battle with the police. Each afternoon, from 400 to 2 000 people, mainly the local women since their husbands were at work, confronted the tear gas, percussion grenades, and bulldozers used by the CRS.[37]

In February, on the actual site, 20 000 people demonstrated and set up a sheep farm.[38] More sheep were also being brought in by farmers from Larzac, who for eight years had successfully resisted being expelled from their area to make way for a military camp.[39] More recently in their region a struggle against uranium mining had also begun.[40] At Plogoff, on the final weekend of the 'consultation' the battle reached a high pitch of intensity. On Friday, 15 March, police and locals exchanged tear gas and Molotov cocktails.[41] Two days later, between 40 000 and 70 000 people demonstrated at the site,[42] and the following day 7 000 demonstrated at nearby Quimper where some protesters were being tried.[43] When the vans finally departed it was clear that the demonstrators had won at least a moral victory. Of the 60 000 entitled to view the EDF plans, only 210 had visited and signed the register.[44]

The strength of the opposition was sufficient to indicate that, at least on the periphery, people were no longer intimidated by the power of the CRS. Whether their strength would be effective was another matter. But, as one local commented, 'They may put the plant up but if they do they are going to need a cop behind every construction pole'.[45] Two months later the force behind that comment was apparent when an estimated 150 000 people gathered to protest against the planned plant. It was by far the largest demonstration ever held against nuclear power in the history of France.[46]

The growth of opposition in regions remote from the central authority of the French and UK governments is not restricted to

CENTRAL CONTROL, REGIONAL REVOLT

the periphery of France. Across the channel, in southern Ireland, a spirit akin to regionalism runs strong. It is not geographically located. Instead it is found in the republican sentiments forged in the country's struggle to become independent. Recently these sentiments have combined with a growing opposition to the hazards posed by nuclear power and uranium mining.

IRELAND   In 1973 the Irish Electricity Supply Board (ESB) announced plans to build a nuclear reactor at Carnsore Point, the most south-easterly tip of the country. Concern over the hazards by some met with scepticism by others that it would fill a need: electricity consumption would have to grow at around 8.5 per cent for ten years before its capacity could be fully used.[47] Locally, a Wexford Nuclear Safety Association was set up to oppose it, and a local FOE was revitalized. Together they began pressing for an inquiry.

By 1978 even the relatively weak parliamentary opposition was calling on the ruling Fianna Fail Party to hold an inquiry. As well the Irish Transport and General Workers' Union (ITGWU) had become concerned about the hazards posed by the proposed reactor. It set in motion a process of debate within other unions which by July 1978 was strong enough for the Irish Congress of Trade Unions to call for an inquiry.[48]

The reaction of the government to this growing opposition was the predictable information campaign, this time in the form of a discussion document called 'Energy-Ireland'. The outcome was equally predictable: a further growth of opposition now encompassing organizations with solid images like the Council for the Status of Women. In August 1978 a festival and rally of 15 000 at the reactor site confirmed the growing mood of opposition and catalyzed the growth of a network of local citizens action groups.

By now the political implications of the project had become more widely appreciated. With perhaps 10 per cent of Ireland's electricity depending on a single reactor for which all fuel and servicing would be provided from other companies, Ireland could find its independence compromised. Sinn Fein (the political wing of the Provisional Irish Republican Army), the Irish Sovereignty movement and, in marked contrast to its UK counterpart, the Communist Party of Ireland, by now had all declared themselves against the reactor.

To the north, in County Donegal, concern over nuclear hazards,

mixed with resentment over foreign intrusion, was welling up this time in the context of uranium mining. In September it had been announced that several companies associated with Northgate (a Canadian multinational) and Rio-Tinto Zinc, were prospecting for commercial reserves of uranium.[49] In particular, residents of Fintown and Glenleighan had become aware that the mining could endanger the local water supply.[50] By March 1980 the occupation of one site, the sabotage of equipment, and demonstrations had forced the companies, at least temporarily, to withdraw their equipment from the area.[51]

Meanwhile, in February 1979, the mounting pressure had led Energy Minister Des O'Mally to announce that an inquiry into Carnsore would be set up, although planning would continue. Pressure continued to build. In June the ITGWU came out formally against the plant.[52] It was joined in September by the Electrical Workers' Union, the largest white collar union in the Electrical Supply Board workforce.[53] Six months later the ITGWU served notice that if the inquiry came out in favour of the plan the union would press for a referendum.[54] The threat proved unnecessary for the foreseeable future. By May 1980 it seemed that a combination of mounting pressure and a new prime minister and energy minister had produced at least a temporary victory. Their statements clearly indicated that plans for the inquiry, and Carnsore, had at least temporarily been shelved.[55]

WALES, SCOTLAND, AND THE ORKNEY ISLANDS

In the UK, in marked contrast to ambitious plans of the Central Electricity Generating Board (CEGB) and the almost unanimous support by English MPs for them, opposition was also building up on the periphery. Its focus covered some reactor plans, test drilling for waste storage, and uranium mining.

The only substantial legal barrier to electricity authorities implementing their nuclear plans in the UK is the necessity to obtain planning permission from the local council. Where the council refuses, the matter is referred to the Secretary of the Environment in England or, where appropriate, to the Secretary of State for Scotland or Wales, who will usually institute an inquiry and then make a final decision. In the UK this process has provided the one legal channel through which opposition on the periphery has been able to disrupt or block the ambitious plans pressed forward from the centre.[56]

CENTRAL CONTROL, REGIONAL REVOLT 191

In Wales, April 1980 saw the people of Portskewett celebrating the abandonment of a reactor that had been planned for their side of the Severn Estuary, only 1.6 kilometres from the centre of the town.[57] Opposition had built up over two years to the point where the local Monmouth District Council had reversed an earlier decision in 1972, and now opposed the plant. The Secretary of State had intervened in mid 1979, but a year later the CEGB announced it had let its planning application lapse.[58] It was a victory for the local people but, illustrating the lack of any substantial formal power at the periphery in the UK, that victory was at the expense of the people of the Lothian region of southern Scotland who had become equally strongly opposed to a reactor at Torness.

The two-reactor complex for Torness had been planned by the South of Scotland Electricity Board (SSEB) since 1973.[59] An opposition coalition, the Scottish Campaign to Resist the Atomic Menace (SCRAM), was set up two years later and by 1978 had developed substantial momentum. So strong had its support become that it was, for the first time in the UK nuclear debate, beginning to explore channels for stopping the project outside those provided by the state.

A first occupation of the site was held in November 1978. It ended two weeks later with arrests and the cottage on the site, which the occupiers had been renovating, being bulldozed into the sea. Six months later more than 7 000 people took part in a three day anti-nuclear festival near the site. Of these, 2 000 crossed the fence on a flight of stairs constructed of hay provided by a local farmer, and occupied the site. The level of action was raised further when fifty crossed into an inner compound and deliberately damaged bulldozers, causing an estimated £2 000 damage.[60]

Both these actions were the subject of heated discussion as to whether they were 'violent'. Whether or not they were, they seemed to do nothing to reduce the legitimacy attached to the occupiers' concerns. A poll published a few days later showed 90 per cent of local residents 'totally opposed' to the reactor construction.[61] A few weeks later a petition with 20 000 signatures opposing the plant, gathered from local residents since the occupation, was handed to the Scottish Secretary of State.[62] The local council came out in opposition to the plant,[63] and fifteen Scottish MPs called for a review of the project.[64] In September 1979, 7 000 people marched in opposition to Torness through Edinburgh – its largest demonstration since the war.[65] The level of opposition was

impressive, but so was the determination of the central government. In March 1980 it announced that it intended the project to go ahead.[66] Whether or not that would finally occur, the struggle over the plant had probably already transformed the 'realm of the possible' in the strategic thinking of nuclear opponents, not only in Scotland, but elsewhere in the UK.

In other parts of the UK, local opposition to nuclear projects had met with more immediate success. One of these was the Orkney Islands. 'NO URANIUM, the posters shout from nearly every shop door in Kirkwall, the capital of the Orkney Islands . . . A poster beside the ferry reads "Welcome to Stromness – Uranium Never, Orkney for Ever".'[67] That was a journalist's report of the atmosphere in Kirkwall, in March 1979, as a public inquiry was launched by the Scottish Secretary of State into plans by the SSEB to mine for uranium on the islands. The proposal offended the islanders' sensitivity to both pollution of their land and the broader nuclear hazards.[68] Opposition to it also resonated with their already strong desire for home rule. More than 3 500 written objections to the proposal were received out of a population of 17 000. Summing up this concern, the council submitted a draft structure plan which contained a clause prohibiting the prospecting, mining, or extraction of uranium on the islands.[69]

Pressure mounted, with almost 20 per cent of the island's population taking part in one demonstration.[70] In December 1979 the islanders gained a partial victory. The Secretary of State announced that he would neither approve nor oppose the clause put forward by the council.[71] At least for the time being the islanders seemed to have made a substantial advance. In this they were not alone. Earlier in 1978, on the Scottish mainland at Banchory, pressure from Aberdeen FOE, the Scottish Nationalist Party, and locals had forced the SSEB to withdraw from even applying for planning permission to test drill for uranium there.[72]

The pressure generated by proposals to mine uranium was impressive. Not surprisingly, equally strong opposition and regionalist sentiment was being aroused in those parts of the UK periphery in which the national Atomic Energy Authority (UKAEA) wished to test drill for possible waste dumping. The drilling proposals emanated in 1975 when the UKAEA joined a European Economic Community study aimed at locating possible underground sites for the disposal of high level radioactive wastes. Suitable hard rock was to be sought which, under the UKAEA's

controversial Harvest plan, could eventually become the place in which the potent highly radioactive residue from reprocessing would be entombed after being cast into glass blocks.[73]

The UKAEA was supposed to produce results by 1979, but by then it had only two places in which it could drill. One was an abandoned mine in Cornwall for which planning permission was not required. But its rocks were deeply cracked and unsuitable. The other was at Altnabreac in Caithness, a county in which one job in five is provided by the Dounreay Nuclear Power Establishment. The UKAEA had first sought to drill in Northumberland, England, right in a national park. The council was not pleased and declined. Remarked one local government official, 'If they want granite, there's plenty in Scotland'.[74] But when the UKAEA began to look in Scotland, elsewhere than Altnabreac, it encountered growing resistance.

As early as 1977, the Scottish National Party had launched a campaign against the drilling with a pamphlet entitled 'Scotland: a nuclear dustbin?' In March 1977 it had collected 15 000 signatures on an anti-dumping petition,[75] and by 1980 the party was pledging itself to 'non-violent disobedience, if necessary'.[76] In addition, a number of trade unions had come out in opposition. Their concern cumulated in a motion against waste dumping being carried by the Scottish Trade Union Congress.

Wherever the UKAEA showed its face the reaction was hostile. A proposal to drill in the remote Mullwarcher Valley in Galloway, a wild picturesque part of the Glentrool Forest Park, produced two opposition groups, Campaign Opposing Nuclear Dumping (COND) and SCRAM-South West (SCRAM-SW), in the nearby towns of Dalmellington and Loch Doon. By mid 1977, SCRAM-SW had collected 10 000 signatures, one quarter of the adult population of Galloway, in opposition to dumping. COND had held a sizeable demonstration at Loch Doon.[77] Others followed in Ayr, the largest city in the area. By then opinion polls were showing 80 per cent of the people of Ayr opposed to drilling. Not surprisingly, when the UKAEA finally sought formal approval from the local Kyle and Carrick councils, in August 1978, it was refused. It then turned to Scourrie, on the north-west tip of Scotland, but every adult in the area except two signed a petition opposing drilling.[78] Permission was again refused. By the end of 1979, permission had been sought, and was meeting opposition from councils at Scourrie, Shin Forest, Corrur, Tornashean,

194   *GLOBAL FISSION*

Mullwarchar, and Northumberland, as well as in the Outer Hebrides and at Harris.[79] The UKAEA had appealed to the Secretary of State, but it also seemed to have made a tacit admission that that might not suffice, by causing five additional sites to be announced in parliament in July. No doubt it hoped that this might throw up one or two more tractable sites. But another result was also possible. As *The Times* had remarked two years earlier:

> The protests over Loch Doon may just be another sign that the honeymoon for nuclear developments in the United Kingdom has ended. Future proponents of nuclear energy may find it increasingly difficult to convince the public that the benefits of nuclear power outweigh the dangers.[80]

In 1980, with inquiries under way and opposition to drilling still mounting not only in Scotland,[81] but also in Wales,[82] there seemed every reason to expect that further escalation of public resistance to nuclear proposals could follow, at least on the periphery of the UK.

FROM PERIPHERY TO CENTRE 1979–80

Enough has been said to make it quite clear that the nuclear battle in France and the UK has two important common features. First, the state and the nuclear industry are particularly closely intertwined. Second, there are few formal barriers (such as a federal structure or divided political parties) to the implementation of the ambitious nuclear programmes that this combination has produced. And third, the reaction against those programmes is not only against their direct hazards, but also against the political power relations they represent. This is clearly demonstrated by the tendency for opposition to those programmes to be most forceful in those parts of the two countries which have a strong regionalist tradition or drive for independence.

The situation in France and the UK is not, of course, identical. On the contrary, the British nuclear programme is much more tentative, and in far greater trouble than its French counterpart. As well, the national economy of the UK is in a deep malaise, while the French economy seems comparatively robust. And in the UK the nuclear programme may still falter in its passage through parliament, to which it is ultimately answerable. In France, the nuclear programme is never in any real sense either affected by, or even

CENTRAL CONTROL, REGIONAL REVOLT 195

discussed in parliament, and is instead implemented under the steadily growing authority of the president.

In both the UK and France, the power of the central government has been sufficient to at least place ambitious nuclear programmes seriously upon the agenda, in marked contrast to the stagnated situation of the nuclear industry in most of the rest of the Western world. Paradoxically, as these central programmes are implemented, they further increase the opposition on the periphery. The outcome of this increasing conflict between opposed but growing forces is not easy to predict. As they become larger it is likely that the equilibrium between them, maintained by what is considered legitimate action, may become less stable.

There are two other possibilities. One is that the opposition on the periphery will die away. At the end of 1980 there were no signs of that. The other is that cracks may be forced into the solidity of the central authority, providing a channel through which opposition may flow from the periphery to the centre. There were already some, admittedly still early, signs that such cracks might develop.

In the UK the defeat of the Labour government at the 1979 elections provided a possibility for the party to begin to re-examine its previously adamant support for nuclear power. The election of a new leadership had already given expression to a growing opposition to nuclear weapons, both in the general community and within the party. As will be seen later, this had led to the largest demonstrations in the UK since the Aldermaston marches. At the same time, the Thatcher government's initial talk of building perhaps as many as twenty new nuclear reactors by the end of the century,[83] and the more realistic announcement of investigation of five sites (for possible construction in the late 1980s and 1990s), had helped fuel the debate.[84]

The decision to effectively scrap twenty-five years of UK development of the advanced gas cooled reactor, and instead buy US-designed boiling water reactors, not only provided further grounds for debate but also stripped away some of the more nationalistic reasons for supporting the programme. But more importantly, the development of strong opposition at the periphery was giving rise to the possibility of new coalitions of forces. One of these was the subject of an editorial in the journal *New Scientist* in November 1979: 'The most significant political event since the general election happened last weekend ... politicians – and

Margaret Thatcher in particular – would be foolish to make the mistake of ignoring the foundation of the Anti-Nuclear Campaign (ANC)'.[85]

The editorial referred to the formation of a new national coalition of anti-nuclear forces. Significantly, it included not only some prominent trade unionists (among them Arthur Scargill of the National Union of Mineworkers), anti-nuclear groups, and the Young Liberals, but also the Scottish and Welsh nationalist parties, and some Labour Party members of parliament. Continued *New Scientist*: 'There is no doubt that the ANC has the potential to grow into an enormous campaign, as influential as the Campaign for Nuclear Disarmament . . . The coalition of forces is fragile. But the potential is undeniable'.[86]

Despite that potential, there was no clear indication that this particular coalition would live up to its promise and establish, with the existing anti-nuclear groups, an increased momentum of opposition. But whether or not it succeeded in this, notice had been served that the time was drawing close when new powerful coalitions might be forged to fracture the solidity of the centre as it attempted to press forward its ambitious nuclear programme in the UK.

In the UK and France another factor would operate to deepen any fractures caused in the centre. This was the steady revelation of incidents accompanying the expanding array of nuclear technology. Thus, in the UK, the public acknowledgement in 1980 that a waste storage tank at Windscale had leaked large amounts of high-level liquid radioactive waste for seven years before it was detected in March 1979, caused alarm not merely in the community, but also among workers at the plant.[87] The General and Municipal Workers' Union, which is one of the main unions at the plant, had previously made little criticism of nuclear power. It now expressed serious concern over the safety hazards at the plant.[88] In France a set of accidents had already led to more significant fractures in the centre's nuclear authority.

It has already been noted that the workers at la Hague had early become concerned by the increasing hazards and accidents associated with the operations of the reprocessing plant. This concern was expressed first in the local branch of the CFDT, and then became part of the process which led to the CFDT calling for a moratorium on nuclear developments nationally. Since many members of the CFDT are Socialist Party sympathizers, this could

hardly fail to cause discussion on the issue there. In October 1978, just before the national elections, the Socialist Party announced that it now supported an eighteen to twenty-four month moratorium on further nuclear orders, and an immediate halt to the construction of the Super Phoenix reactor. With a following of around 20 per cent of the national vote, it was a significant fracture in the centre's nuclear unity.

Later, in April 1979, when the government announced the pending construction of two new reactors at Gravelines,[89] the Socialist Party called for a moratorium on reactor construction and an inquiry into whether nuclear power is safe.[90] Its call was joined by a coalition of ten other organizations representing organized labour, scientists, and consumers, including the PSU and the CFDT. They launched a national drive to collect signatures on a petition calling for a national public debate, a moratorium on nuclear construction, and democratic decisions on major energy choices.[91] In July the Socialist Party announced a grassroots offensive: to organize a public debate even if the government refused to participate.[92]

Equally significant, a new fracture was beginning to show. The CGT, the confederation of unions associated with the Communist Party, announced its intention to initiate comprehensive debate among its own members. Although not yet prepared to condemn nuclear power outright, it joined the CFDT in condemning the pace with which the government's nuclear programme was proceeding.[93] In early October 1979 that fracture was further widened when it was reported that the government had authorized the start-up of two nuclear plants (Gravelines and Tricastin) despite scores of cracks detected by a worker in their tubular base plates.[94] A collapse of the supports could be catastrophic. Worse, a confidential Framatome technical note 'fell' into CFDT hands revealing that such cracks could become serious in three to twelve years.[95] For the first time, the CGT now joined with the CFDT in a strike to prevent the loading of the two reactors. It only ended when EDF announced that the loading would be delayed.[96]

It was still too premature to predict that the alliance would last. In fact, when leaks were found at the Bugey plants two weeks later, the CGT refused to join a further strike.[97] However, CFDT concern was reinforced in April 1980 when a transformer short circuit and fire occurred at the la Hague plant. According to the CFDT, ventilation systems became inoperative for ten hours, causing

substantial contamination of the plant, and possibly exposing the surrounding community to a serious release of radiation.[98] It seemed clear that it was not the end of such accidents. And already, at the 1979 union elections within the EDF, the CFDT had gained 2.5 per cent, while the CGT had lost ground.[99] The CFDT's stand at Gravelines seemed to have been vindicated, and it seemed likely that anti-nuclear pressure within the CGT would continue to mount. At the beginning of 1981, with a national election looming, there were signs that the Communist Party leadership was taking steps to force the dissidents back into line on the issue. But although this might well prove temporarily successful, it seemed likely that the forces set in motion to resist that pressure, in the longer term would not be easily contained either within the CGT or the Communist Party itself.

When the presidential elections were held in May they produced a defeat for the Conservative candidate, then Giscard D'Estaing, for the first time in twenty-three years, and victory for the left's candidate, Socialist Party leader François Mitterand.[1] Soon afterwards, the assembly elections produced a majority for his party in the legislature. There is often a big gap between policy and action. Nevertheless it was clear that the Socialist Party's policy, to complete all reactors already under construction and then pause for a long 'period of reflection' during which time energy-conservation measures would be emphasized, had now produced a potent division at the very centre of French power.[2] The announcement, within weeks of Mitterand's election, that the plans to build the reactor complex at Plogoff had been suspended, together with the new government's decision to suspend construction of five other reactors pending parliamentary and public debate, graphically emphasized the party's potential to substantially reduce the pace of expansion of the French nuclear programme.[3]

In both the UK and France the opposition at the periphery, whether of union branches or local communities, was slowly leading to a realignment of forces at the centre. There could be no guarantees that this process would reduce the pace with which the nuclear programmes were being pushed forward. But it seemed likely that while the centre continued to force the programmes forward, it would meet an increasingly powerful response not only in the periphery but also, perhaps more slowly, on its home ground.

The past success of the UK's and France's central governments

CENTRAL CONTROL, REGIONAL REVOLT

in pushing forward their nuclear programmes is anomalous within the Western world. In only two other countries, Taiwan and South Korea, are similarly ambitious programmes on the drawing board, and in those countries there is still serious doubt whether they will be put into practice. But although they are anomalous, the trend that they demonstrate – of a strong relationship between regionalism and energetic opposition to the nuclear enterprise – can also be observed in many of the other countries which together in the nuclear conflict constitute its international arena.

9

THE INTERNATIONAL ARENA

It was three months after the accident at Three Mile Island, Whitsunday, 3 June 1979. Responding to a call issued by an international gathering in Switzerland a year earlier, in over eighty places in thirteen countries more than 200 000 people took to the streets.[1]

In the city centre of Barcelona, Spain, a sea of 50 000 people swelled through the streets in protest against the ambitious nuclear plans of the Spanish central government. The surging column of people represented just one highlight of three days of continual demonstrations linked by a common opposition to nuclear power and stretching across the world.

Almost simultaneously, demonstrations of opposition to nuclear power flowered in West Germany, the Netherlands, Belgium, Switzerland, Ireland, Scotland, England, France, Luxembourg, Canada, Japan, and at over forty sites in the USA. If it had been in doubt before, it was now clear that the international operations of the nuclear industry are faced by citizens movements capable not only of effective action, but trans-national collaboration.

The potential for such cooperation was demonstrated not just by the coordination of simultaneous actions in a range of different countries. Several actions were aimed at nuclear sites near national borders, drawing demonstrators from adjoining countries. On that Sunday, near the Dutch border at the site of the Kalkar fast-breeder reactor under construction in West Germany, 25 000 people from both West Germany and the Netherlands gathered in protest. Several hundred wooden crosses were erected to symbolize the lethal potential of the plant and the plutonium it is designed to manufacture.

At Doel in Belgium, near Antwerp and the Dutch-Belgian

THE INTERNATIONAL ARENA

border, about 20 000 came together to protest against a proposed nuclear complex. Despite the presence of several thousand police hidden near the site the demonstration proceeded without incident. Organized jointly by citizens groups in the south of the Netherlands and Flanders, it was the largest anti-nuclear event ever held in Belgium. At the French-West German border, police refused 1 000 cyclists permission to enter France with a model of a nuclear power plant mounted on a trailer. After the cyclists blocked the bridge in protest, the police relented.

On Monday the police were more successful in blocking the citizens of Luxembourg – a country which had already rejected nuclear power as an option – and West Germany from joining a demonstration at Thionville in north-eastern France. The French component of the demonstration continued on regardless.

In Basel, Switzerland, the city from which the original call for the international coordination had emanated, 2 500 people gathered to participate in an international forum on nuclear power and strategy to oppose it.[2] By the time it ended it was clear not only that international cooperation was possible within the opposition movement, but that opposition to nuclear power had reached a new level of intensity.

A year later, on Whitsunday 1980 over the weekend 24 to 26 May, 300 000 people in twenty-five regions of Europe and the USA again took part in a variety of activities to show that the tide of opposition to nuclear power still carried plenty of momentum. Most dramatic was the huge demonstration, already described in chapter 8, at Plogoff in France. But in Italy also tens of thousands marched in Rome, Milan, and Venice, and in the Netherlands thousands rallied against waste dumping off the coast near Injmuiden, while at the small reactor at Dodewaard, thousands more gathered and laid down an ultimatum that they would occupy the plant in September if the government refused to close it. Other actions stretched across the UK, Belgium, France, Spain, Luxembourg, West Germany, and the USA.[3]

The phenomenon of large numbers of people taking to the streets simultaneously in a variety of countries, and in common protest against nuclear power, provides an impressive demonstration of the energy and strength which the activist core of the opposition to nuclear power has achieved. It is also a reminder of the importance of the role demonstrations have played in illustrating the growing strength of the opposition movement. But such demonstrations are not the nuclear conflict. They are merely a

symptom of it, and their significance varies greatly from country to country depending on the actors in the conflict and the different cultures, social structures, and histories which comprise its backdrop.

For the opposition to nuclear power, demonstrations may be a sign not only of strength, but also of weakness. The supporters of nuclear power, also comprising a substantial cross-section of the community, need do nothing dramatic to press forward their desire for further nuclear developments. While the nuclear industry and its corporate allies, and the ruling political parties and agencies of the state combine their efforts to push the nuclear programmes forward, all that is required from the community is acquiescence. It is only when opposition from one section of the community seriously threatens the power or unity of this front that it need call for support from the remainder of the community.

Even when the opposition is strong, if a united front is maintained across the state and the industry, the resulting battle of legitimacy may not favour the opposition. This, at least until recently, has been the case in France. But even in that country its state authority is not totally integrated and unbreachable. The local authorities in Plogoff, for example, also part of the state, have taken a clear stance in opposition to the central government's nuclear plans.

In many countries the state is far more fragmented, with regional governments, courts, or agencies possessing substantial autonomy and power. Also there is often a spectrum of opinion within the state over which of its often contradictory objectives should be emphasized, and how they should best be achieved. This is illustrated by the different policy platforms of the various parliamentary parties, which all largely support the present structure and functioning of the state but differ on the question of the best strategy to make it operate most efficiently. The conflict becomes far more than a simple test of strength between the nuclear industry combined with the state, and the opposition. Instead, the state itself becomes an arena for struggle. Nowhere is this more clearly demonstrated than in Austria and Scandinavia, where significant parts of the state such as key political parties within the government have turned against nuclear power, severely limiting or cutting off nuclear programmes. The opposite is illustrated by France and the UK, where the state has proved largely impervious to the opposition.

THE INTERNATIONAL ARENA

The above examples, however, represent opposite extremes. Most countries in the international arena lie between these and provide a more complex picture. But despite the apparent complexity it is still possible to discern the same broad trends – the contest within the state, confrontation with it, and the role of regionalism – all still playing a vital role. To illustrate this three countries which have played an important role on Whitsunday 1979, and more generally in the global conflict over nuclear power, will be examined: Spain, the Netherlands, and West Germany. In each of these three countries the battle over nuclear power became significant in 1975 when government nuclear plans began to be set in motion. In Spain, under the authoritarian rule of General Francisco Franco, the plans seemed particularly ambitious.

In the late 1960s the Franco government was anxious to build its industrial base. Technologists were brought into key positions of authority,[4] and plans were developed to build even further the growth in electrical demand which had been developing at 10 per cent each year for the last decade. Drafting of a National Energy Supply Plan began in 1969. It was completed in 1972 only to be confronted by the OPEC price rises and their implications for energy autonomy. Again amended, it was finally approved in 1975. It proposed a heavy commitment to nuclear power.

Expectations ran high, with proposals for as many as thirty-eight nuclear power stations by 1985. With a land area of only 315 000 square kilometres, Spain was moving towards a high density of nuclear reactors. Already three nuclear reactors were operating and seven more were under construction. In accordance with the plan a further eight were given preliminary construction authorization. However, even the construction of these eighteen nuclear reactors was to prove steadily less and less attainable.

On 20 November 1975, the year these plans were finalized, Franco died. With his death, the iron grip that he and his state apparatus had exerted over his political opponents faltered. Suddenly the government's programmes and its power were more amenable to challenge.

In the Netherlands, opposition was galvanized in the same year when the government announced its intention to build three new large nuclear reactors. They would represent a major addition to the existing small capacity of the reactor at Borssele and the tiny reactor at Dodewaard.

In West Germany also, 1975 brought with it nuclear opposition,

this time in the form of the world's first nuclear occupation at Wyhl. On this occasion the state found it politically impossible to expel the occupiers. Suddenly the government's entire programme of fifty-three reactors for the year 1985 seemed open to challenge.[5]

In each country, the ground was prepared for a national confrontation over the nuclear plans of the central government. But the differences within the fabric of their societies would in each country bring out very different forms of conflict.

SPAIN: REGIONAL REVOLT

Undoubtedly, the conflict in Spain is the most transparent, falling most readily into the earlier picture of 'centre and periphery'. Although emerging from the authoritarian grip of Franco, the Spanish struggle is overwhelmingly characterized by its relation to demands for regional autonomy.

As early as 1976 popular resistance to the hazards that would accompany the nuclear programme had begun to appear. At the level of the central government the newly emerged Socialist and Communist parties responded to this concern, and it was not until August 1979 that the National Energy Plan in a much amended form was finally put before parliament.

In the months leading up to the parliamentary debate, opposition to the plan was highlighted by major demonstrations of over 50 000 people in Barcelona, Madrid, and Valencia.[6] But despite their avowed positions of concern, during the two-day debate the Socialist and Communist parties made little attempt to seriously head off the nuclear part of the plan, and it was adopted.[7]

By now, however, the programme was only a pale shadow of earlier plans. Seven reactors were already under construction and these were to be completed. Of eight more with 'previous authorization', only three were to be completed before 1987. The programme for thirty-eight reactors to operate by 1985 was thus stripped back to fifteen by 1987. Later, the left-wing parties might strengthen their opposition to nuclear power, bringing even the diminished programme under attack. But in the meantime it would not in any case go unchallenged. Already the central government was facing major opposition to it in the regions.

Of the eighteen authorized and operating reactors, three were to be near Madrid. The remainder were confined to three regions of

THE INTERNATIONAL ARENA 205

Spain. Six would be in Extremadura near the border with Portugal, three in Eskuadi, the Basque country in the north near Bilbao, and six on the east coast, five of which would be in Catalonia, near Barcelona.[8] The major part of the burden would be shouldered by Extremadura, the poorest region of Spain, and Eskuadi and Catalonia, the homes of the Basque and Catalan independence movements. They have a history of determined struggle to separate their regions from the control of the central Spanish government.

Under Franco the electricity utilities had been bastions of the dictatorship, effectively responsible to no one but themselves. Sitings of reactors had been made without consultation or public hearings. In at least one case, on the Basque coast, a major utility did not even bother to comply with the permissive Franco legislation.[9] Their attitude, however, was soon to rebound.

As early as July 1976 the local city council of Tortosa in Catalonia had begun proceedings against the Asco reactor planned for the area. It claimed that the plant's cooling water would kill the fish in the Ebro River. In the Basque country the first group aiming to halt a specific nuclear project was formed at Bilbao. A coalition of three citizens groups, it was patterned on an earlier group which had successfully halted a government-supported Dow Chemical plant outside Bilbao.[10] The new group's pledge, to stop the construction and operation of the two Lemoniz reactors in the Basque country, was to resonate powerfully through the region over the forthcoming years.

THE SIEGE OF LEMONIZ   The Lemoniz site lies in a misty cove by the Bay of Biscay surrounded by rolling hills. However it is only 12 kilometres from Bilbao, an industrial city with over one million inhabitants in the heart of the Basque country. Up to twelve reactors have been planned for the 15 kilometre long strip of Basque coastline.[11] Of these, two are already under construction at Lemoniz. The Basques contend that they, not the Madrid government, should determine whether Lemoniz is built. Their argument is simple but persuasive. As one Basque leader put it: 'A radiation accident at Lemoniz would wipe out the entire Basque people'.[12]

The company which owns Lemoniz, Iberduero, is particularly offensive to the Basque separatists. It is controlled by the industrial barons who flourished under Franco. Says lawyer Juan Cruz, an

opponent of the plan: 'The company is dominated by the same oligarchy that turned Bilbao into the most polluted, most contaminated industrial city in the world'.[13]

On 14 July 1977 the first legal demonstration in a year flooded through the streets of Bilbao. The 3 kilometre route was filled with a seemingly endless procession of men, women, and children. An estimated 150 000 to 200 000 people marched that day. It was the largest anti-nuclear demonstration the world had ever seen.[14] It was also a massive demonstration against the authority of the central government. Thousands of Basque flags studded its ranks and chants against nuclear power were mixed with chants in support of the Basque separatist movement, ETA.[15] The physical and political implications of the nuclear project had become totally intertwined. Many Basques are prepared to die for the liberation of their homeland. They soon showed that they are also prepared to risk their lives to stop Lemoniz.

In December 1977, when demonstrators attempted to break through the police lines to reach the Lemoniz reactor, they were violently repulsed and one demonstrator, David Penya, was shot dead.[16] Three months later when 100 000 to 150 000 people again massed together in opposition to the reactor, the mood was angry. Said the communique released by the organizers, 'When after all these years of resistance the authorities' only reaction is shameless passivity, what else is there left for us to do?'.[17]

Five days later ETA answered the question. A bomb, allegedly smuggled in by site workers, exploded in the reactor core.[18] Tragically, little notice had been taken of a warning given shortly before the blast, and two workers were killed.[19] Neither ETA nor Iberduero accepted blame for the deaths, but there was an inevitable backlash against ETA. It enabled the central government to veto the newly created Basque General Council's plans for a referendum on the future of the plant.

Nevertheless, the bombing achieved at least part of its principal aim. It caused $70 million worth of damage to the plant, and delayed its construction by an estimated two years. Not long afterwards, when tens of thousands of Basques paraded through the provincial capitals to celebrate the Basque national day, the most commonly heard slogan was reported to be: 'ETA, Goma-2, Lemoniz, No, No, Lemoniz!'.[20] (Goma-2 is the name of a plastic explosive often used by ETA.)

Over the following year the opposition lost little of its intensity.

THE INTERNATIONAL ARENA 207

As well as a further larger demonstration of around 100 000 people in Bilbao in April 1979, [21] there were increasing protests by local governments. The councils of Lemoniz and Munguia, and the city government of Bilbao called for a halt,[22] and in the Spanish senate, Basque senators set up an investigatory committee.[23] Iberduero took little notice.

Whitsunday 1979 provided an horrific demonstration of the fierceness of the escalating confrontation between the central government and the local people over Lemoniz. In the small town of Tudella, in the south of the Basque country, the police shot and killed one of 5 000 demonstrators. One participant gave this description:

> It must have been 4 o'clock in the afternoon. Perhaps some-one threw a stone, but nothing more. The cops who were regulating traffic asked us to turn back, telling us that some serious things were going to happen.
>
> The activists, a few hundred, were holding a sit-in on the bridge which a group of guardia civil were crossing. There were blows from police clubs, discussions, negotiations: several members of the Tudela municipal council were trying to intervene without success ... The woman was sitting down, she said something to them, they were less than a metre from her. One of them [the police] fired point blank. Gladys del Estal Terreno, a twenty-three year old student from Saint-Sebastien, was dying on the bridge, from a bullet which entered her mouth and exited through the top of her head.[24]

By Monday midday, barricades were burning in the centre of Pamplona. The next day ETA responded by killing two policemen in Madrid. In the Basque country the opposition to nuclear power and the local Tudela municipal council called for a strike in protest against the slaying of Gladys del Estal. The strike spread through the Basque country bringing industry and commerce to a virtual standstill.

Shortly afterwards ETA kidnapped Ignacio Astiz, the Commissioner of Industry and Energy. He was held for five days and questioned about the government's nuclear policies. Two days after his release a bomb exploded in Lemoniz for the second time in fourteen months. Again, tragically, a maintenance worker was killed. For Iberduero, the bombing hardly required further investigation. The company's offices and installations had by now been attacked over a hundred times.[25]

208   *GLOBAL FISSION*

The tally was added to four months later when the Equipos Nucleares factory in Santander, which was repairing the Lemoniz reactor, was attacked. At 9 p.m. on 11 November 1979 five men armed with submachine guns and pistols entered the plant. They removed the workers in the plant to safety, and then blew it up with plastic explosives. It caused an estimated $6 million worth of damage.[26] Iberduero and the central government still seemed prepared to forge ahead and get the plant into operation. But whatever their intentions, it seemed likely that the siege of Lemoniz had only just begun.

IN OTHER REGIONS 1979–80   For the Spanish central government and the power companies the Basque region was an obvious tinder-box for opposition.

However in other regions of Spain, such as the impoverished province of Extremadura in the south-west, no resistance seemed probable. Despite its six planned reactors, its poverty and lack of any large scale independence movement seemed to provide little basis for conflict of the style encountered in the highly organized Basque country. However, although opposition was slow to develop, it has become clear that even here it can reach substantial proportions. The first place where that could be seen was over the proposed Valdecaballeros plant in the western province of Bandajos.

Early in 1979 the authorities were so confident of their position that they began construction of the plant without waiting for permission. When approval was given retrospectively in August, it precipitated an immediate reaction. In the town hall of Villanueva de la Serena, twenty-five mayors of nearby cities and towns went into permanent session. Thirty thousand people defied a ban to demonstrate in the town, while half as many again, blocked by the civil guard, held protests outside the town.[27] A week later the number of mayors occupying the town hall had risen to over a hundred, and they had threatened to resign unless the plant was stopped.[28] The two-week occupation only ended when nuclear power officials announced that approval for the plant's construction would be deferred pending a report from the region's energy advisor.[29]

By early November, four reports from departments of the regional council had been released. All of them emphasized that the dangers of siting the plant in Extremadura exceeded the possible

THE INTERNATIONAL ARENA

advantages. Both the Socialist Party and the Communist Party, which had generally supported the national energy plan, declared their opposition to the plant. But the regional authority, dominated by the governing political party, the Central Democratic Union (UCD), was unable to withstand sustained pressure from the central government. By March 1980 both it and the local council had yielded.[30]

For the central government, however, the events set in motion in Extremadura by the nuclear plans still seemed far from over. Most significantly, as one commentator pointed out, the plans 'have awakened a dormant independence movement in Extremadura' which 'with little industry and backward agriculture is providing sites for power plants whose output will be consumed in the industrial and urban centres of Madrid and Seville'.[31] Action was now evident in Caeceres province, where seventeen towns declared their opposition to the impending start up of the Alamaraz nuclear station. Further north, in the province of Salamanca, Spanish nuclear authorities were facing united opposition to the construction of a proposed plant from local government, townspeople, and peasants.[32] The seeds of an independence movement in Extremadura had still barely germinated but the nuclear plans seemed to be providing essential nutrients for further growth.

Elsewhere in Spain the potential of such movements was being clearly demonstrated. Although in February 1980 active opposition could be seen for example in Catalonia, where 10 000 people marched near the Asco reactor site, for the central government other more worrying political effects were accompanying the nuclear plans. Most alarming was the success that the anti-nuclear movement was having in the electoral struggle for regional control.

The early warnings were observed as early as April 1979 in Asco. There, in the first democratic municipal elections since Franco, an anti-nuclear candidate won 40.9 per cent of the vote. Shortly afterwards, local officials ordered the demolition of the first of two nuclear plants being built there.[33] It was, however, early in 1980 that the interlinked demands of local autonomy and the halting of the central government's nuclear programme showed a new powerful stage of development. Wrote one observer in the *Guardian*:

> Regional elections follow one another in Spain and they all look alike. On March 20, the Catalan nationalists came out on top in the election for their assembly, just as the Basque

autonomists had done 11 days earlier in their province. On February 28, the Andalusian autonomists also scored points, but lost the referendum solely because of a highly controversial constitutional procedure. Three nationalist successes in three weeks. And that is too much for Adolfo Suarez's centrist government, which is beginning to worry over the repercussions.[34]

The immediate implications of the victories were still unclear. The powers that would be offered were still vague and would depend on negotiation. Nevertheless, a political process with substantial momentum was clearly in motion. It had been heralded by the powerful events, scattered through the Spanish countryside, of Whitsunday, 1979. It was now clearly visible in the electoral successes of the movements for local autonomy. With each development the strength of these regional movements seemed to increase, and the grip of the central government and the future of its nuclear programme seemed less and less secure.

THE NETHERLANDS: A DIVIDED PARLIAMENT

As in Spain, the battle over nuclear power in the Netherlands combines some regional opposition to central plans, and some division over those plans within the national parliament. But, in marked contrast, the regional opposition is distinguished by its lack of violence, and the divisions within the parliament have been significant enough to lay open to serious doubt the possibility of any further nuclear developments.

Since the Dutch government's announcement in 1975 that it intended to build three large new reactors, a debate had blossomed across the country.[35] By 1976 it had become so fierce that the government announced it would postpone its decision on the programme until after the next elections, scheduled for May 1977.

By then the two major national groupings of unions, the Catholic based Dutch Catholic Trade Union Council (NKV) and the socialist based Dutch Council of Trade Unions (NVV), had already adopted positions in favour of a five-year moratorium on any decisions to build nuclear facilities.[36] Even more significantly, just before the election the largest political party, the Labour Party, responded to the growing pressure within the community and its own ranks by adopting at its election congress a position calling for an indefinite moratorium on all nuclear activities.[37]

THE INTERNATIONAL ARENA   211

Hopes were high among nuclear opponents that the Labour Party would win the election, although there were some doubts as to whether it would actually carry out its policy platform. After the election opponents of nuclear power were overjoyed. The Labour Party had received the largest vote, gaining forty-three seats. It was thought that the party would form a coalition with the Christian Democrats who, with twenty-three seats, had a policy of 'utmost reluctance on nuclear power', and the Democrats who, with eight seats, favoured uranium enrichment but not the construction of further reactors. However the parties were unable to hold together, and in November it was announced that the pro-nuclear Liberals, with twenty-eight seats, would form a coalition government with the Christian Democrats. Now it was clear that the battle against the nuclear plans would have to be won by pressure from below rather than edict from above.

Debate continued and opposition to nuclear power steadily grew. Over 60 000 signatures were collected in under three months on a petition calling for an indefinite moratorium on nuclear power. The government's spring 'information evenings', in which local citizens could express their views in writing on *where* the nuclear reactors should be built, but not on *whether* they should be built, only helped stimulate the growth of opposition.[38]

Additionally, the debate was fuelled by concern over nuclear weapons proliferation. A proposal by the USA to deploy the neutron bomb in Western Europe led to the formation of a coalition of groups called Stop the Neutron Bomb. Their campaign was supported by over 300 groups, including the influential Dutch Inter-Church Peace Council (IKV), an inter-church organization on which all major churches in the Netherlands are officially represented. In the autumn of 1977, 1 200 000 people – 9 per cent of the country's population – signed a petition against the deployment of the bomb.[39]

On the other side of the nuclear weapons issue lay plans to extend the country's small uranium enrichment plant at Almelo. Owned and run by URENCO, a British, Dutch, and West German consortium, the Dutch government, with considerable prompting from West Germany, had decided to enlarge the facility into a commercially viable enterprise. Sixty per cent of the enriched uranium to be produced would be destined for West Germany to form a vital link in its planned fuel cycle. As the Treaty of Paris prohibits the Federal Republic of Germany from carrying enrich-

ment plants on its own soil, Almelo is important not only for West Germany's own nuclear programme, but also for its plans to export nuclear technology to other countries. But for many people in the Netherlands it was abhorrent that some of the enriched uranium was intended to fuel the massive nuclear complex which West Germany had agreed to supply to Brazil.

The West German agreement with Brazil had generated considerable US opposition because of its military implications.[40] A long history of attempts by various organs of the Brazilian government to obtain enrichment technology suggested that Brazil's interest was not restricted to nuclear power.[41] Given the relationship between both West Germany and Brazil with the USA, neither would have been able to develop a nuclear weapon openly. It was widely suggested that one aim of the proposed provision of reprocessing and enrichment facilities to Brazil was to make available to both Brazil and West Germany the raw materials necessary for the construction of nuclear weapons. In the Netherlands this fear was a major underlying reason for the wave of opposition which greeted the government's proposal to accede to West German pressure and extend the facilities at the Almelo enrichment plant.

On 4 March 1978 the opposition in the Netherlands coalesced in a large demonstration at Almelo. Supported by a coalition of organizations ranging from church groups to the major communist parties, some 50 000 demonstrators massed outside the plant.[42] An increased breadth of opposition had been created by this convergence of opponents of nuclear power and nuclear weapons.

Given the divisions over the issue not only within the community but also the parliament, it was not surprising that the government temporarily faltered in its resolve to extend Almelo. Finally it pressed on with its plans. In doing so it was certain to further strengthen the already powerful current of opposition within the community.

Although much of this opposition was being expressed nationally within churches, unions, and left-wing parties, additional obstacles to the nuclear plans were being thrown up in the outlying regional provinces. These have traditionally considered themselves disadvantaged in relation to the rich cities and provinces in the west. They have little respect for decisions made in the remote halls of The Hague.

In the northern provinces there is a movement for local autonomy, and this is particularly strong near Friesland, where the

THE INTERNATIONAL ARENA 213

people speak their own language. Adjoining it are the provinces of Groningen and Drenthe. Here there is not only a regionalist sentiment but also a strong socialist tradition with substantial support for the left wing of the Labour Party. The ground is thus well prepared for opposition to nuclear power, and that has been strengthened since 1976 by the central government's emphasis on investigating the possibility of using the underground salt domes as repositories for high level nuclear wastes.

The potential of the combination of regionalism, radicalism, and this government plan was demonstrated graphically on the weekend of Whitsunday, 1979. Near Groningen, in the town of Gasselte, some 25 000 people came together to express their opposition. Symbolizing the breadth of the concern, the demonstration was opened by the mayor of Gasselte who gave a strongly anti-nuclear speech on behalf of the council. The mood of the demonstration was non-violent but determined. Said Berend Mensing, one of the organizers, 'Not a single drilling team will be allowed to enter Groningen or Drenthe'. Some of the other organizers warned that this could be the 'last really peaceful demonstration against the Government's plans . . . the opposition will be non-violent but every attempt to start drilling will be met with every means at the disposition of the movement'.[43]

Even if the government wished to confront the local opponents of the waste drillings, there might well be weaknesses within its own forces. The demonstration had been supported by a coalition of all regional parties and town councils. Before the demonstration organizers had been invited to address the police at their regional union meeting. By the time of the demonstration it was clear that the opposition to the waste dumping proposals was shared by many police.

In marked contrast to the events in Spain, the chairman of the Dutch Union of Policemen had declared that his members would not help the government suppress such demonstrations even if civil disobedience was involved. 'We also have children and fear the danger of this toxic waste', he said. 'Our task is to maintain public order and when the government is destroying that order we can no longer obey'.[44]

On the day, not only did they do nothing to obstruct the demonstration but, as reported by the Dutch newspaper *NRC Handelsblad*, 'It was also striking that a number of policemen in uniform joined the demonstration thereby showing their support for the protest'.[45]

It was of course an exception, but it showed that even the police – part of the 'coercive apparatus' of the state – are open not only to challenge over nuclear hazards, but to 'capture'.

During 1979–80, both regionally and nationally, opposition to enrichment, reactors, and waste dumping continued to combine with increasing concern over nuclear weapons. Just after the accident at Three Mile Island, *Nucleonics Week* reported 'Opposition to nuclear power appears to be as strong as ever and Dutch government officials are said to be in despair about ever getting local authority approval for test borings in Dutch salt formations'.[46] November 1979 saw 20 000 people protesting in Utrecht against the proposed installation of new NATO cruise missiles in the Netherlands.[47] One month later the Dutch parliament voted to reject the missiles.[48]

In this atmosphere the government was constrained in pursuing its broader nuclear programme. Unable to move forward, it chose to pursue a two-year information campaign accompanied by 'general public discussions', which would precede any decision on the three proposed nuclear reactors.

The discussions were expected to continue until 1982, but by mid 1980 the government had made it clear that it was in favour of adding a further three nuclear reactors to the nation's electricity grid.[49] Whether it would be allowed to would be determined in the political arena during the early 1980s.

No one could predict with certainty what the outcome of the debate in the Netherlands or the ultimate fate of the much delayed nuclear reactor programme would be. In October 1979 an opinion poll showed that the number of Dutch people opposed to nuclear power had risen from 40 per cent a few years earlier[50] to 60 per cent.[51] Increasingly, either the nuclear programme, or a combination of the programme and the government itself, seemed in danger of being submerged beneath the swelling tide of community opinion.

WEST GERMANY: CONFLICT ON FOUR FRONTS

Of all the nuclear battles illuminated by the Whitsunday demonstrations, the struggle in West Germany is the most complex. Unlike the Netherlands and Spain, West Germany's central government has been relatively untroubled by overt divisions over nuclear

THE INTERNATIONAL ARENA 215

power in the parliament. In contrast to the Dutch Labour Party, the ruling Social Democratic Party (the SPD) in West Germany is an enthusiastic promoter of nuclear power. Its major potential rival and coalition partner, the Free Democratic Party (FDP), has been scarcely less enthusiastic. With no anti-nuclear opposition party waiting in the wings, the ground has seemed well laid for the government's ambitious nuclear programme to be pushed through vigorously, just as in France. However, from the first confrontation at Wyhl in 1975 it was clear that it would not be so easy. Not only would there be a nuclear battle, but it would rage on four fronts: in confrontation directly with the state, in the courts, in the political parties, and within the regional parliaments.

CONFRONTATION AND COURTS 1975–80 The opposition at Wyhl had been both regionalist in character and determined. The government had retreated and a broad citizens movement had spread across the country. This embraced existing conservation organizations, scientists, political groups, and activists, and led to the formation of an array of widely differing anti-nuclear 'citizens initiative' action groups.

From 1975 to 1977, the more militant sections of this movement tested their strength by directly confronting the state. As described in chapter 6, a series of demonstrations had occurred at Brokdorf, Grohnde, and Kalkar. Numbers present swelled from 8 000 in October 1976 to as many as 50 000 in September 1977. In addition, the militancy of the demonstrators – and especially a much smaller core drawn from the radical left 'K Groups' – grew rapidly, only to be confronted by the state's bulwark of police counter measures.

By now it was clear that in direct centralized confrontation with the state the opposition was not strong enough to halt the nuclear programme. Nevertheless it was still possible to obstruct the programme. To do this it would be necessary to manoeuvre within the state.

As in many other countries, West Germany has not only a central government, but also an assortment of regional governments. The historical process by which these regions have been welded into the present federation has left their governments with substantial powers. It has also produced a complex system of courts.

The West German legal system is decentralized, with a three-tier

216   GLOBAL FISSION

hierarchy of administrative and appeal courts. As one analyst puts
it, 'this structure allows dispersal of control through the Lands as
well as the federal authorities'.[52] Further, the West German Atom
Law requires that all safety precautions be as strong as science and
technology allow, and that safety issues must take precedence over
economic or political considerations. Together, the law and the
legal structure can provide important but not insurmountable
hurdles to many nuclear projects. They can create long delays
which have provided vital time for the further growth of
opposition.

The first candidate was the Wyhl reactor, suspended by court
order from early 1976. A flood of writs, court actions, and
planning objections followed. The construction licence for the
Bilbis-C reactor in Hessen was suspended pending a solution to the
nuclear waste problem.[53] Construction of the Kalkar reactor was
held up, the suspended animation of Wyhl reinforced, and a
successful court action launched against the start-up of the
Esensham reactor on the Weser River.[54] By Whitsunday 1979
regional court actions had brought much of the central government's nuclear plans to a standstill.

Twelve plants yielding a total of 9.5 GWe were in operation. A
further eleven were under construction. Of these, three (Wyhl,
Brokdorf, and Grohnde) had had their construction licences
revoked, and a further four (Hamm, Bilbis-C, Neckar-2, and
Neupotz-1) were awaiting construction licences, these having been
delayed by a network of government decisions requiring the
industry to solve safety problems, including the waste problem,
before they could be granted.[55] According to the Electrical Utility
Association, a total of 11 GWe of nuclear power had been held up
by the citizens opposition permeating much of West Germany.[56]
The scale of this reverse to government plans was substantial. In
1978 alone, some $6 billion of investment in nuclear power had
been halted by threatened or actual court orders.[57]

In attempting to cope with the opposition the state and federal
governments had weaved too and fro and left behind a quagmire of
regulational and legal conditions. The licensing of nuclear reactors
had been made conditional on a satisfactory storage system for
nuclear wastes, and the stages of licensing had expanded to many
tiers. The average construction time for a nuclear reactor had
extended from four to five, to eight to nine years.

On average, seven different permits were now required to

THE INTERNATIONAL ARENA

license a reactor, while the nuclear projects were losing some $6 to $9 million (DM 15 to 20 million) each month the projects were delayed. For the Kalkar prototype fast-breeder project, at which 25 000 people had massed in protest on Whitsunday weekend, delays and construction stoppages had been typically paralyzing. The project had only passed through the first three permits, yet already the information on these applications occupied more than 10 000 files filled with 50 million stamps and 3 million signatures. Costs for the project had risen to four times its original $340 million (DM 800 million).[58]

For the nuclear industry the net result of this, combined with intertwined economic factors, was alarming. The government's estimates for West Germany's nuclear capacity for 1985 had been halved − plummeting from 40–50 GWe to 24 GWe.[59] West Germany's reactor manufacturing company KWU, a company of major importance among the handful which make up the world's reactor manufacturing industry, was in an increasingly difficult position. It had not had a single domestic order since June 1975 and five of its earlier orders were blocked by court decisions or licensing delays. Until the situation improved it hoped to limp along on orders from overseas. Said Klaus Bartheld, head of KWU, 'Without orders from abroad we would have been bankrupt long ago'.[60] A major portion of those orders had already caused great concern in the USA and the Netherlands. It was KWU's contract to supply a full nuclear fuel cycle to Brazil.

The contract for the ambitious programme for the construction of nine nuclear reactors, a uranium enrichment plant, and a waste reprocessing plant, had been given to West Germany by the military government of Brazil. Initially costed at $10 billion, it had risen to a price estimated at between $20 to $30 billion in 1979.[61] The contract was a vital component of KWU's survival strategy. But among the Brazilian people, the benefits their government alleged would stem from the project were being questioned.

The original plan, drawn up in the early 1970s in the heyday of the Brazilian 'economic miracle' was spectacularly ambitious. It called for sixty nuclear reactors to produce 75 GWe of electricity by the year 2000.[62] Later in the decade the plan was stripped back to ten. Soon even this target was cut back.

By the beginning of 1979 only the first stage of the programme, to add two new reactors to the existing Angra-1 plant, was actually under construction. Plans for further plants were being widely

attacked. The billion dollar price tag on each reactor alone would have a substantial effect on the country's foreign debt of $55 billion, the largest in the Third World.[63] In a country where an estimated 45 per cent of infant deaths in the first year of life are associated with hunger, and an estimated third of the population live on less than the legal minimum wage of $90 per month, the vast expenditure on nuclear power seemed an outrageous misallocation.[64]

With more than 150 GWe of cheap hydro-electric capacity available and widely conceded to be capable of producing electricity at cheaper prices than nuclear plants, there seemed no requirement for their introduction.[65] Even at the most optimistic growth rates, it is unlikely that the hydro potential could be exhausted before the year 2020.[66] Inside Brazil, within the limits on dissent enforced by the military regime, the questioning of the government's programme was growing in effectiveness.

In June 1979 industry journal *Nucleonics Week* reported 'Forces opposed to Brazil's nuclear power program are rapidly gathering momentum', and that the government was prepared to take part in talks which might lead to 'adjustments' in the programme.[67]

Importantly, opposition now arose within government ranks. Some began to realize that the Brazilian dream of obtaining sufficient control over nuclear technology to become a nuclear exporter in its own right was likely to remain in the world of fantasy. In a world of shrinking nuclear programmes West Germany was hardly likely to relinquish its expertise and create another competitor.

The crisis came in October 1979 when leaked secret documents confirmed that control over reactor construction would lie totally with West Germany. The resultant reaction forced the government to decrease the number of projected reactors from eight to four.[68] But in the community even this programme, amounting to a mere 8 per cent of the original plans, continued to generate mounting opposition.

It is not necessarily safe to openly express opposition in Brazil. Nevertheless, several thousand people demonstrated against the Angra-1 plant in Itaorna in 1979, and a similar number demonstrated at Resende in Rio de Janeiro in April 1980.[69] The Brazilian Society for Scientific Progress, Brazil's principal scientific society, called for a national referendum over nuclear power.[70]

Regional opposition too was becoming more assertive. The state government of Rio Grande do Sul voted to make any permit for the

THE INTERNATIONAL ARENA   219

nuclear facilities planned for its region subject to a referendum of those living around them.[71] In June, when it became known that the fourth and fifth reactors were to be placed near the city of Iguapé, the mayor, after several demonstrations, declared that although a member of the government party, he supported his city's opposition to nuclear power.[72] By the end of 1980, Brazilian officials were acknowledging that the completion of the remaining programme would be delayed at least until the end of the century, and that some reactors might never be built.[73]

For KWU the dissolving Brazilian contract further heightened the pressing need to obtain some reactor sales domestically. But as West Germany entered the 1980s there was no evidence that the court actions which had brought the programme to its state of virtual paralysis, were going to disappear.

The judiciary is of course part of the state. It is always, at least in theory, open to the legislature to change the ground rules on which the legal battles are fought. In particular the whole court battle over nuclear power in West Germany could be dramatically transformed by amendments to the Atom Law to weaken safety requirements and reduce the possibility of citizen intervention. Whether or not this occurs depends not only on whether it is desired by the government, but whether it is politically possible. The battle in the courts is dependent on the broader political struggle within the community and that struggle's effects within other parts of the state. In West Germany in particular, this battle already depended on important skirmishes within the political parties and the regional parliaments.

PARTIES AND PARLIAMENTS 1976–80   As early as June 1976 a delegation of 750 citizens groups had declared they would instigate a campaign to persuade the electorate to vote only for candidates who declared their opposition to nuclear power before the forthcoming general elections.[74] By 1977 such a campaign seemed a real possibility, with more than 8 000 citizens initiative groups estimated to be spread across the country, and with a large public following.[75] According to a survey that year, 12 per cent of the West German population was prepared in principle to work for a political party, while 34 per cent would join a citizens initiative.[76] Although many of these groups had initially been organized around various environmental and community concerns, by 1977 large numbers of them had also taken up the fight against nuclear power.

At the national level the mood of nuclear opposition had already begun to assert itself within the ranks of the two government coalition partners, the SPD and the FDP. By the time of the parties' bi-annual conference in November 1977 there seemed real danger that the rank and file might push through policies binding the parties to a moratorium on any further nuclear construction. The SPD leadership and its union supporters headed this off by organizing a major union demonstration *for* nuclear power. On 7 November about 40 000 unionists were booked off work, paid about $10 (DM 20), and bussed down to the Dortmund stadium.[77] It may not have been spontaneous, but it gave a major weapon to the pronuclear forces at the SPD conference.

Both parties finally adopted policies which, although requiring the nuclear programme to be 'minimal', left the parliamentary sections with virtual carte blanche. But the fierce debate on the issue reflected real political forces in the electorate. Three months later this was demonstrated by the surprising success of environmental candidates at the elections of the regional parliaments of Lower Saxony and Hamburg. The 'green lists' won 3.9 per cent and 3.5 per cent of the vote respectively, taking most of the votes from supporters of the FDP. For the FDP, which had previously been part of the coalition governments in the two regions, the loss was vital, taking its vote below 5 per cent. Under the constitution it was therefore automatically eliminated from all parliamentary representation in the two states.[78]

In both the SPD and the FDP the growing anti-nuclear pressure, if not decisive, was becoming increasingly uncomfortable and difficult to contain. At its November 1978 convention the FDP voted to reject all plans for fast-breeder reactors for West Germany,[79] mirroring earlier resistance to such plans from the left wing of the SPD.[80] The FDP decision, however, seemed to have little effect on the parliamentary wing which in February 1979 used a threatened resignation of all cabinet ministers to force the backdown of a group of their colleagues. These had threatened to vote against a continuation of the country's fast-breeder project.[81] Three months later, in May, the left wing of the SPD and the FDP were reported to be pushing for decisions to abandon nuclear power altogether at the forthcoming party conventions. Theo Somner, chief editor of the *Die Zeit*, was reported as saying that a de facto moratorium of about two years in West Germany seemed likely 'but that for the

first time the demise of nuclear power in West Germany could not be ruled out entirely'.[82]

Despite the growing tension within the two coalition political partners, neither was yet quite ready to carry a nuclear moratorium motion. In July 1979 the FDP national convention defeated two nuclear moratorium motions by the barest majority – 193 votes to 191, and 184 votes to 181. That thin margin was obtained only after leader Hans-Dietrich Genscher had threatened to resign if the motions were carried.[83] SPD party leader and chancellor, Helmut Schmidt, also threatened to resign unless his party's December conference defeated motions for a nuclear moratorium. The motions were duly defeated, but observers noted that the green vote was now becoming strong enough to threaten the party's electoral majority.[84]

Nationally, the parties still seemed committed to an expanding nuclear programme. The programme's stagnated condition in the face of this reflected the fact that opposition at the regional level, both inside and outside the governments, was too great to force the programme through. Within the SPD this was illustrated by the attitude taken by the Baden-Württemberg state branch of the party. The opposition which had germinated in that region around the proposed reactor at Wyhl now encompassed much of the state, penetrating deeply into the SPD. In mid July 1979 the regional party convention had demonstrated this by passing, by a large majority, a resolution stressing the unsolved problems of nuclear waste disposal and calling for a nuclear moratorium until 1984. Elsewhere, in Lower Saxony, the problems of nuclear waste disposal and local opposition were coming together to demonstrate the degree to which the state was being fractured by the nuclear issue.

UNDERMINING GORLEBEN For the West German supporters of nuclear power the problem of what to do with the radioactive wastes accompanying their nuclear programme proved ever more distracting. By 1976 the existing reactors already produced some 1 700 cubic metres of waste annually, and this was expected to rise to 22 000 cubic metres by 1980. Said Dr Frank Haenschke, chairman of the reactor safety subcommittee of the Bundestag, the lower house of parliament: 'The unsolved difficulty of getting rid of this waste will be West Germany's security

problem No. 1 in the 1980s'.[85] With the growing popular concern over the hazards of nuclear power it was politically necessary to find a credible solution to the problem.

The government had already begun to search for possible locations for permanent nuclear waste dumps. In the summer of 1975 a series of fires occurred in the forest areas of Lower Saxony. The region comprises one of the few untouched natural beauty spots of West Germany, with moors of marshes, fields, and forests in which rare birds and plants abound. A few months later the government announced that the salt domes beneath three of the four fire locations were being considered as possible dumps for solidified nuclear waste.[86]

At the first site to be selected, Unter Luhs, opposition broke out immediately with an occupation. Farmers blocked roads to the site with their tractors, demonstrating their support for the occupation. The government retreated and sought a site with a better political climate.[87] The pressure to settle on one was considerable. The small transitional storage area at Asse was fast becoming filled, and political concern was mounting. Editorialized the liberal newspaper *Süddeutsche Zeitung* in 1976: 'we have taken a seat in a rocket whose steering mechanism is faulty and for which a landing place has not yet been found'.[88] Finally the government selected Gorleben, a remote promontory on its north-east border protruding into East Germany, as the site for the waste disposal complex.

The proposed complex was to be built on an ambitious scale. Its estimated cost rose from $3.25 billion in 1976 to $6 billion by 1979.[89] It was to comprise a storage pond capable of holding 3 000 tonnes of spent fuel, a reprocessing plant, fuel fabrication facilities, and underground storage vaults capable of receiving 30 to 40 tonnes of high level waste, and 50 000 drums of low level waste each year.[90] Its sixty or so buildings were to cover some 12 square kilometres.[91]

Opposition to the project first developed locally and then, with the first occupation in 1976, Friends of Gorleben groups sprang up across the country. At one stage it looked as though the local group in Lüchow might be able to halt the project by buying a piece of the land on the site from a local farmer. But although they raised $320 000 (DM 750 000), nearly double the asking price of $171 000 (DM 400 000), he sold it to the government.[92] By 1978 there had been several large occupations of the site, including a

THE INTERNATIONAL ARENA

223

non-violent strategy camp which ran for three weeks and was attended by 8 000 people. In June a day of direct action brought out demonstrations in fifteen cities across West Germany in opposition to the Gorleben plan. In the same month the green lists' success culminated in the region of Gorleben, where the local candidates won 17 per cent of the vote.[93]

At the level of the regional government the leader of the Christian Democratic Party administration, Walter Albrecht, who had first suggested the site, now found the granting of a construction permit politically suicidal.[94] On the other hand, the federal government was pressing him to give permission. He announced that he would make a decision before mid 1979.

The pressure was increasing on both sides. In January 1979 Bonn offered to pay a lump sum of $100 million to the Lower Saxony government for police protection and other services needed for the plant. In all, the government would receive $275 million in compensation payments.[95] But on the other side, a meeting of 1 300 anti-nuclear activists from all over West Germany had announced that action would be started against Gorleben 'without any limitations to the type of protests'.[96]

Drilling to investigate the salt began in March.[97] Two weeks later the accident at Three Mile Island shook the world. Said Chancellor Schmidt, 'Harrisburg will result in wide re-thinking on options concerning nuclear energy'.[98] Three days later the opposition underlined the urgency of that as an estimated 80 000 to 100 000 people converged on Hannover in opposition to the Gorleben complex. It was the largest demonstration against nuclear power ever held in West Germany.[99] For the government of Lower Saxony the pressure against the complex had become too great. On 16 May 1979 it announced its decision to indefinitely postpone permission for the reprocessing plant, the crucial link in the integrated complex which would separate out the high level waste. Said Albrecht in a televised speech, even if all the risks for the population involved were removed, 'the double question remains – if construction of such a facility is indispensable and if it can be carried through politically'.[1]

The abandonment of the planned Gorleben complex highlights the ability of citizens to force part of the state into conflict with the rest over its nuclear programme. Here West Germany's federal structure had been used to place yet another obstacle in the face of

the largely stalled nuclear programme. It was an important achievement for the opposition, but it was certainly not the end of the battle even over waste reprocessing and storage.

Albrecht had already reserved the right to permit both investigative drilling and the temporary storage of medium and low level wastes on the site. In September, a few weeks after a major demonstration in Berlin to counter any federal government move to override Albrecht's decision,[2] Chancellor Schmidt announced a new nuclear waste strategy. The plan to build a single integrated complex was abandoned. The government and nuclear industry would now attempt to build smaller units at different locations.[3] Favoured sites were Karlsruhe, Hanau, and Borken.[4]

This new strategy had the advantage of a simultaneous attack on several perhaps less politicized fronts. But it carried the nuclear controversy into further areas of West Germany. Already in Arhaus, Northrhine Westphalia, the Christian Democrats had in recent weeks lost 20 per cent of their vote at a local election over a decision to build a temporary waste storage pond. By mid 1980 this regional concern was obstructing the implementation of several small scale waste storage and reprocessing proposals for various parts of West Germany.[5]

During 1979 to 1980, the West German nuclear conflict continued to be very visible nationally. Most dramatic was the largest demonstration yet held on the issue in West Germany when, on 4 October 1979, between 100 000 and 150 000 people marched through the streets of Bonn. Significantly, this event was the result of a coming together of opponents of nuclear power and the peace movement. It called not only for a shut-down of all nuclear reactors but also for an end to the deployment of nuclear weapons in West Germany.[6] The same alliance could be seen again at Lingen a year later.[7]

Equally striking was the emergence of the Green Party as a contestant in the national election. As the re-election of the Schmidt government in October 1980 demonstrated, it would be very difficult for such a party to sway peoples' votes at the national level on the nuclear issue, when many other issues loom forcefully. But in local and regional elections the Green Party attracted increasing support. As commentator Roger Boyes of the *Financial Times* concluded after the Green Party candidates in the state election for Baden-Württemberg (the home of the Wyhl reactor site) had gained 5.3 per cent of the vote, and six had been elected:

THE INTERNATIONAL ARENA

'The leaders of West Germany's major political parties . . . agree on one thing. They know now that the small anti-nuclear Ecologist Party has to be treated as a serious political force'.[8]

The beginning of 1981 saw Chancellor Schmidt finding it increasingly difficult to maintain the support of his own party over the twin issues of allowing US tactical nuclear weapons in West Germany, and pressing on with the nuclear programme. On the one hand, the SPD left wing now strongly opposed the missile deployment,[9] with the Green Party adding to the pressure by launching its own signature campaign against it.[10] On the other hand, the SPD was now being torn by a ruinous dispute over the Brokdorf reactor.

In January 1981 a decision was taken by the federal government to press ahead with the construction of Brokdorf. Within three weeks it found itself confronted with the Social Democratic government of Hamburg, a 50 per cent partner in the project, voting to defer the project for three years.[11] As the debate over Brokdorf continued, West Germany was rocked by another announcement. The national Reactor Safety Commission had found that there were faults in the primary cooling circuits of all its boiling water reactors. It ordered them closed for repairs. Initial estimates were that this would put their combined capacity of 3.3 GWe, or 38 per cent of West Germany's entire nuclear electricity production, out of action from between one and three years.[12] Given that the construction standards set in West Germany have in the past been considered the highest in the world, it was not surprising that one commentator described it as 'perhaps the largest single blow to nuclear power in Europe since programmes began'.[13] There could be no doubt that it would add further fuel to the nuclear debate already raging in West Germany.

In examining the nuclear battle in West Germany it is easy to pay too much attention to the more dramatic developments, such as the national elections, and too little to the complex decentralized activity in the regions. Yet it is these which may in the long term prove more important. As West Germany moved into the decade of the 1980s one thing was clear. The conflict over nuclear power in that country will continue to develop, and along more than one front. Looking at the conflict's recent history, it is apparent that, given the extraordinary success at the regional level, which has removed the Gorleben complex from the West German nuclear

programme, the success of the Green Party at the state level, the very large demonstrations, and the still unfolding problems of the stalled nuclear programme, a shroud of uncertainty still surrounds the country's nuclear programme.

In West Germany, as in the Netherlands and Spain, the nuclear struggle remains unresolved although their nuclear programmes have been dramatically obstructed during the last half of the 1970s, contributing substantially to the overall malaise of the world's nuclear industry. The nature of the struggle in the three countries has varied markedly, yet broad common trends are observable.

Where the central government's nuclear plans have touched the community, opposition has arisen, often most intensely in those regions where local history and culture reduces the legitimacy of the central government's authority. In addition, the state has become an arena for struggle. Both regionally and nationally the nuclear debate has penetrated political parties, governments, and other agencies of the state, as well as the broader community. At times courts, regional governments, political parties, and even the police have been diverted from acquiescing to the pro-nuclear stance of the central government, and where this has occurred it has played a vital role in helping to limit or halt the expansion of the nuclear enterprise.

These features of the various conflicts over nuclear power in a variety of countries make it clear that despite the intimate relationship between the state and the nuclear industry their plans are open to challenge and may ultimately be overcome by community groups. That said, it need hardly be noted that the challenge presented by such groups is also not without its weaknesses. In the long term the battle will be determined by the degree to which each side manages to overcome its deficiencies.

For the opposition to nuclear power, one serious weakness which, if overcome, would vastly increase its effectiveness, relates to the role taken to date by the majority of social democratic and left-wing parties and unions.

Although there have been exceptions, such as the Labour Party and many unions in the Netherlands, the PSU in France, and the Communist Party in Sweden, the role of the labour movement has largely been to support nuclear programmes. The SPD in West Germany is joined by the Communist Party (and until recently the Socialist Party) in France, the British Labour Party, the Austrian and Swedish Social Democratic Parties, and a host of others, in

THE INTERNATIONAL ARENA

enthusiastic promotion of nuclear power. Similarly, many unions in Austria, West Germany, the USA, and many other countries have taken positions in favour of nuclear projects. Beyond this, and not unrelated to it, the official positions of countries outside the capitalist world have been almost uniformly in favour of nuclear power. The generally pro-nuclear position of many official bodies within the labour movement has been a constant obstacle to the greater growth and effectiveness of opposition to the nuclear enterprise. Any change in this situation, either way, could play an important role in the future development of the nuclear conflict. For this reason it is important to examine the relationship between nuclear power and the left. To do this, however, it is first necessary to review the attitudes to nuclear power held by the countries which lie outside the capitalist world.

10

COUNTER-CURRENTS WITHIN THE LEFT

Even if Soviet Russia succeeds in carrying out completely its project of general electrification, without introducing any essential change in the system of control and organisation of
the people's economy and production, it would only catch up with the advanced capitalist
countries in the matter of development.

A. Kollontai, 1921[1]

IN THE USSR

Nowhere in the world have nuclear plans been more avidly
pursued than in the USSR. Its plans not only deeply affect the
thinking of many sectors of the left in the West. They also reflect a
common ideological position underlying that thinking. Roberts
summarizes it thus:

> The socialist movement has suffered for many generations
> from the illusion that technology is value free. Adopting a
> misleading schema in which an essentially non-political 'base'
> (the forces of production) is simply to be taken over and
> endowed with a different 'superstructure' (socialist relations
> of production), it has failed to appreciate the political content
> of that technological base.[2]

The view that the technology of capitalism could, and should, be
adopted as the 'base' upon which to build socialism, has been
widely held in the USSR since the crucial formative days of the
1920s following the Bolshevik seizure of power in 1917.

Once they had seized power the Bolsheviks were confronted
with the task of building a socialist society. But how? The expectation that socialist revolution would occur in one of the most indus-

COUNTER-CURRENTS WITHIN THE LEFT 229

trialized capitalist countries had been stood on its head. Instead, it had occurred in Russia, still semi-feudal. Without the advantage of a large educated working class, how would the people build their socialist economy?

Surrounded by capitalist countries whose leadership feared the implications of the revolution, Russia seemed under threat. In addition there was an understandable desire to produce results quickly, both to solve the pressing legacy of economic problems, and also to demonstrate the effectiveness of the new socialist order. There was little time to worry that the adoption of the large-scale factory technology developed in the West and based on an hierarchical organizational structure, might entrench similar power relations in the emerging society. It was easier to accept the technology as ideologically neutral. In the hands of the capitalist class, it had served them well; in the hands of the workers' state, why should it not be equally effective in the service of the people?

This view was quite consistent with the broad thrust of current Marxist thinking, and it was clearly articulated by Lenin: 'We ourselves will organize large-scale production on the basis of what capitalism has already created'.[3] 'We shall not invent the organizational form of work, but take it ready-made from capitalism — we shall take over the banks, syndicates, the best factories.'[4] And if this was the goal, then there was seemingly only one way it could be achieved. 'We, the party of the proletariat, have no other way of acquiring the ability to organize large-scale production . . . except by acquiring it from first-class capitalist experts.'[5] Given the exigencies of the time, the position is entirely understandable, but it helped lay the basis for constricted goals and the emergence of a technocratic elite.

That there was no choice but to adopt the technology offered by capitalism reflected a view that capitalist technology is the only way technology could develop. The ideology underpinning this view is the 'idea of progress',[6] where science and technology are seen as moving forward along a single track, propelled by a politically neutral, autonomous process of discovery. But as writers such as Dickson have demonstrated, this is a very distorted picture of technological development.[7]

Any society selects, over time, from a vast array of possible inventions. The selection is mediated by the need to choose those which are compatible with the society's dominant social and political processes. Even the problems which inventors address are

usually selected with some concern over whether the solution will prove useful, or profitable. In the development and use of inventions the selection process is even clearer. For example, the advanced industrialized part of the capitalist world set out deliberately, and with much help from the state, to develop and use nuclear power. But a different society, with different political and economic needs, and a correspondingly different ideology, might well instead have embarked with similar enthusiasm to build efficient small-scale solar technology.

Suggesting that the direction of technological development adopted was inevitable and politically neutral helps legitimate the social structure that was needed to produce it and now maintains it. In the capitalist countries, nuclear power helps legitimate the centralization of power in large corporations and the state, while in the USSR it acts similarly to legitimate the large technical bureaucracy and the centralization of power now characteristic of its social organization.

These considerations were not totally unknown, even in the 1920s. There were those who did challenge the idea of accepting capitalist technology as the operating principle for the new socialist society. In particular there were those who advocated the development of *Proletkult* – people's art, science, and technology. They argued that rather than build the society to meet the requirements of the existing available technology, 'our task consists in bringing the content and methods of science into line with the requirements of socialist production'.[8] But this was not the dominant view. Instead, the USSR joined the West in equating economic growth with progress, and in believing that this could most efficiently be achieved through massive investment in the most technically sophisticated large-scale technology that a heavily specialized technical élite could devise.

Above all, electrification of industry represented a glittering prospect, encapsulated in Lenin's famous dictum, 'Socialism is electrification plus Soviet Power'.[9] More broadly he expressed this as, 'industry . . . must be rehabilitated on the basis of modern technology, which means the electrification of industry and a higher culture'.[10] And in this project of electrification, what could be 'higher', more daringly advanced along the 'road of progress' than the ultra-sophisticated frontier technology of nuclear power!

In 1949, the USSR exploded its first nuclear weapon. In 1952 the USA exploded its first hydrogen-fusion device, and one year

COUNTER-CURRENTS WITHIN THE LEFT

later the USSR more than matched this with the explosion of what was almost certainly a fully operational hydrogen bomb. The following year, only six months after President Dwight Eisenhower's 'atoms for peace' speech, the USSR set into operation the world's first industrial nuclear power reactor. It was the first step in an ambitious plan to generate nuclear electricity across the USSR. The first targets, however, proved greatly inflated. In the sixth Soviet five-year plan, 2–2.5 GWe of nuclear plant was to be constructed between 1956–60. The actual capacity achieved proved to be only 0.6 by 1960, 0.91 by 1965, and even by 1970 only 1.52 GWe. By 1971, when the programme was ready for serious implementation, the goal was to construct 8 GWe of nuclear power by 1975, and 30 GWe by 1980.[11]

Ambition, however, was not matched by capability. By 1975 only 4.4 GWe of nuclear power had been brought into operation, barely more than half the original target.[12] Three years later, in 1978, some 8 GWe was operating, about that targeted for 1975. Compared with the 47 GWe by then operating in the USA it was still only a slight achievement,[13] representing only 2 per cent of the electricity then generated in the USSR.[14] By then it was clear that the target of 30 GWe set for 1980 seven years earlier would be totally unachievable. That target was steadily revised downward to as low as 13–14 GWe.[15] In June 1980 the actual operating capacity was only 12.7 GWe, little more than one third of the original target.[16] A target of 110 GWe for 1990 was still current, but whether it can be even remotely approached in practice is a different matter.[17]

In order to speed the pace of its reactor construction, the Soviet authorities had embarked on a programme to mass produce nuclear reactors. This Atommash Project covers some 640 hectares at Volgodonsk near Volgograd. With enormous presses set up in a 277 000 square metre main building, the massive plant is intended to produce reactors on an assembly line basis. It was intended to begin operating by 1981, but by 1980 it was clear that it was already two years behind this target.[18] Two of the problems cited were the bureaucracy's difficulties in coordinating the massive project, and the need to redesign the type of reactor being produced.[19]

For planners steeped in traditional engineering practice, mass production has an intoxicating ring of efficiency. However, gains only follow if the production line can be used to make a substantial

number of reactors without major design changes. In the USA, the steady stream of incidents, accidents, and revelations of hidden hazards has forced a continual series of design modifications. In the USSR the success of the mass production technique would depend at least partly on the prevention of increased pressures for redesign. In the past, the strength of Soviet state authority and the control over the flow of information has made the growth of serious grass-roots pressure for such redesign somewhat limited.

Vitaly K. Sedov, director of the Novovoronezh nuclear power station, made this clear when he proudly stated that his country had never been bothered by anti-nuclear demonstrations like those that have besieged reactors in the USA.[20] But even though overt opposition of this type has been avoided, it is not for lack of what would in most other countries be considered good reasons for concern. In many ways, the Soviet approach to nuclear power hazards has been among the most cavalier in the world.

The number 5 reactor under construction at Novovoronezh is a Soviet reactor with a difference. It is the first to be built with any form of containment structure. In the past they have been built in standard factory buildings. This, says Yuri Svintsev, director of the Kurchatov Institute's nuclear safety laboratory, is because 'the plants are so safe'.[21] But apparently no longer. It is now conceded that future plants will be built with the thick concrete containment domes mandatory everywhere in the West. Perhaps, eventually, they will also be provided with emergency core cooling systems similar to that which only just prevented the ultimate reactor catastrophe from occurring at Three Mile Island.

Until recently the authorities simply ruled out core meltdowns as a credible occurrence. A. Petrosyants, director of the State Committee on Atomic Energy, is quoted as saying, 'We consider a containment vessel unnecessary. However we will install one when our customers want one'.[22] This lack of concern over the possibility of meltdowns was not restricted to conventional reactors. Even in the case of the much more dangerous and sensitive proto-type fast-breeder reactors, no backups have been provided for the possibility of a core meltdown, which could lead to a nuclear explosion, nor has any special containment been provided apart from the steel reactor vessel.[23]

Unlike most other countries, the USSR has considered that there is no special requirement to keep reactors away from populated areas. Writes one Soviet reactor expert:

COUNTER-CURRENTS WITHIN THE LEFT

The successes in the area of atomic power plant safety as a whole have been so great that the selection of locations for atomic power plants at the present time is not limited by safety requirements, being determined only by technical and economic factors.[24]

This being so, it might be expected that Soviet regulations would be designed to keep reactor emissions of radiation to a much lower level than is permitted in the USA. However this is not the case. At the time of the above statement, a Soviet pressurized water reactor was permitted to release twelve times as much radiation in the form of noble gases, and a hundred times as much radioactive Iodine-131, as in the USA. The resulting dose to the thyroid can be over four times that permitted by US regulations.[25] Exposure to other forms of radiation may be difficult to determine since workers often do not wear radiation-monitoring film badges.[26]

At the 'back end' of the nuclear fuel cycle, although the use of plutonium has been admitted as being hazardous and capable of leading to proliferation of nuclear weapons, the desire to use nuclear energy is considered sufficiently strong to make its use 'obligatory'.[27] The Soviet strategy for handling reactor waste is based on reprocessing it and recycling the plutonium. The highly radioactive liquid residues from this process have been pumped under pressure into deep bore holes in the ground, a practice that would be considered highly controversial elsewhere.[28]

It was almost certainly in the context of reprocessing nuclear wastes that the USSR incurred what is generally believed to have been the worst nuclear accident to date. Although Soviet authorities have never admitted it, and until recently even Western nuclear agencies were doubtful, it now seems highly likely that a major nuclear accident occurred at Kyshtym, a town on the eastern side of the southern Ural mountains, in December 1957. The accident, which is believed to have killed hundreds of people and covered an area of more than 1 600 square kilometres with a mixture of radio-active materials, seems to have been common knowledge among nuclear scientists in the USSR. Its occurrence, however, had not been revealed to the public in either the capitalist or non-capitalist world. It was inadvertently revealed in the West by an exiled Soviet scientist, Dr Zhores Medvedev, who mentioned it in passing in an article published by *New Scientist* in November 1976.[29] Medvedev was surprised to learn that the nuclear establishment in the West

234   *GLOBAL FISSION*

claimed no knowledge of the accident and was less than interested in news of it. In the USSR the accident was known to have been related to the reprocessing and disposal of waste from military reactors in the area. With a waste controversy burgeoning in the West, nuclear administrators such as Sir John Hill, chairman of the UK AEC, attempted to dismiss the story as 'science-fiction', 'rubbish', or a 'figment of the imagination'.[30]

A month after Medvedev's story was released it was independently confirmed by Professor Lev Tummerman who had emigrated from the USSR to Israel in 1972. A firm believer in the benefits of nuclear power, Tummerman was concerned that anti-nuclear groups in Israel would take up the story as an argument against nuclear reactors. He was therefore intent to make it clear that the accident was caused by waste. Anxious to show that reactors had been blameless, he described how he had driven through hundreds of square kilometres of deserted, contaminated countryside in which all the villages and towns had been destroyed to make them uninhabitable. Later Medvedev was to amass an imposing body of evidence from Soviet scientific journals which provided convincing proof that lakes, land, and the ecology of the area had been badly contaminated with radioactive isotopes.[31]

Later interviews, by British journalist Andrew Cockburn, with former inhabitants of the area now living in Israel, confirmed that many, possibly even 'thousands' had died. Women had had to have abortions, and people had become victims of an 'unknown mysterious disease'.[32] This was further confirmed by a Central Intelligence Agency report obtained by Ralph Nader's Critical Mass group under the US Freedom of Information Act.[33] A belated study from within the US nuclear establishment, released by Oak Ridge National Laboratory in January 1980, confirmed Medvedev's findings and pinpointed the location of the accident at a site east of Kyshtym. It showed that the names of more than thirty small communities in the area had been deleted from Soviet maps published since the accident, and proposed a credible mechanism to explain the explosion that had distributed radioactive waste over Kyshtym so catastrophically.[34]

The reluctance of the Soviet authorities to publicly admit the disaster in the Urals was matched by a reluctance to admit any nuclear accidents whatsoever. However, possibly through change of heart or a slip of the tongue, the Power and Electrification Minister, Petr Neporozhniy, in April 1979 admitted that at least

COUNTER-CURRENTS WITHIN THE LEFT

two serious accidents had occurred in Soviet plants. These involved fractured cooling pipes and an exploding steam generator. Other sources have reported different accidents, one of which involved earthquake damage to a pressurized water reactor.[35] In October 1979 Soviet scientists publicly admitted that in all their experimental fast-breeder reactors the liquid sodium coolant had at one time or another caught fire. Unofficial reports vividly describe a major fire which occurred in the Beloyarsk fast-breeder reactor at the beginning of that year and almost led to an evacuation of nearby villages.[36]

The secrecy in which the Soviet nuclear industry is veiled makes it extremely difficult for the general public to form an opinion on the comparative value and danger of nuclear power. Nevertheless, there has been an occasional indication that at least within the administration there is not total unanimity on the value and safety of the nuclear programme.

Criticism of the utilization of nuclear power is subject to censorship in all Warsaw Pact countries. Controversy over nuclear safety therefore practically never occurs in the mass media. Where concern exists it is only likely to be expressed at meetings of experts – in scientific conferences, nuclear management meetings, and nuclear industry journals. One of the earliest public criticisms of nuclear power in the USSR was voiced by Pjotr Kapiza in the Academy of Science at its 250th anniversary in October 1975.

Known as the 'grand old man of Soviet Physics', a member of the Presidium of the Academy, director of the Academy of Physical Problems, and the former director of the laboratory which developed the Soviet atomic and hydrogen bombs, Pjotr Kapiza described in his speech three major problems of reactor safety. 'A big atom station . . . is a big danger for surrounding nature, and especially for humanity', he said. Waste disposal, too, is a problem, and proliferation 'can lead to the point that an atom bomb can serve as a weapon of blackmail of even a group of gangsters'.[37]

The speech was greeted with loud applause and was later reprinted in full in the Academy's January bulletin. The inevitable official reply came four months later in the 3 May issue of the national newspaper *Izvestia*. Written by Anatoly Alexandrov, recently elected president of the Academy, it praised nuclear power and attacked unnamed opponents. It was not clear whether this reprimand would necessarily end the debate. As one Western diplomat observed, 'you can never be sure if we will hear the debate continue

or if we will even recognize it when it happens'. But he noted that it took the authority of the Academy president writing in a widely-read paper, and thus risking the debate being raised more widely, to counter the Kapiza line.[38]

Later reports suggest that the debate did not die with Kapiza's article and its rebuttal. 'Leading scientists' were reported to have admitted that a flood of protests came when a decision was announced to build a nuclear power plant at Novovoronezh.[39] In December 1979 Alexandrov himself admitted for the first time that 'some Russians' were worried about the effects of nuclear energy, although he hastened to add that their fears were unfounded.[40] Much more significant, however, was an article that appeared a few months earlier in *Kommunist*, the Communist Party's leading theoretical journal.[41] Written by nuclear scientist Academician Nikolai Dollezhal, and economist Dr Yury Koryakin, the article had, after making the expected bow to the necessity of nuclear power for Soviet growth, plunged into a biting series of criticisms of the hazards posed by the Soviet programme.

Concerned about the 'ecological capacity' of the environment to absorb the heat and nuclear wastes the reactors produce, the authors stress that the dangers from released radiation are rapidly increasing. They note that the waste problem is still unsolved. 'We cannot yet say that guaranteed, reliable, economic and effective processes have been developed for all stages in the production of the external fuel cycle', they add.[42] Further, because reprocessing plants have been built in isolated areas while reactors have been built near population centres, waste products will have to be carried long distances through populated areas. And with fifty to seventy reactors operating in the year 2000, large amounts of water and land will be put out of action in order to ensure operating safety, removing from production grain-growing areas capable of providing bread for several million people.

As a *New Scientist* report put it:

> Dollezhal and Koryakin's article is clearly representative of a powerful nuclear safety lobby in the Soviet Union . . . there is clearly much private debate on nuclear issues in Russia and the environmental wing in this debate is gaining strength.[43]

Even though the article was widely cited and discussed on Moscow Radio, the problems it raised were almost certainly a pale reflection

COUNTER-CURRENTS WITHIN THE LEFT   237

of a more direct debate going on among those managers, technicians, and scientists who had access to nuclear information. This
inference could be drawn simply from the fact that the article was
published at all. It was given added support by Soviet participants
at a 1979 British–Soviet energy seminar when the dangers of
nuclear waste disposal were mentioned. They were reported to
have replied with an unusual lack of confidence, 'doesn't it worry us
all?'[44]

Even if restricted to the bureaucracy, it is evident that a debate
over nuclear power is developing in the USSR. In part it can be
understood as a spontaneous development of concern by the
scientists themselves as the problems of the growing programme
increase. But that growth of concern must also be stimulated by
their absorption of the arguments being raised with such depth of
feeling in the West. It seems likely that that process will further
intensify the Soviet debate, perhaps even eventually moving it into
a more public arena. If so, it may well add further barriers to the
implementation of the already greatly retarded Soviet nuclear
programme.

EASTERN EUROPE

In other communist countries similar disagreements exist over
nuclear power, accompanying a similar setback in nuclear expectations. In those countries within the Soviet sphere of influence
which are members of COMECON, the Council for Mutual
Economic Aid, a typically ambitious nuclear target had been set in
1974. According to the Joint Forecast of Development of the
Nuclear Power Industry, the total nuclear capacity in the Eastern
bloc (outside the USSR) would reach 30 GWe in 1980, and
between 90 and 120 GWe by 1990.[45] By 1979, however, it was
clear that expectations would not be met, and the target for 1990
was reduced to 37 GWe.[46]

This substantial decrease could not be attributed to the effects of
opposition. As in the USSR, the official confidence in nuclear
power had been pushed forward in the mass media with little
suggestion that there might be any problems attached. Where
problems were mentioned, it was in the context of disparaging
remarks about the opposition to nuclear power in the West.[47] For
example, the East German periodical *Kernenergie* reported on delays

in the West caused by 'so-called citizen's initiatives, in which discussions are prolonged by new, mostly unfounded objections'.[48]

Where grounds for concern are conceded, this is usually qualified by the suggestion that these problems could only occur under capitalism. For example, Günter Leuschner, a radio commentator on 'Voice of the GDR', says that anti-nuclear campaigns 'show mistrust towards the ruling powers as to whether the protection of man and his interests are more important than the interests of entrepreneurs'.[49] But Leuschner did not go on to seek parallels with the East German situation, despite evidence that such parallels exist.

One example was the rejection in 1968 by the East German Radiation Safety Institute of the proposed Nord nuclear plant near Greifswald, reportedly because it was not well enough designed against radiation leakage. Despite this recommendation the government authorized construction of the plant.[50] Outside these technical and administrative circles, the first clear sign of public opposition to nuclear power in East Germany was a demand in 1979 by the synod of the Protestant church of Mecklenburg for a public discussion 'on the possibilities and dangers of the peaceful use of nuclear power'.[51] Further church involvement may provide an important new channel through which debate could develop.

In Czechoslovakia there is already evidence of a debate over nuclear power, at least among one section of the general public. In December 1978 the Czech human rights group Charter-77 began distributing *Document 22*, an exposé of conditions at the Jaslovske Bohunice nuclear power station.[52] According to the document, employees at the station had been compelled (under threat of loss of premium payments) to expose themselves to radiation levels well above the safety standard. Even more seriously, in an accident at the station in 1976, two workers had been killed.

Over the past three years the workers on the 110 MWe gas-cooled, Soviet-built reactor had been working under great stress. Originally intended to be fuelled automatically, the relevant machinery had never been constructed, and workers had had to place the fuel rods in position manually. According to *Document 22* workers often worked sixteen-hour shifts.

On 5 January 1976 an error occurred in the mounting process. A fuel element shot out of the reactor under a pressure of 60 atmospheres, together with a large quantity of radioactive carbon dioxide. Because an escape door had been locked, apparently to

COUNTER-CURRENTS WITHIN THE LEFT

deter petty thefts, two workers were suffocated. Six weeks later another accident led to radioactive material entering the drainage system, contaminating a stream.

The anonymous authors of *Document 22*, unrestrained by the problem of censorship, urged open discussion on Czechoslovakia's plans to expand nuclear power capacity by 10 GWe by 1992, and demanded local referenda on whether nuclear power stations should be constructed at all.

In other COMECON countries, evidence of any organized opposition to nuclear power at any level is hard to find, save for a few isolated items such as a letter to the editor of *Polityka*, a Polish journal, in which the author replies to a pro-nuclear interview by stressing the biological consequences of radiation.[53] This lack of published evidence does not of course mean that there is no local opposition. Indeed, there have been reports of an illegal citizens action group operating against further nuclear expansion in the Soviet Baltic republics.[54] However it is in Yugoslavia that clear local opposition has operated openly and effectively to produce changes in government nuclear intentions.

Yugoslavia is completely independent of the USSR politically. Unlike any of the other Eastern European countries it is not a member of COMECON, and its social organization is more democratic, stressing principles of workers' self-management. It is also considerably more open to public discussion and interchange of information with the West than the COMECON countries. But one of the things which it does have in common with them is that its central government has ambitious nuclear plans.

In official circles it is assumed that Yugoslavia will have about 12 GWe of nuclear power by the year 2005.[55] So far, however, most of this lies in the realm of intent. The country's first nuclear plant, a 640 MWe Westinghouse reactor, is under construction at Krsko, and was to go into operation in mid 1981. There has been little resistance to this first project. But when the Union of Yugoslav Electrical Enterprises in 1974 announced its intention to build a second reactor on the island of Vir, just off the town of Zadar and the holiday beaches of the Adriatic, it was surprised to find itself facing local opposition.[56]

The authorities were forced to take account of this when in 1978 the Local Assembly of Zadar removed the nuclear power plant from its long-term energy development plan. Strong local pressure had built up against the plant and the reasons were carefully argued

in the Assembly meeting. The people cared deeply for their unpolluted environment, clean sea, fishing industry, and booming tourist trade, and not at all for the future benefits that might be drawn from the plant. As Gabriel Ronay noted in the *New Statesman*:

> Fear of radiation and a nuclear mishap mobilized public opinion long before the Harrisburg accident, and since thousands of local people working in West Germany as Gastarbeiter saw with their own eyes that anti-nuclear protests can be effective, they began putting increased pressure on their representatives in the local legislature.[57]

The strong regional identity of the local Croatian people is backed by the Yugoslavian constitution whose provisions allow regions to exercise significant powers of decision making. Often the tension between regional and central views in Yugoslavia is overcome by the League of Communists' persuasion. This time, however, despite some pressure, the local people would not be moved.

The Croatian Assembly, which exercises authority along the entire Dalmatian coast, had fallen in with the plan to build the reactor. In the event of a deadlock, the constitution provides for a referendum, but neither side was anxious to force the issue, which was already developing into a test case over who makes the important decisions in Yugoslavia.

Early in 1979, the Zadar Assembly formally demanded that the Croatian republican government and the electrical union drop the Zadar plan. The federal government in Belgrade responded by sending a delegation of senior scientists to reassure the local people that the reactor would be totally safe, but by then the Harrisburg accident had happened, and they were not in the least convinced.

Despite a series of power cuts, brought about by a drought affecting the country's hydroelectric system, the Zadar Assembly unanimously rejected the Croatian government's plan and took the unprecedented step of stating that Zadar would refuse to cooperate any further on the project. On 27 June 1979, the head of the Croatian Electricity Board announced that the proposed plant would be built on the Sava River in Prevlaka, far to the east.

Whether or not the people of Prevlaka would accept this gift, it was clear that an important precedent had been set by the Zadar Assembly's fight. It had struck an important blow for regional

COUNTER-CURRENTS WITHIN THE LEFT

autonomy, and the right of ordinary citizens of Yugoslavia to question and reject a sophisticated technical project. The sound of that blow could reverberate in Yugoslavia for many years to come.

CHINA

Even more elusive than the strands of debate over nuclear power in the USSR and the countries within its sphere of influence is the debate over nuclear power in China. By 1980 the country had built only military reactors producing plutonium for its nuclear weapons programme. Although industrial reactors had occasionally been considered, this had not resulted in any being constructed.

It is hard enough to track the changing winds of Chinese policy on national objectives. Attempting to isolate inner attitudes to the future use of nuclear power is even more perilous. In the 1950s China had followed much the same development model as the USSR and had been considerably assisted in this by the presence of some 10 000 Soviet technical experts. However when the tension between the two giant countries came to a head in 1960 and the experts were withdrawn, Chinese policy shifted.[58] Under the pervasive influence of Chairman Mao Tse-tung, China entered a period of experimentation in which great emphasis was placed on the achievement of national independence through self-sufficiency. The purchase of nuclear reactors from another country would have seriously conflicted with this policy.

The construction of nuclear reactors in China in any case seemed largely unnecessary. The country has enormous reserves of coal and oil, as well as a very substantial hydroelectric capacity.[59] Some 36 per cent of China's 30 GWe of electricity production is provided by hydroelectric stations[60] and, in line with the emphasis on regional self-sufficiency, the Chinese have been building about 5 000 new small-scale hydroelectric stations each year. Although each had only an average capacity of 45 kWe, the 75 000 operating by 1978 were making a substantial contribution to electricity production. With total hydroelectric potential estimated at 580 GWe, large coal reserves, and the two sources covering complementary areas of the country, there seemed little need for nuclear reactors.[61] However, as was shown earlier for other countries, 'need' is never an essential prerequisite for a government to embark on a nuclear energy programme.

With the death of Chairman Mao and Premier Chou Enlai in

242   GLOBAL FISSION

1976, and the subsequent displacement of the 'gang of four' by
Deng Xiaoping, the policy of self-sufficiency was overshadowed by
the new objective of the 'four modernizations': agriculture,
industry, science and technology, and national defence. The
ideology underpinning this plan was clear from the way it was to be
achieved – in no small part by substantially increased trade with the
West. In that sense 'modernization' seemed closely intertwined
with the direction 'progress' had taken in the West. Whether necessary or not, it was now possible that nuclear power would seem a
valued item in China's modern technology portfolio.

In early 1978 the *Journal of Commerce* reported that several
Chinese officials visiting France had indicated that in the context of
the new economic policy Chinese authorities were considering a
nuclear energy programme. Apparently Framatome would be
involved in the arrangement.[62] France subsequently approached
the USA for permission for Framatome to sell one or two
Westinghouse-designed nuclear reactors to China. The USA gave
approval in November 1978, subject to a pledge by France that it
would have Peking approve of an inspection system to establish
that the reactor was not being used for military purposes.[63]

By mid 1979 the situation had once again changed. A draining
military excursion into Vietnam had placed strain on the Chinese
economy. In addition, it was becoming clear that China did not
have enough foreign reserves to pay for the ambitious ten-year plan
outlined one year earlier. Finally, there was reported to be pressure
from within the Communist Party to modify the emphasis on the
palliative powers of new technology. The result was a slowing
down of the whole plan.

In May it became clear that China was reconsidering the reactor
purchases.[64] An immediate observer reaction was to attribute the
hesitation totally to the deceleration of the ten-year plan. But when
the cancellation of the contract was confirmed in July 1979, Vice-
Minister for Commerce, Cui Qun, made it clear to French Industry
Minister André Giraud that at least one motivation for the cancellation had been the accident at Three Mile Island.[65]

That there had been some discussion over the hazards accompanying the use of nuclear reactors was clear. However, limited as
it was to the inner circles of government, the weight given to the
various considerations remained totally obscure. As shifts had
occurred in political fortunes and policies in the past, there was no
guarantee that the present course would be maintained. By late

COUNTER-CURRENTS WITHIN THE LEFT   243

1980 reports were again appearing that China might attempt to build two 'prototype' power reactors during the 1980s,[66] and that it would share in the joint construction of a nuclear plant with the China Light Company in Hong Kong, to be completed in about 1990. Whether or not these projects would proceed, only one thing could be said with any certainty. That is that China had so far managed to steer clear of locking itself into any substantial nuclear power programme. By doing so, it had at least kept its future options open.

One thing common not only to China but to the USSR and the other countries of Eastern Europe is that the debate over nuclear power is coming very late compared with the battle over the issue raging in the West.

Inside the communist countries the development of a nuclear debate is beginning to occur partly through the stimulus brought by news of the debate continually flaring in the West. Even though information filters into the communist countries, other than Yugoslavia, far less readily, it does slowly diffuse. In particular, the desire to sell nuclear technology in the West, and the interchange of technical information, is slowly forcing the technical community to face some of the problems raised outside.

While consideration of the problems raised by nuclear power remains largely restricted to the technical community in the communist world the debate is likely to be constrained, since there, as in the West, technical experts often view the issues through a different paradigm than the rest of the community. But even with the tendency of experts to emphasize the progressive nature of sophisticated technology, the slow process of diffusion is raising the prospect of a steady reassessment of the relative value of nuclear power outside the capitalist world. That process, too, may be largely obscured from Western eyes. But as news of it begins to percolate out of the communist countries it carries the potential to play an important role. It will help further develop a process of re-examination of attitudes to nuclear power already occurring in the labour movement in the West.

RE-EXAMINATION ON THE LEFT

In most countries of the Western world there are a bewildering array of socialist parties and groupings. While all are bound

together by common opposition to the great disparities in wealth and opportunity which capitalism creates, they differ greatly on many other key considerations. Some, including a number of anarchist groups, owe their intellectual heritage to socialist contemporaries of Karl Marx, who placed a different emphasis on the sort of society which should be striven for. Others, while purporting to walk in Marx's footsteps, differ on whether socialism can be achieved by a strategy of social democracy in which the workers would vote themselves into power, or by the revolutionary seizure of state power. Others again, characterized by varying emphases on the writings of Lenin and Trotsky, concede that power is unlikely to be surrendered peacefully by the class which benefits most from the existing social arrangements, but differ on how a revolutionary transition can be achieved.

In many countries there exists at least one party, often the Communist Party, which developed during the second quarter of the twentieth century when Soviet leader Joseph Stalin, and Stalinism, dominated world Marxist practice and theory. The leadership of some of these parties still sees the role of the USSR as crucial to the attainment of socialism world-wide, and substantial deviations from Soviet policy are not entertained. However, by no means all communist parties, or other socialist groups slavishly follow the policies of the USSR.

In the early 1960s the profound split in policy that emerged between the USSR and China rang the death knell of any dreams of a unified communist movement and opened the way for the emergence of 'Maoist' parties to join an already large array of other groups which had broken away from the official Moscow line. This assembly was later joined by a number of communist parties which split from Soviet policy after the invasion of Czechoslovakia in August 1968. In addition, a diffuse process of rethinking of socialist theory had begun under the broad title of the 'new left'.

In this way the single goal of socialism had led to the development of a broad spectrum of political groups, parties, and their adherents in the West, characterized very loosely as 'the left'. Those which still hold close to identical positions to the USSR will usually reflect its official position on nuclear power. However, to characterize even these parties' usually fervent support for nuclear power as mere obedience to Moscow's dictate would be to grossly oversimplify. More importantly, the ideology leading to the use of nuclear technology in the USSR is shared by the leadership of many

COUNTER-CURRENTS WITHIN THE LEFT

of these parties. In other parties and groups which have departed from Soviet policy a similar attitude to nuclear power still often prevails.

In Western Europe, where the communist parties do have policies which substantially differ from each other and from those of the USSR,[67] their leaderships nevertheless still tend to give enthusiastic support to nuclear power. Even in Italy, whose Communist Party has played a leading role in the 'Eurocommunist' debate by developing some novel strategies for challenging capitalism, the party's leadership is less willing to question nuclear power. It has vacillated between urging a revitalization of the government's stalled nuclear programme (for two reactors, in addition to the four presently operating, by 1986, and a further six thereafter),[68] and expressing cautious concern when opposition within the party's membership intensifies. This happened, for example, in the wake of the accident at Three Mile Island.[69] Similarly, the leaderships of the social democratic parties of Western Europe, which often have little truck with the USSR, nevertheless are often consistent proponents of nuclear power.

Despite the staunchly pro-nuclear positions taken by the leaderships of many parties on the left, there has been a steadily increasing division and debate over nuclear power within their parties' membership. In some parties this has already been reflected in a complete change of policy. The communist parties of Australia, Sweden, and Ireland, and the labour parties of Australia and the Netherlands have already reversed their previously held positions of general support for nuclear power. They are joined by a host of much smaller left-wing parties, such as the Radical Party in Italy, and the PSU in France, which have played important roles in opposing nuclear power. Others, such as the French Socialist Party, the Irish Labour Party, and the Spanish Communist Party, have begun to modify their positions in favour of limitations to nuclear expansion and the establishment of wide-ranging inquiries into the nuclear issue. Even some Soviet leaning parties, such as the Socialist Party of Australia, which broke from the Communist Party of Australia after it condemned the invasion of Czechoslovakia, have adopted positions opposing nuclear power, at least in capitalist hands.[70] This may seem a strange position to take, but in political reality it allows the party to throw its support behind the anti-nuclear movement even though it is not yet fully prepared to question developments in the USSR.

246   GLOBAL FISSION

In other parties of the left, such as the West German SPD, the French Communist Party, or the Swedish Social Democrats only marginal changes have yet been made to their official policies, but a debate rages within the rank and file which is increasingly divided on the issue. Overall, the picture is clear. Within the left parties and groups there is a steady erosion of support for nuclear power, and in many groups this has already produced shifts in policy.

The effects of this debate are not limited to the parties and groups of the left, for to these are linked by common membership, organizational affiliation, or simply shared ideology, the other organizations of the labour movement. Most particularly, this includes the unions. In the unions there is now growing evidence of steadily increasing debate over nuclear power.

THE UNIONS AND NUCLEAR POWER

In most countries the debate over nuclear power within the parties and organizations of the left has followed the debate in the broader community. This has been true of many unions as well, although it would be misleading to suggest that all unions may be typified as belonging solely or even partly within the left. In some countries many unions have right-wing leaderships, and in all countries some unions are extremely conservative in their policies and actions. But whether of the right or the left, and with a few important exceptions, the early position of almost all trade union leaderships with a policy on nuclear power, was in favour of it. For the conservative trade unions the force of government and corporate support for the nuclear enterprise was sufficient reason for them also to support it. For the trade union with more left-wing leadership, nuclear power represented progress, a technical development necessary in any forward looking society whether socialist or capitalist. But irrespective of political alignment, the arguments put forward by unions in favour of nuclear power usually reflect two key themes.[71] Hans Bethe, a US physicist and Nobel Laureate who supports nuclear power, put it this way in Austria in 1977: 'In some of our (US) States we defeated the anti-nuclear movement only because we had trade unions on our side. And we had them on our side because we made it clear that nuclear energy means job security and national independence'.[72]

These two themes have been pressed forward by pro-nuclear union leaderships around the world. They were taken up by the

COUNTER-CURRENTS WITHIN THE LEFT 247

president of the Austrian Trade Union Confederation before the referendum on Zwentendorf. In that case, the rejection of nuclear power by the heavily unionized population of Austria demonstrated that the rank and file are not certain to be convinced that the issue is as simple as this formula would suggest.

In some unions the pro-nuclear pressure from leadership has not only been very strong but has extended beyond the mere use of argument. In the USA, where little more than 20 per cent of the workforce is unionized, many of the unions are conservative 'business unions' and have often collaborated with industry for immediate economic gains. Their support for nuclear power is usually strong.[73] In 1976 the American Federation of Labour–Congress of Industrial Organizations (AFL–CIO) conference on Nuclear Energy and America's Energy Needs concluded that 'rapid development of nuclear power is a "must" without which the nation's economy would falter'.[74] In the same year, organized labour joined forces with the utility companies in the USA to defeat the anti-nuclear referendum 'Proposition 15' in California. The following year 3 000 workers from the building and construction trades demonstrated in favour of nuclear power at the Seabrook reactor site,[75] and the next year the same unions signed a national agreement with key nuclear companies, which included no-strike clauses aimed at speeding the construction of nuclear reactors.[76]

Especially within the USA, but not only in that country, the gap between the labour movement and the opposition to nuclear power can be enormous. For their part, those sections of the anti-nuclear movement which have grown out of environmental groups have often been suspicious of working-class organizations whether they be unions or left-wing political groups or parties. On the other side, many on the left have also tended to view the environmental movement with suspicion. As A. Roberts has observed:

> this suspicion leads to a dismissal of the anti-nuclear struggle – indeed of environmentalist issues in general – as a fashionable middle-class phenomenon that does not interest the working class, and hence is no concern of the true revolutionary, who will concentrate on the real issues: those at the point of production in the realm of state power.[77]

As Roberts goes on to establish, such an attitude is extremely shortsighted, relying on a dogmatic belief that the present concerns of the working class are static, and on a doctrinaire adherence to a

248 GLOBAL FISSION

selection of Marx's writing in a manner which fails to emulate the highly creative thinking of Marx himself.

Ironically, the tentative attitude of many left groups towards environmental issues, and the avowedly pro-nuclear attitudes of some, has not prevented right-wing trade union leaderships from accusing those within their own ranks who oppose nuclear power of being 'left-wing'. When Richard Ostroweski, a welder of thirteen years experience with Consolidated Edison and a shop steward in the Utility Workers Local 1–2 of the AFL–CIO, organized a meeting of members to explain some of the hazards they face, he found himself charged by the union with 'disloyalty'. During the proceedings, New York's anti-nuclear Shad Alliance, with which he was involved, was described by union officials as a 'communist organization' trying to 'take away our jobs'. When the union announced he would be removed from his position, he was greeted with cries of 'Commie' and 'Hang him!'[78]

The point is not so much that the union had produced what a court would subsequently describe as a 'chilling effect' on Ostroweski's freedom of speech and association,[79] as that this action illustrates the yawning gulf that may separate unionists in many countries from the anti-nuclear movement. Whether the union is on the right or the left, there are ideological factors which may make it difficult for it to view the anti-nuclear movement with anything but grave suspicion. But although that gulf is at first sight intimidatingly deep, there are both theoretical reasons and concrete evidence to suggest that it is far from unbridgeable.

BRIDGING THE GAP

The union is not the only social organization to which its members belong. They also watch television and encounter the debate over nuclear power in just the same ways as the rest of the community. Like the rest of the community, they begin to form opinions on the issues raised. And in the USA, where the official union positions have generally been strongly pro-nuclear, a national Harris poll in April 1979 found 42 per cent of union members and their families opposed to building more nuclear plants.[80]

The developing debate within the community thus provides a base on which unions may begin to build anti-nuclear policy. But for this to happen there must be ways in which that debate can be raised within the union structure. One channel is provided by the

COUNTER-CURRENTS WITHIN THE LEFT

divisions which already exist between unions on the issue. Even within the USA, not all unions are in favour of nuclear power. The United Mine Workers, for example, concerned about the future of the coal industry, opposed nuclear power from the start. It has expelled members for publicly supporting nuclear power, and actively campaigned against the extension of the Price-Anderson Act and the construction of the Clinch River breeder reactor at Oak Ridge, Tennessee. The United Auto Workers has also opposed some nuclear projects, including mounting a court challenge to the Fermi prototype fast-breeder reactor in 1956. At its 1977 convention, while still not opposing nuclear power outright, it expressed deep reservations over nuclear power's 'many unanswered questions'.[81]

There are also emerging differences within union branches. One of these is that some union branches may lie in regions with an important local identity. In this sense they may lie in the union's periphery. In France this was illustrated by the position of opposition to extensions to la Hague taken by the local Manche branch of the CGT[82] despite the much more pro-nuclear position of the national leadership. In the USA, as will be seen in more detail in chapter 11, it was illustrated by the actions of the local branches of the International Longshoremen's and Warehousemen's Union and the United Public Workers Union in refusing to service a transport ship carrying radioactive waste when it attempted to refuel at Hawaii.

In addition, sections of some unions are associated with the nuclear fuel cycle to differing degrees. In some unions, even though some members might have to change their jobs if nuclear programmes were truncated, this does not prevent other members from taking an anti-nuclear position. Thus a debate over nuclear power has continued within the US Oil, Chemical, and Atomic Workers Union (OCAW) for several years, although the anti-nuclear position did not reach sufficient strength in the 1979 elections to give anti-nuclear candidate Anthony Mazzocchi the presidency. In particular, the debate in the OCAW was inflamed in 1974 when one member, Karen Silkwood, was contaminated with plutonium. She was already engaged in documenting alleged health and safety violations at the Kerr-McGee plant in Sequoyah, Oklahoma, where she was employed. Shortly afterwards she died in a car accident. Although police have ruled her death an accident, some believe she may have been deliberately run off the road. In

1979 opposition to nuclear power within the OCAW was strengthened when a federal court found Kerr-McGee liable for the contamination and awarded $10.5 million damages to Silkwood's relatives. Kerr-McGee is appealing against the decision.[83]

The familiarity with the hazards of nuclear power through some workers' personal experience can lead to a policy of at least restricted opposition to nuclear power. This seems to have been a relevant factor for the CGT in France with their policy of a limited moratorium on further nuclear expansion, and for the opposition to nuclear power by the largest trade union confederation in Japan, SOHYO.[84] In both cases workers' exposures to radiation through accidents caused considerable discussion of nuclear issues.

All these events provide entry points for a dialogue to develop between the anti-nuclear movement and trade unions and, within trade unions, over nuclear power. The ground for that dialogue is greatly improved by the dialogue already occurring in many of the political parties, especially those of the left, to which many of the more active trade union members belong. As the examples already given suggest, in many unions such a dialogue is under way. It revolves around several key issues which fall into two different categories: the physical hazards of nuclear power, and its social implications.

PHYSICAL HAZARDS

At almost all stages of the nuclear fuel cycle workers are exposed to radiation. Levels typically may be ten times larger than permitted for the general public, and sometimes when accidents occur exposures can be much larger. The effects of such exposures, as noted earlier, are still controversial. Working in 'hot' areas is becoming a subject of increasing concern. For example, in early 1980, workers at Tihange in Belgium went on strike demanding an early retiring age for those exposed to heavy radiation doeses. They belonged to the moderate Gas and Electricity Employees Union, and it was the first occasion on which they had expressed concern over nuclear hazards.[85]

The practice of 'burning out', where workers are exposed to as much as a year's allowed dosage in one short operation taking as little time as a minute, is a particular source of increasing concern. As installations age there is evidence that they become more radioactive. At West Valley, New York, the reprocessing plant became so radioactive that it had to be permanently closed. As plants age, the need for 'burn out' doses and the pressure for

COUNTER-CURRENTS WITHIN THE LEFT

workers to submit themselves to larger radiation doses becomes greater. Such pressure is not necessarily gentle. In New York a worker was sacked for refusing to work on a 'hot' job at the Buchanan reactor. He said, 'All of us used to joke about going up there because no-one put pressure on us. Now they do and my life might be at stake'.[86] In France, at the la Hague, Saclay, and Comhurex-PierrLatte plants repeated exposure accidents resulting in the contamination of workers have been an important factor in hardening CFDT opposition to the government's nuclear programme. In the UK, evidence of contamination of workers at Windscale highlights the same issue. Over 1979–80, British Nuclear Fuels opted to make out-of-court settlements totalling nearly £100 000 to three families of Windscale workers who died from cancer-related illnesses.[87] Adding to this pressure, some seventy workers at the Aldermaston nuclear weapons plant have lodged claims for damages due to alleged over-exposure to radiation.[88] Evidence that union concern is growing was provided in 1980 when the General and Municipal Workers' Union began to express concern over safety at Windscale.[89] In the UK the concern is only beginning to surface, while in other countries such as France, Japan, and even the USA, it has developed further. Undoubtedly there is almost everywhere a growing awareness of the scientific controversy over radiation hazards, and increasing concern by workers over their own safety. Equally significant, however, is the growth of discussion over the broader social implications.

SOCIAL ISSUES The nuclear industry was quick to argue that nuclear power means energy, and that means growth and jobs. This alleged nexus was taken up and used with great effect by nuclear proponents who dismissed the opponents of nuclear power as anti-innovation, anti-jobs, and therefore anti-worker. The opposition was at first slow to meet this argument, and then unsure how to intrude its reply into the unions. Thus, Frank Chapple, a pro-nuclear speaker for the Electrical, Telecommunications and Plumbing Union, was able to argue persuasively at the UK Labour Party Conference in 1977 that the position of the nuclear opponents is 'we have suffered enough from growth and it is time that we called a halt to the technological threat'.[90] This tactic is slowly becoming less effective as the opponents' arguments begin to be raised within the trade union movement iself. These call not

for a halt to innovation, but for innovation in a different direction – for the development of energy-saving techniques, and the use of often elegant modern technologies ranging from heat pumps to photo-electric cells, wind-electric generators, solar collectors, and fluidized-bed gas turbines.[91]

The consistent feature of these technologies is not that they are 'regressive' but that they are capable of being used and often manufactured in a decentralized, flexible way to provide electricity or eliminate waste of energy. As the thrust of these arguments begins to penetrate so do their implications for workers. As John Carroll, vice-president of the Irish Transport and General Workers' Union argues, 'The solar industry is job intensive and again can provide two and a half jobs for every job that the nuclear industry can provide . . . if we had a mandatory programme of efficient home insulation, many many jobs could be provided'.[92] As already stated, his union has since taken a leading role in opposing the reactor proposed for Carnsore.

This position may now be found in policies of other unions. In 1979 the British Union of Construction, Allied Trades and Technicians adopted a policy advocating 'the rapid development of alternative and conservation technologies in preference to nuclear power . . . providing jobs for construction workers while reducing the cost of heat to the consumer'.[93] Similar arguments can be found surfacing in trade unions from Austria to the USA.[94] Some unions, most notably those within the French CFDT, have even launched their own alternative energy plans.[95]

With the breakdown of the typical 'back to the cave' arguments against opponents to nuclear power, the alleged relationship between quality of life, employment, and energy consumption also comes under more critical scrutiny. Slowly, unionists are beginning to inquire whether the high level of energy consumption in, say, the USA, rather than being the cause of its wealth, is instead a symptom of the wastage which results from wealth-induced carelessness. In addition, as Trade Unionists Against Nuclear Power in Austria point out, 'In the USA, in spite of its nuclear industry and the highest per capita energy consumption, which is three times higher than Austria, there is massive unemployment'.[96] Trade unionists are beginning to realize that there are choices to be made between different goods to be produced, different ways of producing them, and different ways of providing electrical power. When the criteria

COUNTER-CURRENTS WITHIN THE LEFT

traditionally most dear to unionists are applied they are finding that further expansion of the heavily subsidized and capital intensive nuclear industry does not necessarily provide the best answer.

There is also a growing awareness that the increased centralization of power in corporations and the state which accompanies nuclear technology, is not in the long term interests of organized labour. Thus Jim Roulston, state president of the Amalgamated Metal Workers and Shipwrights Union (Australia), and vice-president of the Australian Council of Trade Unions, writes:

> To achieve their long term global strategies of growth and expansion, multinational corporations need the 'political stability' of a society more authoritarian than our liberal democracy. They also need to suppress opposition of people concerned about the social and environmental costs of these vast technological undertakings. Especially they fear common action by community groups and trade unions. To prevent social protest from hindering their projects the large corporations have forged closer links with government bureaucracies to secure the trend towards repressive legislation which protects their interests.[97]

In Australia, in addition to the specifically nuclear legislation discussed earlier, amendments to the Conciliation and Arbitration Act and the enacting of the Commonwealth Employees Act (1977) have brought into law provisions to severely limit the unions' right to strike or boycott work they consider socially undesirable. With fines of up to $A1 000 per day, and powers to seize union property and dismiss union officials, the implications are wide-ranging.[98] Similarly in France, in July 1980, the parliament passed a law empowering EDF to fire any person who intentionally disrupts the normal operations of any establishment that 'holds nuclear materials'. Significantly, it precipitated protest from not only the CFDT but also the CGT, and a joint action by the Communist and Socialist parties challenged the law in the Supreme Court.[99]

Whether based on the physical or social implications of nuclear power, the development of such concerns provides a firm basis for a dialogue between the anti-nuclear movement and the trade unions. In some countries a stage beyond dialogue, where unions have become a significant factor in the development of opposition to nuclear power, has already been reached.

THE GROWTH OF THE ANTI-NUCLEAR UNION

Included among unions that have now begun to play an important role in opposing nuclear power are the Irish Transport and General Workers' Union, the largest trade union in that country; one of two major French union federations, the CFDT; and one of the two major Japanese union federations, SOHYO.[1] In Canada, the British Columbian Council of the Confederation of Canadian Unions is opposed to uranium mining and the development of nuclear energy,[2] while in the Netherlands the national union federation, the FNV, has called for a moratorium on nuclear developments, as has the trade union confederation in Norway. In Scotland, the Transport and General Workers' Union called first for a 'review of the further development of nuclear power' and then, in September 1979, this became the policy of the union throughout the UK. With 2 million members it is the UK's largest union.[3]

Although for many unions the process of re-examination of the nuclear issue is still at a relatively early stage, its effects are unmistakable. In the UK in mid 1980 four trade unions, including the National Union of Mine Workers, passed anti-nuclear resolutions. Another three had tough debates over the issue.[4] In the USA, June 1980 saw unionists from twenty-nine unions in sixteen different states come together to form the Labour Committee for Safe Energy and Full Employment.[5] Within three months it held its first anti-nuclear conference. It was endorsed by nine unions and attended by 800 unionists from ninety unions.[6] Even in the countries where trade unions traditionally have been strongly in favour of nuclear power, debate on the issue is beginning to well up from the rank and file and shift union policy. The potential of such a shift has already been demonstrated. But nowhere has the role of anti-nuclear unionism been better demonstrated than in Australia.

The first Australian unions were formed in the 1840s in the tradition of their British counterparts. But within fifty years they had seen the economy, fuelled by the gold rush, expand on a bubble of land speculation, only to collapse in the 1890s into a ruinous depression. The subsequent widespread unemployment and the intense period of class struggle which it precipitated was an important factor in the development of unions with a strongly independent spirit and an overtly political outlook.[7] In addition, some long periods of relatively full employment, especially in the 1950s and

COUNTER-CURRENTS WITHIN THE LEFT

1960s, gave them, in contrast to many European unions, sufficient confidence to exercise this independent spirit.

At one count, over thirty 'political' strikes occurred over national issues alone between 1916 and 1978 in Australia.[8] Among the more notable were the refusal of workers in 1946 to load Indonesia-bound ships carrying materials which might be used by the Dutch to suppress the nationalist uprising, the placing of 'black bans' on ships carrying arms to Vietnam, and a national strike over changes to the national medical health insurance scheme in 1976. In the late 1960s the unions began to form policies over environmental issues. These included a resolution by the national Australian Council of Trade Unions to boycott mineral exploration on the Great Barrier Reef. Not long afterwards the Builders' Labourers' Federation began to institute so-called 'Green Bans' prohibiting projects endangering areas or buildings of historic or environmental significance.[9]

It was this background which made possible the leading role that some unions have played in the development of opposition to the nuclear fuel-cycle in Australia. Prominent among these have been the workers of Port Kembla in New South Wales. In the summer of 1938–39 they made an historic decision to refuse to load pig-iron bound for Japan. They argued that it was likely to return in the form of bombs.[10] The fact that they were proved right in 1942 when Japanese bombs did fall on Australia helped reinforce the confidence to take similar action in the future. Forty years later it was the workers of the same port who declared it a Nuclear-Free Port, and undertook to block the passage of any nuclear-related materials through it. By then they were just one group of workers around the country who were concerned over the hazards posed by nuclear power world-wide. They added their efforts to a developing campaign to obstruct plans by the government and mining companies to mine and export Australia's uranium.

11
THE AUSTRALIAN EXPERIENCE

Australia, separated from all other countries by the expanses of the Pacific and Indian Oceans, is atypical. It is a country with a vast land area but few people. Of the 14 million inhabitants, most are clustered in a few large coastal cities, the majority of which lie in the south. No commercial nuclear reactors are established inside or outside these cities, and none is ordered or under construction. It might seem the last place to expect the nuclear debate to penetrate, yet for five years that debate has ebbed and flowed across the country.

The catalyst has been uranium. In the Northern Territory, the sparsely inhabited northern 'top end' of the continent, alternately baked by tropical sun and drenched by torrential monsoons, lies an estimated 18 per cent of the world's uranium, and a much larger proportion of the uranium not already committed to nuclear projects. As early as 1972 contracts were signed for the sale of some of these reserves and plans were under way for rapid development and sale of the rest. In 1975 the conservative Liberal Party government came to power with a firm ideological commitment to see the reserves developed and sold as rapidly as possible. Yet it was only in 1979 that the first two new mines were able to begin even preliminary operations. As late as 1981, over 70 per cent of Australia's reserves still remained untouched.

The delays to the export and mining of Australia's uranium, and the implications of this, provide a sharp illustration of the degree to which the debate and battle over nuclear power is international in character. In Australia, international considerations underpin the delays to uranium mining and the debate within the community to

THE AUSTRALIAN EXPERIENCE

257

which they are deeply connected. A powerful debate over the hazards of nuclear power has rippled through Australia severely affecting the timing and extent of uranium mining. On the one hand, the motivation has not been merely the effects that the use of that uranium would have on Australians, but also on people in other parts of the world now confronted with nuclear technology. On the other, the debate in Australia, and the energy future of Australians will finally depend on the outcome of the international debate over nuclear power and other energy options.

The importance of international considerations is just one of several unusual features of the nuclear debate in Australia. Others include the development of a national anti-nuclear movement even before there were any nuclear projects under way, the speed with which it developed political force, and the central role played in this by the labour movement. Because these features stand out so clearly in Australia, they provide a good illustration of the potential of the slower and less visible development of similar trends in other countries. This can be clearly seen in an examination of the obstacles that have delayed and constricted the flow of Australia's uranium into the international nuclear fuel cycle.

THE HURDLES

The Liberal Party is committed to the belief that rapidly opening up the country's resources to overseas customers is in the best interests of the Australian economy and Australians. Other considerations, whether they be environmental, concern for the effect on Australia's ailing manufacturing industry, or the desire to conserve resources for the future, tend to take second place in the party's priorities. While the opposition social democratic Australian Labor Party (ALP) places some emphasis on the benefits of preserving Australian ownership of key resources, the Liberal Party, when it assumed office in November 1975, was committed to the rapid and broadest possible exploitation of Australia's uranium reserves. However, the path towards that goal was already strewn with some formidable obstacles.

CITIZEN OPPOSITION   Since the early 1970s the international nuclear debate drew the attention of a handful of environmental groups in Australia. First to take up the issue were the

dynamic FOE groups. They were soon joined by other conservation groups, sections of the women's movement, student organizations, Aboriginal groups, church groups, ALP branches, and the small Australia Party and Communist Party of Australia. By the beginning of 1976 representatives of a number of these groups were meeting to plan joint action. Umbrella organizations had been set up in most states under titles like the Campaign Against Nuclear Energy (CANE) and the Movement Against Uranium Mining (MAUM).

The movement was still small, the bulk of the population being unaware of the issues associated with nuclear power. Nevertheless, the fledgling movement had already gained a substantial victory. Actions had been set in motion within the labour movement which would cause great hurdles along the road to uranium mining.

THE ALP AND THE RANGER INQUIRY   Before the Liberal Party assumed office the ALP had held government for three years. Early in 1975 a growing number of ALP branch members and parliamentarians led by the Minister for the Environment, Dr Moss Cass, began to express concern over the local and global effects of uranium mining. Said Dr Cass, nuclear energy creates 'the most dangerous, insidious and persistent waste products, ever experienced on the planet'.[1]

In addition to these developments at the national level, in the state of Victoria an anti-nuclear report was produced for the 1975 state ALP conference. As a result the conference passed a motion opposing uranium exports, with only one of the 600 delegates dissenting. In that state, the battle over uranium mining within the ALP was over almost before it had begun, since none of the leadership would be prepared to go against such an overwhelmingly accepted policy.

For the ALP as a whole the problems associated with uranium did not yet feature strongly. It had already approved several contracts for some 9 000 tonnes of uranium, destined mainly for Japan, which had been signed hastily by the mining companies in the dying days of the previous Liberal government. Nevertheless, there were questions in the minds of some as to whether the uranium should not be handled by Australian interests rather than foreign companies. Additionally, the price of uranium seemed to be rising overseas, and the ambitious nuclear programmes then current seemed to promise further heady gains.[2] So, unperturbed

by the prospect of delays, the Labor government fell in with the concern emanating from some members.

In May 1975 the ALP had invoked its new environmental legislation and launched a public inquiry into the proposed uranium mining. The test case would be a proposal by the Ranger Uranium Mines company to mine out its uranium prospect lying in the remote north of Australia. Six months later, the ALP was suddenly and unexpectedly removed from office. The Ranger Uranium Environmental Inquiry which it left behind was to prove a most unwelcome legacy for the incoming Liberal government, a legacy that would raise a substantial hurdle to its mining plans.

ABORIGINAL LAND RIGHTS   One of the key concerns of those who opposed uranium mining in the ALP was the effect the mining would have on the Aboriginal population of the north. The area has been inhabited by Aboriginals for at least 25 000 years, and probably much longer. Throughout Australia, the last two centuries saw the colonization of the country by whites who had ruthlessly hunted and exterminated the Aboriginal people. Later, diseases and disruption to their traditional patterns of living had an equally decimating effect. The Aboriginal people were reduced to a fraction of their former numbers, often living in desperately impoverished circumstances on the fringe of white society. But even in the 1970s, the region where the uranium mining was to take place was remote enough to still be inhabited by some 800 Aboriginals who had managed to maintain much of their traditional land and culture.[3]

The 1970s were marked by an increasing recognition by whites of the plight of Aboriginals. In particular, there was a gradual recognition of the desperate need of Aboriginals for 'land rights' – the right to unspoiled land still related to the culture of surviving Aboriginals. The Labor government responded to this by enacting an Aboriginal Land Rights Act. Under that act no new uranium mining could take place until agreement had been obtained from the 'traditional Aboriginal owners' of the land in which the uranium deposits lay.[4]

THE UNIONS   The ALP was not the only section of the labour movement which had already felt some rumblings over the nuclear issue. In the 1960s Australia narrowly escaped having a commercial nuclear reactor on its doorstep. At that time, Philip

260   *GLOBAL FISSION*

Baxter, then chairman of the Australian Atomic Energy Commission (AAEC), lobbied hard for a nuclear power station to be built at Jervis Bay in the state of New South Wales. Significantly, Baxter, who at times has advocated that Australia develop its own nuclear weapons, pressed for a heavy water reactor, a design that is ideal for producing plutonium for nuclear weapons.[5] A local opposition campaign began, and the South Coast Trades and Labour Council covering the workers in the region announced that it would refuse to build the reactor. Not long afterwards, the concept was dropped, partly because it made little economic sense.[6] Nevertheless, it had set some unions thinking.

In September 1974 concern erupted at a conference of major unions involved in the State Electricity Commission of New South Wales. The six unions involved covered engine drivers, and metal, electrical, and building workers. They resolved:

> That the power workers represented at this conference inform the Electricity Commission that under no circumstances will they man any nuclear power station until the problem of nuclear pollution is completely eradicated by new technical breakthrough, and we call on the ACTU to ban construction of nuclear power.[7]

The Australian Council of Trade Unions (ACTU) is the national body to which all Australian trade unions belong. In December it convened a meeting attended by federal unions covering relevant construction, transport, and manufacturing workers. It called on the ACTU executive to take action:

> Pending adoption of safe procedures acceptable to the trade union movement for the treatment of radioactive wastes [to] place a black ban on the development, mining, treatment and export of uranium excepting that quantity needed for medical research and/or treatment.[8]

In the following September the 1975 biennial ACTU congress, on the basis of a resolution put up by the left-wing unions, declared that all uranium mining should be halted pending the completion of a 'thorough-going' public inquiry. It added that uranium exports should be refused to all countries involved either in the manufacture of nuclear weapons or the use of nuclear reactors.[9] The call was supported by a similar resolution from the white-collar Australian Council of Salaried and Professional Associations. It was a

THE AUSTRALIAN EXPERIENCE 261

strong stand, the strongest taken by a national trade union body anywhere in the world. It was not long before action followed.

OPENING MANOEUVRES

At the beginning of 1976, as the new Liberal government surveyed the path towards beginning uranium mining, it was not without obstacles. One opening was provided by a uranium company which already owned established mining facilities. Between 1944 and 1963 Australia provided uranium from some small mines for the British nuclear weapons programme.[10] One of these was Mary Kathleen in Queensland. Uranium demand dropped in the late 1960s, and the last mine to close was Rum Jungle in the Northern Territory. It left a legacy of some stockpiled uranium oxide, and so much contamination of the local Finniss River that it remains unsuitable for consumption by humans or beasts for 32 river kilometres.[11]

In 1974 Mary Kathleen quietly reopened its mine. It now commenced production of uranium. The new Liberal government was happy to permit exports which would provide a useful precedent for the new mines. However, before the government could move the unions acted.

On 24 May 1976 all trains throughout Australia were brought to a halt. A yard foreman in Townsville, Queensland, in compliance with the policies of the ACTU and his own union, the Australian Railways Union (the ARU), had refused to load sulphur bound for Mary Kathleen. He was stood down, and his workmates walked off the job. The executive of the ARU called a twenty-four hour strike which was joined by the Australian Federated Union of Locomotive Enginemen. It was the first national strike over the hazards of nuclear power anywhere in the world. The threat of escalating union action had suddenly become tangible. Feeling the pressure, the Liberal government announced that no decision would be made to export uranium at least until the Ranger Inquiry reported.

Governments, it has been said, seldom appoint inquiries without knowing the conclusions they will reach. In the case of the Ranger Inquiry, however, the Labor government had appointed commissioners who proved both independent and resistant to pressure. The commissioners' report appeared in two volumes; the first in October 1976, the second on 30 May 1977. In the early press reports following the release of the First Report, it was suggested

that it gave a 'green light' for uranium mining. Although this suggestion was enthusiastically embraced by some government ministers, more careful reading soon revealed that although there were some ambiguities, it gave nothing of the sort. Instead, the commissioners had affirmed many of the key concerns of the opponents of uranium mining who testified before them.

As with the opponents, the commissioners did not restrict their attention to the local impact of uranium mining. Instead, they couched their conclusions in terms of concern for all humanity. The wide-ranging consequences of contributing to the nuclear fuel cycle that they emphasized included the unsolved problem of radioactive waste disposal, the dangers of sabotage of nuclear installations, the possibilities of serious reactor accidents, and the diversion of the plutonium produced by reactors for use as a weapon in nuclear blackmail.[12] In particular they concluded that 'The nuclear industry is unintentionally contributing to an increased risk of nuclear war. This is the most serious hazard associated with the industry'.[13]

In Australia, concern over nuclear weapons has robust political roots. In October 1956, eleven years after the bombing of Hiroshima, the searing light and violent blast of a nuclear explosion had briefly shattered the calm of Maralinga, South Australia. It was one of a series of twelve nuclear test bombs detonated by the British in Australia. Shrouded in military secrecy, even the fallout that descended on Adelaide, Australia's third largest city, remained largely hidden from public consciousness.[14] However, in the early 1970s, when the French exploded their test bombs over the atolls of French Polynesia, public attention was rapidly drawn to the hazards of the radioactive fallout.

In response, the then Labor government took France to the International Court and obtained an injunction against the tests. The public, the unions, and the ACTU enthusiastically instigated a boycott on trade with France. Finally, the French bowed to the pressure and moved to underground testing. But in Australia, the public, and particularly the unions, did not forget the issue.

The unions had links with opposition to nuclear bombs stretching back to the nuclear disarmament campaigns of the 1950s. Many of the more left-wing unions were still active participants in the peace movement, taking strong stands against nuclear weapons proliferation and the arms race. For those unions, the Ranger Inquiry placed a seal of authority over the connections they were

THE AUSTRALIAN EXPERIENCE 263

beginning to see between nuclear power, nuclear weapons, and uranium mining. For the general public too, the Ranger reports were a milestone. Now the opponents of uranium mining had authority beyond their arguments.

Even during the hearings, the Ranger Inquiry had acted as a focal point, galvanizing the opposition to mining into submissions, media work, and an information campaign that put it on the political map. But the release of the First Report marked a turning point in the campaign. It added to the arguments of the opposition the legitimacy attached to an organ of the state.

The legitimacy extended by the commissioners went beyond the concerns expressed by the opposition. The commissioners had concluded not only that because of the 'hazards, dangers and problems associated with the production of nuclear energy' Australia should 'seek to limit or restrict expansion of that production',[15] they had also endorsed the need for, and legitimacy of, public debate. Rather than make explicit recommendations they had accepted the view that many of the questions are ultimately 'social and ethical'. Therefore the report points out that 'the final decisions should rest with the ordinary man'.[16] It concludes that ample time be made available for debate to take place.[17]

Although the inquiry had thrown its weight behind the initiation of public participation, the recommendation was anathema to the Liberal government. Within fourteen days of the release of the First Report, the government authorized the export of uranium to fulfil existing contracts. For the time being it would be drawn from the stockpile remaining from the operations at Rum Jungle. On 25 August 1977, less than a month after the release of the Second Report, which dealt with the impact of mining on the Aboriginals and local environment, the government announced its 'decision': that it would permit new uranium mines to go ahead.[18]

In order to head off public reaction to its decision the government tried to meet some of the concerns raised by the Ranger commissioners. The prime minister, Malcolm Fraser, announced, contrary to the findings of the inquiry and many subsequent reports around the world, that the waste disposal problem was solved and the necessary technology already being developed to commercial scale. In answer to the proliferation problem, the government made a belated conversion to the need for 'safeguards'.[19] However, the government's proposals, in essence requiring customers to place the uranium under the scrutiny of the IAEA according to the

terms of the Nuclear Non-Proliferation Treaty, satisfied few analysts.[20] These safeguards are well known to be severely limited by the technical problems of detecting diversions of fissionable material, and a lack of any credible means of enforcing IAEA regulations if by chance diversion is detected. The Ranger commissioners had commented that these measures could provide 'only an illusion of protection'.[21]

The only substantial innovation made by the government was a requirement that customer countries conclude special bilateral agreements with Australia, undertaking that the uranium would at least remain covered by the international agreements. Few, however, were impressed. As one of the Ranger commissioners put it, the measures were 'virtually useless' and full of 'loopholes'.[22] As time passed any illusion of safety created by the measures was steadily eroded as the government weakened them in a scramble to obtain contracts.

However, whatever the merit of its rationalizations, the government had taken a decisive step. It had swept aside the hurdle presented by the Ranger Inquiry's recommendations. Now it faced the remaining hurdles, reinforced by the opposition to uranium mining during the intervening period.

CITIZENS AND THE LABOUR MOVEMENT

The first Ranger report, in October 1976, had provided powerful support for a rapidly developing citizens opposition movement. Wherever possible this movement engaged in debates and took action to attract the media's attention to the issue. In a small way also, the opposition began to be seen in the streets.

In November and December 1976, 7 000 people marched through the streets of Australian cities.[23] The most active centre was Melbourne where 3 000 demonstrated their opposition. It was a small beginning, but it encouraged the formation of a national coordinating coalition. Named the Uranium Moratorium, it called for a five-year moratorium on uranium mining. It was an important step in a year of unparalleled expansion for the citizens movement.

In April 1977 the first national demonstration coordinated through the Uranium Moratorium brought around 15 000 demonstrators out into the streets of Melbourne, 5 000 in Sydney, and smaller numbers elsewhere.[24] A national signature campaign

THE AUSTRALIAN EXPERIENCE 265

launched at the same time, within four months attracted over 250 000 signatures calling for a five-year moratorium. The act of taking the issue to the people in this way gave rise to a network of locally based citizens groups opposing uranium mining. By the end of the year over 100 such groups had been set up in Victoria alone. In August, four months after the first national demonstration, a second brought 10 000 people out in Sydney, 20 000 out in Melbourne, and a total of 50 000 nationally. By then the opposition was beginning to look like a potential political force.

Additionally, after the announcement of the government's decision in August, a credibility gap opened up between the government and substantial sections of the community over the reasons given for uranium mining to proceed. Opinion polls which in June 1975 had shown 25 per cent opposing uranium mining,[25] swung to 32 per cent in June 1977,[26] and, after the decision was announced, to 42 per cent in September 1977.[27]

From early 1977, the winds of opposition could be felt rising in the ALP from its own membership. The Victorian State Branch had already passed a motion strongly opposing uranium mining and nuclear power. Then on 31 March, the South Australian Branch, which not only favoured uranium mining but was also investigating the possibility of establishing uranium enrichment, did an abrupt about turn. It announced that despite the presence of substantial deposits at Roxby Downs, no uranium mining would be permitted.

In the unions, despite the opposition to uranium mining in the community, there was far from consensus on the issue even among the leaderships. At the head of the national body, the ACTU, the executive was not anxious to confront the government over the issue. The president of the ACTU, Robert (Bob) Hawke, a charismatic and extremely influential figure, was a supporter of uranium mining. When the first Ranger report came out the executive seized on the opportunity to announce that it would not stand in the way of the government's intention to allow uranium mining and exports to meet existing contracts. The national executives of key unions such as the Waterside Workers' Federation had fallen in with this position. Steadily, the official position was being whittled away.

The erosion seemed marked when on 20 June 1977 a 2 kilometre long convoy of trucks, moving at high speed and escorted by large

numbers of police, swept through the suburbs of Sydney in the early hours of the morning.[28] Aboard were containers of 'yellowcake' (uranium oxide) from the government stockpile, bound for loading onto a ship at the docks. Despite militant attempts by demonstrators to prevent the loading, and one crane operator walking off the job in sympathy, most of the waterside workers abided by their executive's decision and the ship sailed with its load of uranium two days later. For opponents of uranium mining there was only one sign of hope: a telegram received from workers at Port Kembla who had decided in principle not to load uranium, despite the views of their union's national executive. However, a few weeks later the prospects changed dramatically.

On Friday, 1 July 1977, the *Columbus Australia*, a ship carrying yellowcake and other cargo, arrived at Melbourne's Swanston Dock for loading with general cargo. The next morning a small group of some 300 men, women, and children assembled outside the dock gates in a hastily arranged protest organized by Melbourne's MAUM. As the dock gates were being opened to allow workers through the assembled people made a snap decision and streamed through also. These small beginnings were to produce powerful repercussions.

On reaching the wharf they found the ship being loaded. Now came discussions between individual demonstrators and workers. For some of the workers it may well have been the first time they had been forced to confront the uranium issue. Finally the waterside workers reached a decision. Despite the directive of their national executive they would not load the ship. Police were called in but were reluctant to act for fear of provoking further union action. The demonstrators resolved to stay. Successive shifts of workers came on, discussed the issue, and each time voted to refuse to continue loading.

That evening the atmosphere hardened as police announced they would clear demonstrators from the vicinity of the ship's gangplank so that loading could continue. This time, after a long discussion, the workers reached a new decision: they would not walk through police lines to go to work. Faced with a seeming impasse, the police consulted headquarters. Soon it was clear they had decided to act. A line of mounted and foot police formed up and charged the by now sitting and singing demonstrators. Some demonstrators were injured and all were arrested. Included was the state secretary of the Waterside Workers' Federation, Ted Bull,

THE AUSTRALIAN EXPERIENCE

who had joined the line of demonstrators. The police action was not a shrewd move.

The next day the media reported the demonstration. Included were photos of battered demonstrators and one reporter's recording of the screams of children and other demonstrators as the police horses charged into them. It was the impetus the waterside workers needed. They immediately declared the entire wharf 'black' in protest, and all work on it was suspended. Two days later, on 6 July 1977, an all-day meeting of the rank and file of the Waterside Workers' Federation reached two historic decisions. First, Victoria's waterside workers would never again be prepared to walk to work through police lines. Second, in defiance of the federal executive, they decided by 2 000 votes to six that they would never again service any ship carrying uranium. That night the *Columbus Australia* set sail leaving over $A1 million worth of cargo on the wharf.

The victory had a galvanizing effect on both the citizens and labour movements. Union action had been shown to be possible. Soon afterwards the Waterside Workers' Federation's Queensland and Northern Territory Branches passed similar motions to the Melbourne resolution. Other waterfront unions fell in alongside.

For those lobbying against uranium mining within the ALP the victory could not have come at a better time. Even as the demonstration was occurring, delegates from all over Australia were meeting in Perth for the biennial ALP conference. The most controversial motion on the agenda dealt with the party's attitude to uranium mining. Finally, although key figures such as former prime minister Gough Whitlam, and Bob Hawke were strongly in favour of uranium mining, the policy supported by the anti-uranium mining forces proved invincible. On the day the *Columbus Australia* sailed, the party adopted a new policy on uranium mining. Moved by the premier of South Australia and the leader of the Victorian ALP it declared that because of the unresolved environmental, technical, and social problems associated with nuclear power and uranium mining:

> * Labor declares a moratorium on uranium mining and treatment in Australia.
> * Labor will repudiate any commitments of a non-Labor government to the mining, processing or export of Australia's uranium, and

268   *GLOBAL FISSION*

> * Labor will not permit the mining, processing or export of
> uranium pursuant to agreements entered into contrary to
> ALP policy.[29]

Even though the resolution still permitted exports to fulfil
existing contracts, it was a significant achievement. For the first
time in the world, a major political party had taken a principled
stand against nuclear power, not merely because of the effects on
the people who would vote at the next national election, but
because of the effects world-wide. Nor was it all talk. Carrying the
threat that a future Labor government would annul any new
contracts, it was bound to cause potential customers to think twice
before entering into them.

The success for the anti-uranium forces in the ALP was not
matched by any such clear victory at the subsequent ACTU biennial
congress two months later. There, ACTU President Hawke's will
prevailed and a complex motion was passed on the uranium
issue.[30] On the one hand it called on the government to hold a
referendum on the issue. On the other it required unions to 'consult' with their rank and file before any action would be taken. In
the end the government showed no interest in the referendum
proposal, and key unions refused to take part in the executive's
proposed 'consultation' procedures. In February 1978, a special
meeting of unions reaffirmed that uranium would be supplied to
meet existing contracts. No labour would be made available for
new mines until the hazards of waste disposal and nuclear proliferation had been contained and the 'legitimate demands' of the
Aboriginal people met. It was left to the executive to decide when
these conditions were fulfilled.[31]

For the government the ACTU motion was a setback and the
ALP resolution a total loss. It had hoped for bipartisan support for
uranium mining from the ALP and a total retreat from any threats
of union action from the ACTU. Community support for uranium
mining also had been steadily eroded, and the pace of this had
quickened after the government's decision six months before.
However, none of these represented an immediate problem. At the
national elections two months earlier, in December 1977, the fact
that the parties with policies opposing uranium mining had
received as many first preference votes as the pro-mining parties
had not prevented the Liberal Party easily being returned to office
for a further three years.[32] In 1978, by far the most concrete and

THE AUSTRALIAN EXPERIENCE   269

immediate obstruction would be Aboriginal rights to the land
under which the uranium rested.

ABORIGINALS BAR THE WAY

The Aboriginals living in the region of the uranium deposits have
repeatedly stated that if given the opportunity they would choose to
have no uranium mining at all. The second Ranger report had
confirmed this, but added: 'There can be no compromise with the
Aboriginal position; either it is treated as conclusive, or it is set
aside. We are a tribunal of white men . . . In the end, we form the
conclusion that their opposition should not be allowed to
prevail'.[33]

The Aboriginal Land Rights Act ultimately gave no choice to the
Aboriginal owners over whether or not mining would be allowed.
Either they must agree to some negotiated level of compensation or
an arbitrator would be appointed by the government to fix a level
of compensation. The act did not make clear what process of
negotiation was to be followed. With the trade unions as well as the
broader population demanding that justice be done, the government at least had to move cautiously enough to allow it to appear to
be done.[34]

A Northern Lands Council (NLC) had already been set up under
the Aboriginal Land Rights Act to represent the Aboriginals in the
area. Although a white creation, it was with this body that the
government would negotiate. The act provided that the Lands
Council could not enter into an agreement until the traditional
Aboriginal owners of a proposed mining area had given their
consent. Any Aboriginal groups or communities affected must be
consulted and given an adequate opportunity to express their
views.[35]

Initially the NLC sought to obtain agreement from the government that at least the recommendations of the Ranger Inquiry, that
mining be done sequentially, would be implemented. This would
reduce the impact on the Aboriginal community and their environment. The government refused.

For the NLC and the communities it represented this presented a
tragic prospect. It is one of the great distortions of Aboriginal
culture to suggest that Aboriginal concern is solely over the
disturbance of sacred sites. More importantly, their concern is with
the disturbance of any land they have traditionally occupied. The

270   GLOBAL FISSION

Aboriginal people and their ancestors have occupied the land for so
many thousands of years that its past merges into the mists of the
'dreamtime'. Their land forms the framework on which their traditions, beliefs, and tribal memories are constructed, and with each
succeeding generation reconstructed.

> The land is my backbone. I only stand straight, happy, proud
> and not ashamed about my colour because I still have land.
> The land is the art. I can paint, dance, create and sing as my
> ancestors did before me. My people recorded these things
> about the land this way, so that I and all others like me do the
> same.[36]

To destroy the integrity of their land is to destroy the memories,
and rape and pillage the culture of the Aboriginal people. Nowhere
is the fear of this greater than among the Aboriginal communities
in the Northern Territory. As Silas Roberts, former chairman of
the NLC, said:

> In my travels throughout Australia, I have met many
> Aborigines from other parts who have lost their culture. They
> have always lost their land and by losing their land, they have
> lost part of themselves . . . We in the Northern Territory seem
> to be the only ones who have kept our culture.[37]

Beyond loss of land, the Aboriginal communities faced an equally
devastating prospect from white intrusion – the alcoholism and
prostitution it inevitably seems to create. In particular, the Ranger
mining town of over 3 000 whites terrified them. In the words of
Silas Roberts, 'It will do nothing for us – only hurt us. Drink and
men looking for girls and everything. We want to keep this city a
long way from our land'.[38]

Despite their opposition to the proposed uranium mining, there
was no choice allowed over whether it would take place. Forced to
negotiate under duress, the NLC employed a US negotiator to act
on their behalf, and on 25 August 1978 the Minister for Aboriginal
Affairs, Ian Viner, announced that agreement had been reached.
The uranium miners were overjoyed. A speedy conclusion would
mean that they could bring in construction machinery and set up
the Ranger project before the torrential rains of the 'wet season'
began. If this set in, site works would be impossible for another six
months. However, Viner's announcement proved premature.

In fact, the agreement had only been accepted in principle and

THE AUSTRALIAN EXPERIENCE   271

initialled by the negotiator. There had been far from sufficient consultation with other Aboriginals to guarantee that the terms of the agreement would be accepted. Nevertheless, the full force of government threats and pressures was now applied to compel the NLC to speedily endorse the agreement. According to reports, one powerful threat was that the government would dissolve the NLC if the agreement was not signed.[39]

On 14 September 1978, after a three-day meeting at Red Lillies Lagoon, the NLC delegates succumbed to the pressure. The following day NLC chairman Galarrwuy Yunupingu informed the government that the NLC had agreed to sign. Four days later an application in the Northern Territory Supreme Court by Dick Malwagu and Johnny Marali No.1, both NLC members, for an injunction to prevent the NLC signing the agreement, was granted.

Even as the meeting had been taking place at Red Lillies Lagoon, telegrams flooded into the office of Bob Collins, the ALP member for Arnhem in the Northern Territory parliament. They came from black communities around the Territory opposing the signing of the agreement and complaining that they had not been allowed to send representatives and had not been consulted. The injunction was granted. After two days of extensive meetings, the NLC on 22 September decided to send the agreement to the communities for more consideration. This discharged the injunction. It was only a matter of time before the avenues for delay would be exhausted and the Aboriginals would crumble before the inexorable pressure from the mining companies and the government.

The agreement was formally signed on 1 November 1978. It provided the Aboriginals with royalties of 4.25 per cent, and made some concessions towards improved environmental monitoring and eventual land reclamation. It fell far short of the NLC's original proposals.[40]

For the Aboriginals, even though they would continue to attempt to prevent further mining, it was clear that it was not within their power to achieve this. Isolated by several thousand kilometres from the southern cities in which the bulk of Australia's population and the majority of the opposition to uranium mining was located, and despite an almost total ignorance of the events in the south, they had struggled desperately to halt mining. In doing so they had prevented any possibility of uranium mining beginning until the end of the wet season. That would occur in the early months of 1979. If there was to be further resistance it would have

to come from the south. With the last formal barriers down, a new stage had been reached. In the south, the opposition to uranium mining, the government, and the uranium companies were all preparing for the beginning of mining.

PREPARING FOR MINING

From the time of the government's decision to mine in August 1977 until the signing of the Land Rights Agreement at the end of 1978, the nature of the confrontation over uranium mining had changed. The period of debate within the community had been supplanted by the opposition actively confronting the state and the mining industry as they attempted to implement their mining plans. Many opponents of uranium mining had hoped that one of the hurdles, Aboriginal land rights, an ALP victory, or decisive union action, would prevent mining. Those hopes had been artificially raised even further by some opponents who had wrongly believed that the isolated and down-trodden Aboriginal communities in the north would be able to resist the massive pressure the government could exert. But now, with the agreement signed and neither an ALP victory nor decisive union action likely in the short term, it was becoming clear that there were no hurdles left which could speedily prevent mining. Instead there was the prospect of a long and difficult campaign of containment in which, to the maximum extent possible, the uranium industry would be laid under siege.

The government's announcement of its decision to permit mining had not in itself dampened opposition. One month after the announcement, the largest demonstrations over the issue to date, and probably the largest since the Vietnam War, moved through the streets of the major cities. Over 20 000 people took part in both Sydney and Melbourne, and over 80 000 nationally. But these demonstrations were still motivated by the hope that mining could be stopped. This was the peak of the mass demonstrations over the issue at least for several years to come. Although very large demonstrations were held in April and August of 1978 and 1979, during those two years the numbers present nationally declined. It was proving hard to maintain the pitch of enthusiasm available during the heat of the battle once the battle had moved into a state of siege.

The reason for the decline was simple enough. With the government decision made, and the Aboriginals having succumbed, the manoeuvrings of the ACTU leadership, the success of the Fraser

THE AUSTRALIAN EXPERIENCE

government at the polls, and the remoteness of the uranium deposits, many in the community who were concerned about uranium mining felt a growing sense of frustration and impotence.

Viewed from a distance, the feeling of impotence was much more intense than warranted. Both the ALP and the unions had strong principled positions of opposition to uranium mining. There would be other elections. With a swelling citizens movement, the unions could be given the strength to take action. However paradoxically, the thought that 'others' would be necessary to implement community concern was itself disabling the development of the very citizens pressure that would produce further obstructions to uranium mining. With the opposition at least temporarily contained by uncertainty, the government and mining companies now sought to entrench their position.

During 1977 the mining companies had been remarkably unsuccessful in swaying public opinion, despite a media campaign reputed to have cost over $A2 million. Despite full-page newspaper and television advertisements featuring famous experts, and participation in many debates and interviews, it gave the familiar result of information campaigns – an expensive reverse for the nuclear proponents. Opinion polls showed a steady decline in support for uranium mining, and a steady growth in opposition. But after the government's announcement that mining would go ahead, the uranium industry changed its tactics. It adopted a strategy of silence.[41] The advertisements stopped. So did its participation in public debates. Government ministers found themselves suddenly too busy to debate the issue. Simultaneously the media, and in particular the newspapers, now found the nuclear debate much less newsworthy.

Although this strategy was successful in helping to contain further growth in large-scale public action on the issue, it was much less effective at preventing the processes of discussion, lobbying, and education over the issue which continued to simmer in the background. Although the size of demonstrations declined, this was not matched by any corresponding reduction in community opposition to uranium mining as measured by the opinion polls. A Morgan Gallup Poll in November 1979 found that the level of opposition had marginally increased since June 1978, while a previous poll using the same question showed no change between June 1977 and June 1978.[42] In 1979 it became clear that the opposition still carried considerable political weight.

274   *GLOBAL FISSION*

In South Australia, an attempt by Premier Don Dunstan to re-examine his state Labor government's opposition to uranium mining was diverted by a hastily formed committee of eighty party members, including two cabinet ministers, members of parliament, and union members.[43] So strong was the party membership's feeling over the issue that headlines of 'Dunstan Changes Uranium Tune' while he was overseas on his fact-finding mission quickly changed to 'No Solutions to N-Problems' when he returned.[44] Although he later retired and a snap state election led to a surprise Liberal Party victory, a process that could have led to an erosion of the federal ALP policy had been averted. At the ALP federal conference in mid July, despite early moves to amend it, the existing policy for an indefinite moratorium on uranium mining was re-endorsed unopposed. It was supported by a petition of 15 000 members.[45] For frustrated opponents of uranium mining it was a reminder of their movement's potential strength. Two months later came evidence that that strength had further consolidated.

It was 14 September 1979 and the atmosphere in the Dallas Brooks Hall, Melbourne, was tense. Packed with union delegates, a biennial congress of the ACTU was for the third time about to vote on the mining and export of Australian uranium. The debate had raged fiercely through the morning and previous afternoon. Now ACTU President Bob Hawke, with the support of a majority of the ACTU executive, sought to have the congress accept uranium mining at the two large new mines – Ranger and Narbarlek – 'as a reality'. Against this, an amendment from the floor proposed that the ACTU support an indefinite moratorium on the construction and operation of new uranium mines. It was supported by left-wing unions as well as numerous others.

As Hawke rose to deliver his closing speech the atmosphere hardened. Addressing the meeting passionately for over an hour he attacked those unionists who had spoken in favour of the amendment. For Hawke a great deal seemed personally at stake. If the amendment was endorsed it would be the first time in twenty-one years that he would have been defeated over a major issue. As the vote was called and the 800 delegates assembled for the division it was clear that the opposition to uranium mining had overwhelmed Hawke's previously unshakeable authority. By 512 votes to 318 the delegates had voted for the amendment and its position of hardened opposition to uranium mining. Noting the associated risks of nuclear war, the absence of safe procedures for storing

THE AUSTRALIAN EXPERIENCE

waste, and the international and Australian environmental problems accompanying the use of Australian uranium in the international fuel cycle, the motion concluded:

> Congress reaffirms continuing opposition to the mining and export of uranium and the present program of development including the proposed establishment of a uranium enrichment plant in Australia initiated by the Federal Liberal–National Country Party Government.
>
> Congress supports the Federal ALP policy of a moratorium on the mining, processing and export of uranium, and the repudiation of non-labor government commitments until satisfactory safeguards are met.
>
> Congress calls on the incoming Executive to immediately embark on a propaganda campaign including leaflets, use of the media and statements by the officers to convince the Australian public and those presently working in the industry of the dangers and consequences of mining uranium.[46]

Symbolizing the change within the ACTU, Hawke soon made his expected announcement and retired from the ACTU leadership to stand for parliament as an ALP candidate. His successor was Senior Vice-President Cliff Dolan, the mover of the amendment. The passage of the amendment was a victory not only for the Australian opponents of uranium mining but, by demonstrating the potential for union intervention over the international hazards of nuclear power, also for opponents of nuclear power around the world. Important though it was, however, it did not contain an explicit programme for action nor did it guarantee that any effective action would follow. It was also not binding on individual unions, and they could not act without substantial support from their members.

Although, in terms of adopted policies, great strides against uranium mining had been taken by both the ACTU and ALP, the process whereby this had been achieved possessed some hauntingly familiar features. As in the case of the SPD and FDP in West Germany or the Communist Party in France, each time there had been a push against nuclear power within the rank and file there had been key sections of the leadership who had worked hard to deflect it. True, in first the ALP and then the ACTU, sufficient divisions right up to leadership level had developed to finally allow the opposition to express itself. Indeed, in the Victorian Branch,

the rank and file opposition had so shifted the balance against uranium mining that the state leader, Clyde Holding, could find it politically attractive to play a crucial role in ushering a similar policy through the national party conference against the wishes of many in the national leadership. But it would be foolish to forget that those who had worked hard in support of uranium mining were also still present and active at all levels. It had been a battle to force the policies through. It would be a continual battle for those opposed to uranium mining in both the membership and leadership, not only to prevent the pro-nuclear forces from undermining the policies, but to force them to assist in implementing them.

Although in principle an important step had been taken towards serious action by the labour movement to obstruct uranium mining, the matter was not therefore settled. In particular, during 1979 and 1980 several serious barriers would retard effective union action. One would be that the unions with the most determined positions would be only marginally involved in the start of mine development.

The first stages of site works, and the digging of ore out of the ground, would be carried out by non-union labour in some areas, or by the Australian Workers' Union (AWU) which had a staunchly pro-uranium mining federal executive. Another relevant union, the Miscellaneous Workers' Union, which had earlier sided against uranium mining, was locked in a demarcation dispute with the AWU over who would cover the mine workers. Since miners would be provided by the AWU, the Miscellaneous Workers' Union decided to take part in the mining in the interests of the union's survival. Other unions found it hard to enforce their policies in the remote north at a time of high unemployment.

The early stages of mine development could therefore be carried out with little hindrance. It was later that union action, if applied, could begin to bite. By mid 1980, a small project at Narbarlek to the east had proceeded relatively unimpeded since the start of works in late April 1979. The ore at Narbarlek, a small volume of 8 900 truck loads of high grade material, had been mined and stockpiled.[47] The treatment of the ore over the next nine years to extract the 12 000 tonnes of uranium, and its transport, might not be so simple.[48] Already key unions needed in the construction of the treatment plant and in transport had strong policies against participating.

During 1979 preliminary site works also proceeded at the much

THE AUSTRALIAN EXPERIENCE

bigger project at Ranger. But already some union action was surfacing. By 11 February 1980, workers at two Queensland steel companies had banned the handling of 4 000 tonnes of fabricated steel bound for the Ranger uranium project. The Queensland president of the Amalgamated Metal Workers and Shipwrights Union (AMWSU) acknowledged that the decision to ban the steel had been made by the workers on the jobs.[49] It was the first strike over the issue since the ACTU congress. It was unlikely to be the last. The ACTU national campaign against uranium mining had also begun, with a tour by President Cliff Dolan and the release of the first of a series of booklets explaining its position of opposition to uranium mining.[50] In addition, the ARU had announced that it hoped to institute a campaign against uranium mining, including bans on its tranport, from early 1981.

The two years from 1978 to 1980 were marked by the beginning of uranium mining and the exploration of strategies to stop or slow it. It was also marked by a transformation in the nature of the conflict. Up to the end of 1978 the debate had centred on the hazard presented by Australian uranium mining to the Aboriginals in the Northern Territory and, through the effects of the nuclear fuel cycle, to the people of the world. These arguments had been sufficient to create the citizens opposition which had delayed uranium mining for some six years. But now a new factor was beginning to enter the debate. Still at an early stage, but nevertheless clearly perceptible, the international issues were beginning to come home.

THE INTERNATIONAL ISSUES COME HOME

If there is one lesson to be learnt from the last two years of the 1970s in Australia, it is that the pattern of events could only be evaluated in an international context. Although in some senses isolated from the rest of the world's nuclear activities, uranium mining is simply part of the world-wide nuclear industry. But as noted in chapter 10, during the last five years of the decade, a combination of the nuclear conflict and other problems had forced the nuclear industry into a period of deep stagnation.

In the early days of the uranium debate the mining companies had spoken with great confidence about Australia's moral obligations to provide uranium to an energy hungry world. Threatening pictures were painted of countries taking Australia's uranium by

278 GLOBAL FISSION

Figure 3 **Variation of AAEC estimates of Western world demand for uranium against year in which estimates were made**

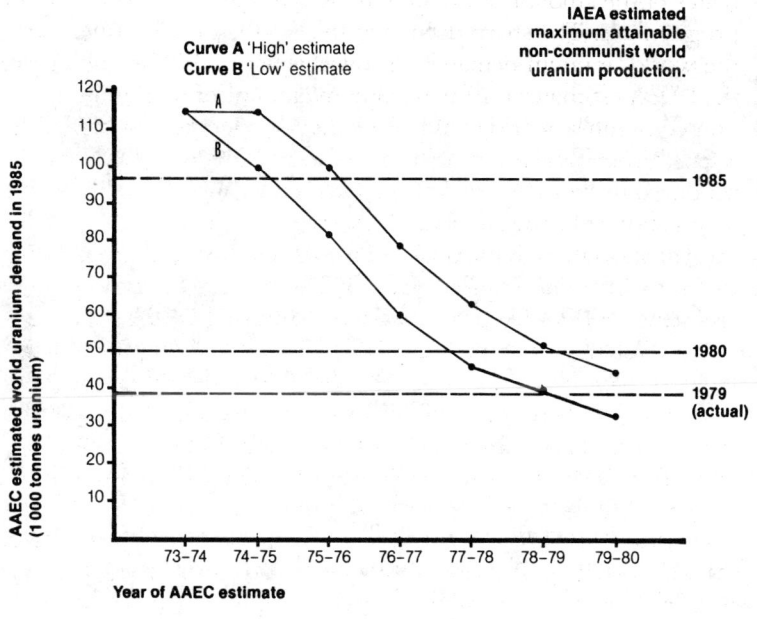

Source: Derived from Australian Atomic Energy Commission, *Annual Report,* 1973-74 to 1979-80, Australian Government Publishing Service, Canberra; and OECD Nuclear Energy Agency, *Eighth Activity Report*, OECD, Paris, 1979, p.68, Table 2.

THE AUSTRALIAN EXPERIENCE
279

force if it was not provided on the market. Threat had been tempered with the promise of the economic bonanza uranium would bring Australia. However, now that uranium was being mined there was little evidence of any stampede to buy the uranium, let alone take it by force. Around the world, the slump in orders for nuclear reactors led to a rapid downward revision in estimates of the amount of uranium needed to fuel the world's reactor programmes. The sharp decline in the AAEC's own estimates for the world uranium demand is shown in figure 3. Also shown are the IAEA estimates for the maximum amount of uranium that the non-communist world could be producing over the first half of the 1980s. From these it can be seen that some 98 000 tonnes could be produced during 1985, more than twice the AAEC's high estimate of the amount of uranium that the world will want. Should the present stagnation in the nuclear industry continue, then according to the AAEC the highest quantity of uranium the world could require in 1985 would be less than was produced in 1980.

For Australia, the problems caused by the slumping nuclear industry were exacerbated by the discovery of large amounts of uranium elsewhere, most notably in Saskatchewan, Canada. As a result, Australia's uranium (at prices less than $A30 per 0.5 kg) had fallen from 21 per cent of the Western world's reserves in 1976 to 17.6 per cent in 1979.[51]

During 1978 and 1979 government ministers travelled around the world seeking contracts with Iran, South Korea, the UK, Italy, Belgium, Switzerland, France, Sweden, Finland, Japan, the USA, and the Philippines.[52] The first problem was to persuade possible customer countries to sign safeguard agreements. Countries such as Japan and Iran proved highly resistant to signing agreements which contained even token limitations on the way that the uranium might be used. The government was in a dilemma. Either it could have token safeguards but no contracts, or a few contracts but no safeguards. It finally chose the latter, steadily weakening the initial provisions. One of the most important clauses that was quietly dropped, along with the requirements for 75 per cent Australian ownership, was that the uranium should remain in Australian hands until placed under IAEA supervision.[53] Finally, it also effectively surrendered control over whether the uranium could be reprocessed. However, even with the much weakened provisions, uranium orders were slow to be placed.

In July 1978, the general manager of Ranger Uranium predicted

280 GLOBAL FISSION

that Australia would be exporting 16 000 tonnes of uranium oxide per year in 1985.[54] But in October 1980 Mr Justice Fox, Australia's ambassador at large for nuclear non-proliferation and safeguards, was reported as saying that Australia was unlikely to be able to sell more than 3 000 to 4 000 tonnes per year in the next five years or so.[55] This estimate equalled the contracts already signed by Ranger and which covered its output, and Narbarlek, which had covered 60 per cent of its output.

By the end of the decade, the uranium market had shown no sign of easing. Indeed, it was tightening further as some countries moved to protect their own domestic uranium industries. In the USA, the country with the largest uranium consumption, a legislative amendment that would cut uranium imports from 30 per cent to 10 per cent had already been placed before congress.[56] Given these factors, it seemed likely that uranium contracts could be hard to find by those who wished to mine the bulk of Australia's uranium reserves, 70 per cent of which still lay untouched.

For the opposition to uranium mining the implications seemed to be that they had delayed the commencement of mining long enough for the effects of opposition elsewhere to have produced a slump in the market. The result was cautious development of uranium mining in Australia. This presented the opposition with a breathing space in which they might reorganize and consolidate. They needed it if they were to successfully confront not only uranium mining, but other nuclear plans looming for Australia in the 1980s.

From the very beginnings of the campaign against uranium mining in Australia the opponents of uranium mining had argued that uranium mining would be only the thin edge of the nuclear wedge.[57] Almost inevitably there would follow proposals for uranium enrichment, fuel fabrication, and even waste storage and disposal. Enrichment plants would require large amounts of electricity, and would it then not be likely that pressure would increase for the use of nuclear reactors? In addition, if enrichment was being carried out in Australia, right-wing groups might well lobby for the building of nuclear weapons in Australia. Clearly uranium enrichment was an important stepping stone.

The possibility of enrichment facilities in South Australia was investigated as early as 1973. This idea was shelved by Premier Dunstan in 1977 but remained under review and consideration, and was subsequently raised by the new Tonkin Liberal govern-

THE AUSTRALIAN EXPERIENCE

ment. In the meantime, the federal government has enthusiastically commissioned its own studies.

In July 1978 detailed proposals were released by the URENCO consortium for a $A250 million pilot plant which could be operating within six years.[58] That proposal was followed five months later by leaked details of an Australian–Japanese study of a cooperative enrichment venture which had been initiated in 1974.[59]

Paradoxically, although the nuclear market had slumped, pressure for nuclear projects increased in Australia. Both the national and state governments seemed gripped by a 'develop at all costs' philosophy. In particular, they competed strongly with each other to attract overseas purchasers of coal and other resources. For example, over the period 1979–80 major increments of Australia's electricity supply were promised for the powering of huge new aluminium smelters in Victoria, New South Wales, Queensland, and Western Australia.[60] The price at which the electricity was offered was extremely cheap and possibly represented a substantial subsidy.[61] Even with its prospects considerably diminished the nuclear industry might find a similar agreement to power an enrichment plant, together with the readily available uranium in Australia, too good to refuse.

By the end of 1980, the Victorian minister for Minerals and Energy, J. Balfour, had advocated placing a plant in his state,[62] and the South Australian premier had released a report indicating that a plant would be economically viable in his state.[63] The governments of Western Australia, Queensland, and the Northern Territory had also indicated interest.[64] But undoubtedly, as Federal Minister for Trade and Resources J. D. Anthony had commented, 'virtually all state governments are interested'.[65]

There was also evidence that the state electricity commissions liked the idea of having nuclear power stations. It seemed they might well succeed in obtaining government approval. Most advanced was the planning in Western Australia. There a proposed aluminium smelter would use up all the state's estimated coal reserves from the Collie field over the smelter's lifetime.[66] The state's premier, Sir Charles Court, had already announced that additional electricity would be made available from a nuclear power plant, and the State Energy Commission intends that it be installed by 1995.[67]

Additionally, in October 1978, Dr G. L. Miles, Deputy Director of Operations of the AAEC, had testified to the senate (the federal

upper house), that he expected one or more Australian states to have nuclear power by 1995, and one year later the AAEC's director had predicted that a reactor would be established in Victoria or South Australia within seventeen years.[68] In Victoria, as early as 1978, a nuclear reactor was mooted for Portland to supply the western half of the state.[69] Since then, despite some protestations to the contrary,[70] a considerable body of evidence has emerged from government reports and statements of government and State Electricity Commission of Victoria officials, that a nuclear plant is intended in the mid 1990s.[71] In the Northern Territory there were no firm plans, but nuclear power was under serious consideration. Said Energy Minister Paul Everingham, 'Nuclear power is the obvious answer. But the political complications are a bit daunting'.[72]

That there were plans looming was one thing. Whether they could be implemented was quite another. Already the community had been educated about the hazards of nuclear power. An opinion poll taken in June 1979 showed 56 per cent of all Australians opposed to the building of nuclear reactors in Australia. Only 34 per cent were in favour.[73] Labour would be required to implement the plans. Given the unions' policy stand against the impact of uranium mining overseas which was already beginning to produce action, the stand taken against construction of nuclear reactors in Australia could well prove invincible. At the ACTU congress, although divided on the issue of uranium mining, unionists had been overwhelmingly opposed to nuclear power in Australia.

The end of the decade, when more general nuclear plans began to become visible, was marked by a resurgence in the opposition movement. From a trough at the end of 1978, the victory at the ACTU congress and the ALP conferences in mid 1979 helped raise confidence. At the 1980 national elections, despite a large swing to the ALP, it was not quite elected. Nevertheless, sufficient seats were won in the senate to give the two parties with anti-nuclear policy platforms, the ALP and the Australian Democrats, enough votes to veto any pro-nuclear legislation after June 1981. In addition, the beginnings of union action, a growing appreciation of the morass which the international nuclear industry faced, and the problems of selling uranium to it, also helped the movement to reconsolidate.

In several key cities the movement was also stimulated by the realization that nuclear projects were on the agenda closer to home.

THE AUSTRALIAN EXPERIENCE   283

This was evident in the progress of a campaign which by the end of 1980 had seen twenty-one city councils in four different states prohibiting the passage of any nuclear related materials (except for medical uses) through their cities. As mentioned before, the workers at Port Kembla had also set a trend by declaring the harbour a 'nuclear free port'.[74]

The desperate lengths to which the government was prepared to go in order to try to attract new contracts was underscored in January 1981 when it dropped the last of its major safeguards. It signed a treaty with France which allowed France to purchase Australian uranium and to reprocess the resultant spent fuel to produce plutonium.[75] Not long afterwards a confidential dispatch from Australia's ambassador to Argentina was leaked to the press. It warned that the existing safeguards were inadequate to ensure that Argentina's 'peaceful' nuclear programme was not being used to create a military nuclear capability.[76]

As the debate became more heated the possibility of union action strengthened. In February 1981 a special meeting of unions involved in the uranium industry, called together by the ACTU, voted to support the decision by the ARU to ban the transport of uranium. The meeting was also told that the Seamen's Union of Australia had banned all shipments of uranium out of the country.[77]

Perhaps equally significantly, a few months earlier, in October 1980, Melbourne had once again seen a large demonstration of around 15 000 people march through the streets in opposition to uranium mining. There was now another major theme: support for a policy prohibiting any parts of the nuclear fuel cycle in the state. In early 1981, as proposed new federal legislation to enable a full nuclear industry in Australia was leaked to the press,[78] the movement announced a new coalition of anti-nuclear groups, church groups, unions, and political parties, to meet the threat.[79] Far from being over, the nuclear conflict in Australia seemed likely to be entering a new stage.

In the 1970s many Australians had opposed uranium mining in order to aid the opposition overseas. In the 1980s the opposition overseas had helped create a situation where it was becoming much more difficult for uranium mining to proceed rapidly. At the same time, the issues which had been a cause for concern overseas were coming home. The course of the conflict would depend on many

GLOBAL FISSION

factors, including electoral, union, and community developments. Certainly its outcome could not yet be predicted.

From the beginning, the purpose of opposing uranium mining in Australia had been to slow down nuclear developments everywhere. The success of that could only be judged from an international perspective. So too, events in Australia will depend not only on the actions within the Australian community, but also on the battle over nuclear power world-wide.

No matter how seemingly remote, the nuclear industry and thus the nuclear conflict are global in their reach. Each ebb and flow of the conflict in each country interacts with others in other countries. The net effect rebounds to interact with individual confrontations over nuclear power. Those involved in the conflict are often unable to see beyond their own local confrontation to the overall development of the battle. Without this overall perspective contestants may allow unrealistic levels of either pessimism or optimism, generated by the transient fortunes of their own particular engagement, to distort their judgement and render their strategy and tactics less relevant. Undoubtedly it has been the opposition to nuclear power with its comparatively fragmented structure, both within and between countries, which has been most vulnerable to this deficiency. Overcoming it might enhance the opposition's effectiveness considerably. By the beginning of the 1980s there was mounting evidence that the opposition to nuclear power worldwide was beginning to confront this problem.

12
STRATEGY AND STRUCTURE

The corporations constituting the nuclear industry are highly centralized. They are organized not merely nationally but trans-nationally. Because of their economic power, hierarchical structures, and carefully constructed lines of command and communication, they have little trouble interacting with government and the media. They are able to plan their programmes on a grand scale. The power focused by these immense structures is obvious. The centralization also helps explain the apparent ease with which the industry can single-mindedly pursue its siting of nuclear reactors in communities desperate not to have them. The long communication lines and chains of command ensure that those making the decisions have only the most remote social links with the people who will be affected.

The pro-nuclear forces consist of the nuclear industry and the section of the community which either actively supports it or acquiesces to its plans, other organizations with compatible objectives, and at least to some degree in all countries, the state and its agencies. Some of the organizations from which the industry draws active support are of the same type as those from which the opposition draws its membership: unions, political parties, and even churches. But the organizational form of the pro-nuclear forces is nevertheless very different from that of the opposition.

The nuclear enterprise is organized in the form of an octopus. Its tentacles grow outwards as it hires new workers. They obey the commands issued in the brain by the executives of the key nuclear companies and government planners. The tentacles embrace, influence, and take strength from community groupings and organizations. But if the nuclear industry dies, the brain dies also and the

GLOBAL FISSION

heart stops. The alignment of forces loses its vitality and the whole disintegrates.

Typically, the life blood of the pro-nuclear forces flows from the nuclear industry to the community. That of the anti-nuclear movement flows in the opposite direction — from the community to the movement. In the movement, formal and informal community groupings with different structures are bound together by a common desire to stop nuclear power. Ultimately the strength of this alignment depends on the strength of concern in the broader community. Political parties, churches, unions, and peace groups may play crucial roles, but if the opposition to nuclear power fades in the broader community, then eventually the ability of these organizations to act is weakened.

The nuclear industry came about as a deliberate, major planning exercise by both industry and the state. This first occurred largely insulated from public scrutiny and then with a high level of public acquiescence which was only much later eroded. But the opposition arose in a quite different way. Its birth process varied from initiation from 'above', as in Sweden, to a largely spontaneous development which kindled and spread wherever the temperature of opposition to particular projects reached flash point, as in Switzerland, France, and many other countries. Elsewhere, such as Australia which had no specific projects at hand, loose state or national coalitions were set up almost from the beginning.

The variety of ways in which opposition groups were formed has led to a rich diversity of forms of organization. These range from elected hierarchies using voting procedures, to open meetings making decisions by processes of consensus. However, as some have found to their dismay, apparently 'structureless' forms of coordination can lead to as great a tyranny as the most structured hierarchy. In the latter, power can be wielded by the people 'with the numbers', while in the former it can be wielded by the people prepared to exercise the most persistent voice and hold out for their objectives over the greatest period of time.

Faced with the expansive yet relatively cohesive organization of the nuclear enterprise, a problem is presented to the opposition: how can the diverse array of groups which have arisen to confront the industry best coordinate their activities to counter the threat it presents? Undoubtedly there would be great advantages for the opposition too, in coordination and decision making at the national and international levels. With organizational structures paralleling

STRATEGY AND STRUCTURE

those of industry and government, access to the media and the institutions of national government could be enhanced. The industry could also be tracked world-wide, and strategy formed on a coordinated basis around the world to hit it at weak spots.

There are serious obstacles to such an approach. The very characteristic which makes such structures strong, the long chains of command and communication, and the centralization of power, render those who accrue authority further and further removed from those on whose success the movement ultimately depends. This may not appear to matter in an organization in which monetary reward and threat of unemployment constrain people to accept dictates from above. But in organizations built around other objectives such a system will only be accepted if it is seen as essential to the achievement of those objectives. There has been a remarkably wide and growing rejection of any such system in the groups which comprise the opposition movement. So widespread has this rejection become that it seems very likely that it arises, at least in part, because the opponents of nuclear power have come to resist not only the physical hazards which accompany the nuclear enterprise but also the related pressure to allow major decisions which could substantially affect their future to be made by any remote or unaccountable authority.

The opposition groups therefore find themselves in a bind. There is a clear necessity to understand the total context in which they are fighting, and to plan and execute coordinated strategy. At the same time there is enormous emphasis on local autonomy, and a great resistance to adopting the conventional forms of organization that would enable them to do so. Any attempts to overcome this by imposing a centralization of authority is likely to dampen the enthusiasm and dedication of the groups and constrain their initiative. Since these are the source of much of the movement's activity, this could prove ruinous. This is not to say that there has been absolutely no role for leadership in the anti-nuclear movement. But the forms which it may take and the extent to which it may intrude have been greatly restricted.

Instead, the opposition has been forced to experiment with an organizational structure which differs markedly in concept from the structure of the industry. It is the 'network', a structure based on maximum autonomy for the local groups of which it is comprised, and tied together by means of diverse forms of communication channels. Although this has limitations, it does not mean the

288 GLOBAL FISSION

Figure 4 **Developing targets of concern**

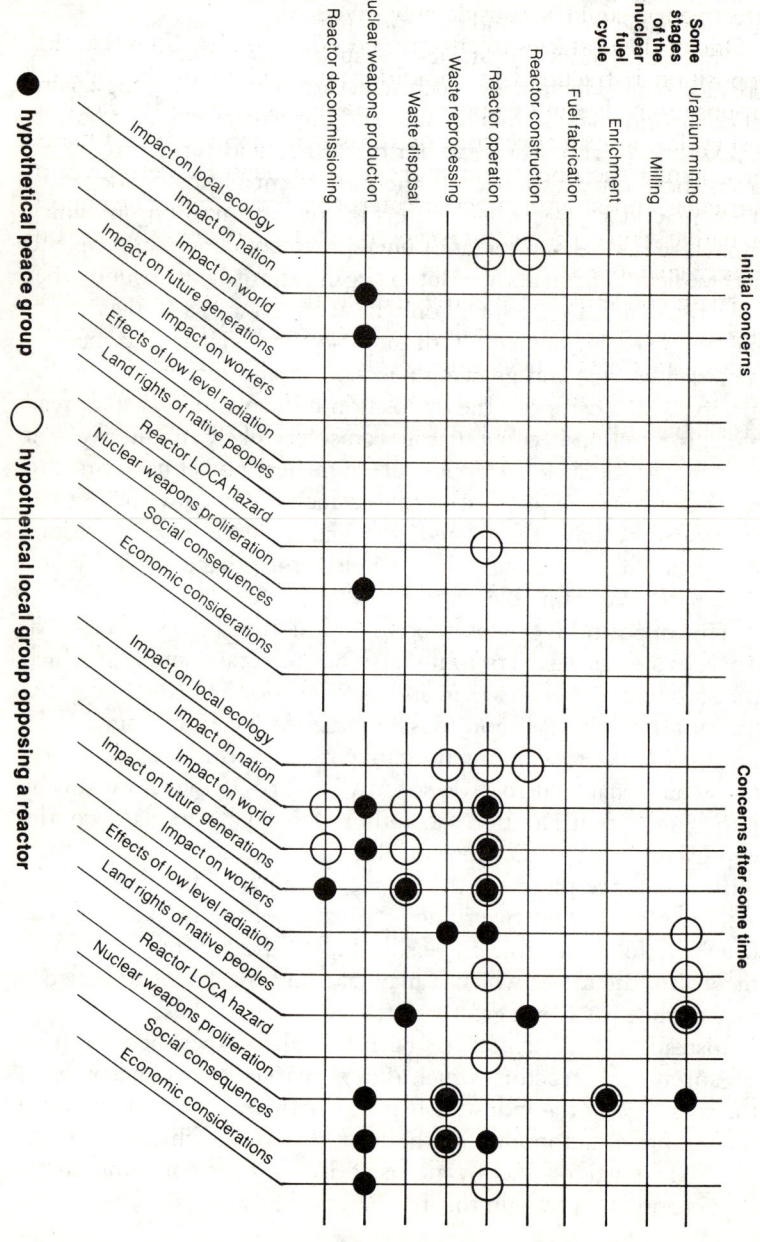

• hypothetical peace group

○ hypothetical local group opposing a reactor

STRATEGY AND STRUCTURE

movement is ineffective, as the stagnated state of the nuclear industry to a large part attests. But if means could be found for more effective coordination and strategy planning, the movement's effectiveness could be considerably enhanced.

One of the obstacles to effective coordination has been that the opposition is fractured by the differing concerns of the various groups. Some began by opposing a particular stage of the nuclear fuel cycle – a nuclear reactor, or a waste dumping project. Others have joined the opposition movement through concern over a particular impact of the nuclear enterprise, for example the impact on native people, on nuclear weapons proliferation, or of wastes on future generations.

These concerns can be envisaged as defining the borders of a grid as shown in figure 4. Along the vertical border are set out the various stages of the nuclear fuel cycle. Along the horizontal border are the areas of impact. For any organization or group its initial targets of concern may appear as a few dots representing particular areas of impact of a stage or stages of the nuclear fuel cycle. However, while this may represent the initial situation, the trend has been for these concerns to spread both horizontally and vertically across the grid. This tends to occur whether groups initially are concerned with a specific nuclear proposal (as with the 'reactor group'), or with a particular area of impact (as with the 'peace group'). Examples include the people of Wyhl who at first only opposed a proposed reactor, but now oppose all nuclear power everywhere, and Aboriginals whose concern has moved beyond land rights to embrace the problems of nuclear power.

This process of spreading concern first occurs because tackling a problem forces groups to educate themselves about its context. Second, it occurs through the spread of news about the actions of other groups facing related problems. As the concerns of the fragmented groups begin to spread and overlap, it becomes more possible for the groups to find a common basis for coordinated action. The need becomes clearer, but the problem remains: how are they to coordinate their activities?

The difficulties of any form of collaboration become greater with the distance between groups. Many attempts to solve this problem therefore have been restricted to within countries or regions. But since 1975 a series of coordination networks, although still mainly within countries, have begun to link the various components making up the movement against nuclear power.

In countries such as the USA, where nuclear power and the opposition developed early, large numbers of lobbyists, citizens action groups, activist students, scientists, and professionals opposed nuclear power for a long time before any formal coordination was attempted. The beginnings were in November 1974 when Ralph Nader arranged the first national conference of nuclear opponents, 'Critical Mass '74', in Washington. By the end of the decade his Critical Mass organization was but one of a multitude of alliances spanning the country. However, there was still no generally supported national umbrella organization. An important step in that direction occurred in August 1978, when a four-day national strategy conference in Louisville was attended by over 150 local, regional, and national organizations and alliances.[1]

Elsewhere the process was easier. In Denmark the movement began as a national alliance. In other countries, such as West Germany, an existing network of environmental groups was forged into a national alliance. Whatever the process, by the beginning of the 1980s national networks of opposition with steadily improving communication channels, including journals and newsletters, and regular national meetings, could be found in many countries of the Western world. But while this was an important development it represented only one of the ways in which cooperation and collaboration had begun to develop.

THE ROLE OF EXAMPLE

The remarkable successes achieved by such a loosely linked assembly of diverse groups and organizations as the nuclear opposition is partly attributable to a process much more subtle than the action of formal organization. It is the stimulation of many groups by the success of one: that is, the confidence gained from example.

The successes of one group can provide inspiration to others tackling similar problems even in different countries. The opponents of uranium mining throughout Australia, and in the USA and Canada, France, Italy, the Black Forest in West Germany, Spain, Scotland, and Sweden are encouraged by each other's successes. The actions of the Australian community raise the possibility of opposing uranium mining even when the use of the uranium is in another country. The actions of its unions change the realm of what is possible again by showing that unions may take responsibility for the consequences of their work. The actions of municipal councils

STRATEGY AND STRUCTURE

in the north of Sweden, which have stopped uranium mining at Billingen, and the similar successes of the Orkney community, demonstrate that uranium mining can be stopped, heartening the opposition everywhere else. At the same time the successes of the opposition to uranium mining elsewhere reinforce the ability to obstruct future moves to reactivate mining in the Orkney Islands. As the struggle moves into the 1980s, the success of one group likely to strengthen others tackling similar problems is the opposition's victory over uranium mining in British Columbia, Canada.

Opposition to nuclear power in Canada has spread rapidly since 1977. Some has been directed at the 5.4 GWe of operating nuclear power and 6 GWe of planned capacity largely located in Ontario.[2] In other states opposition has been directed at the mining and export of Canada's vast uranium reserves. With approximately 10 per cent of the world's estimated additional reserves, at present prices, Canadian uranium plays an important part in the world uranium supply market.[3]

In 1977 exploration for uranium intensified all over the country. Provinces previously untouched by the nuclear debate suddenly became intensely involved. Rich deposits of ore with an average grade four times that in Australia and ten times the world average were found in Saskatchewan.[4] British Columbia swarmed with exploration companies. The emerging debate already had some powerful examples of nuclear problems on which to draw.

As early as 1975 it had become known that homes and schools in Port Hope, Ontario, had been built with radioactive materials. At the nearby Eldorado Nuclear Company's dump the radiation level was several hundred times the permitted level. On one public beach, 10 000 cubic yards of contaminated fill were removed. Levels of radiation at the beach ranged up to forty times the permitted levels. By April 1976, 240 of 652 places surveyed exhibited abnormal radiation levels, and thirty of these were public places.[5] At about the same time it became publicly known that the Elliot Lake area, the site of uranium mining in Ontario, was badly contaminated with radioactivity.[6] Concern began to grow and then escalate. There were allegations of bribes having been accepted by agents of the government, and the Public Accounts Committee of the House of Commons found evidence of corruption and severe mismanagement. Next the government sold a CANDU reactor to Argentina, a transaction alleged to have cost the taxpayers some $130 million. In addition, a suit brought by Westinghouse in the

292   *GLOBAL FISSION*

USA implicated the Canadian government in secret international price-fixing arrangements.[7] A cloud of question marks hung over not only the industry, but also the federal cabinet.

Together these helped increase citizens opposition. Local groups sprang up, and nationally the Canadian Coalition for Nuclear Responsibility, formed in 1975, was greatly strengthened. It gained support from all sections of the community: trade unions, churches, medical associations, politicians, and members of the public.[8] An early test of the movement's potential strength was in Madoc, Ontario, when local people forced the cancellation of a plan to build a nuclear waste repository in their area. Another gain for the opposition was the conclusion of the Porter Commission, whose interim report on nuclear power in Ontario recommended a moratorium on further nuclear licensing if, by 1985, suitable waste disposal methods had not been clearly demonstrated. However another royal commission, into uranium mining in Saskatchewan, brought less welcome news to the opposition. It recommended that the massive uranium reserves in that province be mined.

The reverse for the opposition movement in Saskatchewan undoubtedly dampened the optimism of groups within Canada. There and elsewhere it was groups opposing uranium mining who particularly felt the sting. A victory was needed to restore morale. In 1980 it came, in British Columbia.

The opposition to uranium mining surfaced in the state at Christmas, 1977.[9] Consolidated Rexspar had unveiled plans to mine near Clearwater in eastern British Columbia. But when the company called a public information meeting angry citizens packed the auditorium and the meeting lasted long into the night. Rexspar postponed the proposal. The fear was the same as that voiced at all the other mining proposals for the heavy exploration of the next two years. The mines would pose severe water pollution, dust, and land-use hazards to an area dependent on wine-making, ranching, fruit growing, and tourism for its economy. The uranium would be exported in a few years but it would leave a dangerous legacy of millions of tonnes of radioactive tailings for thousands of years to come.

Opposition was so intense that in February 1979 the government appointed a royal commission of inquiry. However, opposition continued to mount. During the year, with the commission still sitting, more than $6.5 million was spent on uranium prospecting throughout the state. Lobbies such as the Medical Association,

STRATEGY AND STRUCTURE   293

the United Church, the British Columbia Cattlemen, and even the Teachers Federation, passed resolutions calling for a moratorium on uranium mining and exploration. Plans were under way for a demonstration in February which promised to be the largest in the history of nuclear opposition in British Columbia. But two days before the demonstration the premier capitulated. He announced a seven-year moratorium on uranium mining and the cancellation of the royal commission. For the first time in the history of North America a major project from the 'front end' (the mining and processing of uranium) of the nuclear fuel cycle was brought to a long-term standstill by citizens opposition.

The success in British Columbia had the potential to raise hopes and inspire similar groups elsewhere. But there is nothing inevitable about this potential being transformed into actuality. For this to occur, the news of the success, and an understanding of its significance, must reach other groups.

IMPROVING COMMUNICATION

The nuclear industry has ready access to air-travel and the use of the most modern communication facilities. It has its own intra-industry publications, and holds frequent formal and informal conferences of key executives, as well as making full use of its access to the services of government agencies. The strategists and planners of the major nuclear companies therefore have little trouble keeping up with the latest achievements, and setbacks, in their nuclear programmes. In contrast, the nuclear opposition network has until recently relied on the commercial media for much of its information. But this source, at best, bases its reportings on what it or the international agencies consider 'newsworthy'. At worst, the use of editorial discretion to suppress or reduce the impact of stories which would affect local struggles over nuclear power is not unknown. This was illustrated in Australia when a visit to an editor of a major newspaper by a delegation of representatives of political parties, citizens groups, and other community organizations, including representatives of the unions controlling the newspaper's access to paper supplies, at least temporarily accelerated the coverage of the nuclear issue marvellously.

But there is no guarantee that the news of the Canadian success will reach the people in Greenland who are fighting uranium mining proposals. Nor is there any certainty that a knowledge of

294 GLOBAL FISSION

the Greenlander's achievement in obtaining a temporary postpone-
ment of uranium mining in their country, mainly due to opposition
by their majority party Simut[10] would reach those in the Black Hills
of South Dakota, USA, also working hard to stop uranium mining
there.[11] This places a serious limitation on the degree to which
successes can increase horizontally, and is true at all levels of the
nuclear fuel cycle. Recently some steps have been taken to rectify
this problem.

By the late 1970s national newsletters, journals, newspapers, and
even radio stations through which the anti-nuclear movement
could communicate had been established.[12] In November 1977 a
further important step was taken when the Boston Clamshell
Alliance began to produce a monthly international clipping service
called No Nuke News.[13] The following year an even more ambitious
project was begun in the Netherlands to produce an international
newspaper known as WISE or World Information Service on
Energy. Its first issue was in May 1978. Since then it has provided
an invaluable source of information on developments in the battle
over nuclear power.[14] In 1979 these two were joined by an international clipping service produced in Australia.[15]

In addition, communication is being enhanced by international
conferences and exchanges of activists together with coordinated
actions aimed at particular segments of the nuclear fuel cycle. These
have in part been facilitated by the existence of a few environmental groups already organized internationally. Thus it was
possible for the international group Greenpeace to follow the waste
transport ship Pacific Swan. In February 1980 the ship attempted to
dock at Cherbourg, France. The socialist trade union organization
CFDT was able to stop the docking. The ship was then diverted to
Barrow, England, with its cargo. There, with the help of Greenpeace, local residents formed an action group to stop future
shipments and within three days collected over 3 000 signatures on
a petition expressing concern at the ship's presence.[16]

The international scale on which the nuclear industry operates is
also an important catalyst in creating the links needed to oppose it.
The waste ship Pacific Fisher, moving from port to port, has created
a backwash of opposition within some sections of the labour movement. When it arrived at Hawaii on 5 June 1979 carrying 70 tonnes
of spent nuclear fuel rods from Japan and bound for the Windscale
reprocessing plant in the UK, it found itself unable to dock for
refuelling. Local 142 of the ILWU (a dock-workers' union) had

refused to handle the ship. 'We are not going to participate in any activity that jeopardises the lives, health and safety of hundreds of Hawaii residents', said Carl Damaso, president of the union.[17] The local United Public Workers Union joined in saying 'no port in our State should be used'.[18] The state's governor, George Ariyoshi, then announced that the ship would not be allowed to enter the harbour. It eventually retreated to be fuelled at the Pearl Harbour naval base.[19]

Similarly, the ship *Covadonga*, carrying cargo to the Spanish nuclear reactor being constructed at Lemoniz, found itself confronted by a total boycott by Spanish dock workers in 1979. Because the boycott applied to every port in the country the ship was diverted to Bordeaux in southern France. There, in solidarity with their fellow workers in Spain, the dockers also refused to work the ship.[20] It was a significant development because their national union organization, the CGT, supports nuclear power.

These actions illustrate the potential strength of such coordinated action. But its full possibilities are still far from having been explored. For example, although the possibility exists for coordinated action to be taken against the same stage of the nuclear fuel cycle in different countries, for example by those fighting waste dumping in Gorleben, West Germany; the Netherlands; Barnwell, USA; and in the UK, such action has rarely been attempted.

The above actions all represent moves to link groups 'horizontally' across the same stage of the nuclear fuel cycle operating in different places. Important steps have also been taken to link 'vertically' groups which in different places are concerned about the same impact of the nuclear enterprise. One of these has been the growing cooperation between those whose land rights and culture are under threat from the nuclear enterprise.

By a quirk of fate a substantial proportion of the non-communist world's uranium resources lie in land for whose owners the uranium's energy potential represents virtually no benefit, while its extraction represents a devastating threat. In the USA some 500 000 hectares of native land is under lease for uranium exploration.[21] Such projects are threatening native land in a wide belt cutting down the west of North America including the Canadian states of British Columbia and Saskatchewan, and the US states of Washington, Wyoming, South Dakota, Colorado, California, and New Mexico.

As in Australia, tailings dams have leaked and spilled, contamin-

296   *GLOBAL FISSION*

ating water supplies with their radioactive contents. Elliot Lake, Ontario; Shiprock, New Mexico; and Black Hills, South Dakota, are just a few of the many sites on Indian land that have suffered severe contamination. Additionally, the native Americans have been used as an impoverished and therefore compliant workforce for uranium mining operations, producing some 13 per cent of the world's uranium overall.[22] The result, a statistically significant excess of lung cancers and other health problems, is now beginning to surface. The court cases for compensation, and the congressional hearings, are unlikely either to adequately repay the damage or convince the native people against resisting further nuclear intrusions.[23]

Steadily, affected peoples such as the North American Indians are beginning to recognize the need for international collaboration to protect their health and the integrity of their land and culture. Thus the National Indian Youth Council writes:

> The same corporations which are exploiting the uranium resources of the Aboriginal people of Australia and the indigenous peoples of Namibia in southwestern Africa also exploit Indian peoples in the Western Hemisphere. We will work with these native peoples to fight against a common enemy in order to protect our land, way of life, and future.[24]

Much of the struggle by indigenous peoples is not for an improvement in their conditions, but against the steady and seemingly inexorable erosion of their lands, culture, and pattern of life. Generally constituting minority groups, they struggle, often with little success, against the white Western tide which threatens to sweep the remaining ground from under them. In the USA, in May 1978, some eleven bills were introduced to further reduce the rights of native Americans. If passed they could terminate treaties, close native American schools and hospitals, and cancel land and water rights.[25] To fight these trends, indigenous peoples must recruit sufficient support to act, on at least some issues, as the majority. The nuclear issue has potential to catalyze some of the necessary additional strength.

Initially united by their common opposition to uranium mining, the diffusion of targets of concern described earlier can lead white anti-nuclear groups to take up the interests of indigenous peoples, while at the same time those peoples gain an awareness of the problems of nuclear power. In Australia the links between the

Aboriginal Land Rights movement and the movements against uranium mining have at times greatly reinforced each other's effectiveness. To a lesser extent the same is true of North America.

The value of building alliances both between indigenous peoples, and between them and others with common causes, is being grasped. In the USA the 'Longest Walk', a trek by native Americans from California across the USA to Washington in 1978, demonstrated an emerging indigenous national solidarity.[26] A year later a tour of Europe by native Americans was organized by Native Americans against Uranium Mining with the support of the Danish OOA.[27] In October 1979 the possibilities for further cooperation were explored at an international conference held in Copenhagen. It was attended by representatives of the Australian Aboriginals, the native American people, the South West African People's Organization of Namibia, and Greenland.[28] Organized by the Danish anti-nuclear groups, it was a first step in an important new direction.

The developing links between indigenous peoples are symptomatic of the beginnings of 'vertical' links being forged throughout the different groups associated with the anti-nuclear movement. One such grouping already outstanding in its level of international organization is the movement against nuclear weapons. As the concerns which motivate it, and those which motivate the opposition to nuclear power spread and begin to overlap, their objectives begin to converge and the possibility of much closer collaboration increases.

CONVERGING CONCERN: NUCLEAR POWER AND NUCLEAR WEAPONS

The threat of nuclear war represents a far graver and more ominous prospect than even the most catastrophic nuclear accident. During the 1980s, in a world in which a single Trident submarine will carry up to 408 nuclear warheads, each capable of landing within 30 metres of its target, and exploding with a power equal to fifty Hiroshima bombs, the task of preventing the further escalation of the arms race is urgent in the extreme. It is also daunting.

For groups opposed to nuclear power there was initially a strong tendency, especially in the USA, to cast their attention away from the hazards posed by the proliferation of nuclear weapons. Since

many of these groups were formed around opposition to particular nuclear projects the local environmental threat posed was usually the first target of concern and the focus of action. The hazards of nuclear weapons were seen as 'too political' and diversionary. On the other side, the church groups, peace organizations, trade unions, and social democratic parties which have at times given great attention to the problems of proliferation, often harbour an attachment to 'peaceful' nuclear power. Often this stems from either a sympathy for the policies of the USSR or an unquestioning continuation of beliefs (such as that nuclear power represents 'progress') which were popular when groups first took up the nuclear weapons issue.

Nevertheless, the connection between the proliferation of nuclear weapons and nuclear reactors remains inextricable. With an ordinary commercial reactor producing over 250 kilograms of fissile plutonium per year,[29] enough for over forty Hiroshima size bombs, there is, as the Indian government demonstrated in 1974, the real possibility of some of the waste fuel being diverted to the construction of nuclear weapons.[30] Alternatively, the enrichment technology from the front end of the nuclear fuel cycle may be used to provide fissile uranium for the same purposes. As the Australian Ranger Inquiry concluded, the greatest threat posed by the nuclear industry is that it leads to an increased risk of nuclear war.[31]

Slowly the intimate relation between the two issues is being recognized and taken up by the two movements. The targets of concern are spreading, and a basis of common objectives is being laid. The emerging area of common purpose between the two movements enhances the opposition to nuclear power among previously resistant sections of the community, including, in many countries, the labour movement. At the same time it stimulates a growing opposition to nuclear weapons among young activists, without whose efforts the peace movement is relatively impotent. As the peace movement and the movement against nuclear power slowly form links and develop an area of unified concern and action, the effectiveness of both is immeasurably enhanced.

In the USA the peace movement has resurged under the umbrella of the Mobilisation for Survival. Originally the organization seemed to resist placing too much emphasis on the hazards of nuclear power, while the local anti-nuclear energy groups were suspicious of the Mobilisation. However, the Mobilisation now takes a position of strong opposition to nuclear power. With a

STRATEGY AND STRUCTURE
299

membership of over 250 peace organizations,[32] and a demonstrated capacity to organize teach-ins and actions across the country, its declared opposition to nuclear power and actions already taken in support of that policy contribute significantly to the nuclear debate within the community. On the other side, the groups organizing the month-long sit-in on the rail tracks leading into the nuclear weapons plant at Rocky Flats outside Denver, Colorado, the actions against the Trident submarine at Groton, Connecticut, the brave 285 demonstrators who scaled the fence of the US nuclear submarine base at Bangor, Washington, and the 10 000 who marched for disarmament in New York, have all become accepted by opponents of nuclear power as part of their movement.[33]

In Europe it was the USA's proposed deployment of the neutron bomb which led to the issue of nuclear weapons being taken up actively by the anti-nuclear movement. Clearly intended for use against ground troops invading Western Europe, for many it seemed to be opening Pandora's box. In many countries on the Continent the proposed deployment was opposed vehemently. In the Netherlands, by December 1979, more than 10 per cent of the population had signed a petition opposing the bomb.[34] The demonstrations and other actions for a time eclipsed actions against nuclear power, until finally the US government avoided placing its allies in the embarrassing position of having to refuse to take the weapon by deferring its production.[35] In August 1980 the opposition to nuclear weapons was further accelerated by an announcement by the US Secretary for Defense that the USA no longer ruled out the possibility of being the first to use nuclear weapons. He made it clear that his announcement was a warning that the USA was now prepared to deploy tactical nuclear weapons against a large-scale conventional arms invasion of Western Europe.[36]

Naturally enough those who would form the battlefield for such a 'limited' nuclear struggle between the USA and the USSR are not overjoyed at the prospect. There is clear evidence of a resurging mood of opposition to nuclear weapons. This has been particularly exacerbated by the NATO decision, first made in 1979 and reaffirmed in May 1981, to add more than 570 Cruise and Pershing Two missiles to its armoury. In the Netherlands, surveys in early 1981 showed only one in every three Dutch voters in favour of allowing the missiles to be stationed on Dutch soil, and the elector-

ate rated this as the most important issue for the forthcoming national election. Opposition has even penetrated the armed forces in both the Netherlands and West Germany. In the Netherlands a group of fifty officers has been formed to oppose nuclear weapons, and many of these officers have said publicly that they would refuse any order to fire them.[37] A Dutch survey has shown that 60 per cent of conscripts, 20 per cent of non-commissioned officers, and 10 per cent of officers would be in 'moral difficulty' if ordered to use nuclear weapons.[38]

In the UK, the birthplace of the Campaign for Nuclear Disarmament (CND), that organization is for the first time in years beginning to revive. In the first six months of 1980, membership inquiries poured in. In July, the entire year's fund-raising target of £30 000 had already been exceeded by £12 000.[39] The revival was confirmed in October 1980 when about 50 000 people, the largest demonstration in the UK since the ban-the-bomb protests of the 1960s, marched through central London calling for the dismantling of the UK's nuclear arsenal.[40] That demand was given great weight in the same month when a motion for unilateral nuclear disarmament was adopted by the national party conference of the British Labour Party.[41]

In the movement against nuclear power, opposition to nuclear weapons proliferation has become a central organizing theme in several countries. In the Netherlands, opposition to extensions to the Almelo enrichment plant has focused on the proliferative potential in supplying enriched uranium to the major nuclear complex planned for Brazil. In Canada, similar concern has been voiced at the proposed supply of a nuclear reactor to the military junta of Argentina.[42] In Australia and Canada the proliferation of nuclear weapons has been a key argument against the mining and export of uranium.

In Australia the convergence of the two movements was formally recognized in January 1981 when the national body of the movement against uranium mining, the Uranium Moratorium, changed its name to the Coalition for a Nuclear Free Australia, including in its aims opposition to all nuclear weapons systems. In nearby New Zealand, the connection between the two movements has also been unmistakable.

In New Zealand, any expectations for nuclear power in the near future were shattered when first moves towards it were stopped by substantial community opposition.[43] The New Zealand Depart-

STRATEGY AND STRUCTURE   301

ment of Energy's plans to build one or more nuclear power stations had been under way since 1967.[44] In 1974 the department's report to parliament already recommended the construction of a nuclear plant in 1982–83, unless suitable alternatives were discovered. With the department's manager eulogizing the 'safety' of nuclear power, it was also the year that demands for a public inquiry began to swell.

The New Zealand Department of Energy made several miscalculations. First, it worked on the assumption that electricity consumption would grow at an extraordinarily high rate, whereas in reality it grew much more slowly. Thus in 1973–74 it was expected to grow at 9.6 per cent yet the actual growth was only a little over 2 per cent.[45] Second, and more importantly, it underestimated the anti-nuclear feeling already existing in the community. The actions by the previous New Zealand and Australian labour governments in opposing the French atmospheric nuclear tests had reflected a growing awareness of nuclear hazards, and had set the scene for opposition first to nuclear weapons and then to nuclear power. In 1976 the strength of concern was demonstrated with the arrival of two US nuclear warships, the *Truxton* and the *Longbeach*. Apart from being obstructed by the 'Peace Squadron', a flotilla of small craft, the *Truxton*'s visit brought Wellington harbour to a standstill as the waterfront unions walked out on a protest strike.[46] The strike paralysed commerce between the North and South Islands of New Zealand.

Those concerned over nuclear weapons were well prepared to grasp the hazards associated with nuclear power. By mid 1976 a broad coalition including people from the peace movement, the small but active Values Party,[47] and environmentalists – the Campaign for Non-Nuclear Futures – had been formed. By then, also, the official Planning Committee on Electric Power Development was projecting the introduction of nuclear power from 1990, and had reported that a firm government decision 'in principle' to go ahead with the nuclear programme would be needed by 1977.[48]

Protest grew rapidly, and in September the government set up a Royal Commission of Inquiry to determine whether it should proceed with a nuclear programme. At the same time, the citizens coalition began collecting signatures for a petition. By November, in little over three months, more than 300 000 signatures opposing both nuclear power and nuclear powered warships in New Zealand had been collected.[49] From outside New Zealand, a series of official

reports from Australia (the Ranger Inquiry), the UK (the Flowers Commission), and Ontario (the Porter Commission), through their own profound reservations about aspects of nuclear power, helped add legitimacy to the opponents' case.[50] By the time the commission reported in April 1978 the demonstrated lack of need for nuclear power, and the considerable opposition that had been generated, left their conclusion, that in the short-term no decision need be made for a New Zealand nuclear programme, largely a matter of form.[51] By then it was clear that while protests against nuclear warships would need to continue, the political possibility of a nuclear power programme in New Zealand had vanished at least for the foreseeable future.

The countries surrounding the Pacific Ocean – to the west the coasts of North and South America, and to the east the coasts of Asia, Australia, and New Zealand – form the 'Pacific Rim'. In the non-communist part of the eastern half of this rim, the three rich industrialized countries of Australia, New Zealand, and Japan constitute the centres of economic and political power. Of these, it is Japan in which the movement against nuclear weapons has attained greatest political force. However, it is also in Japan that the divisions between the peace movement and the movement against nuclear power seemed most unbridgeable. It is also in this country that a convergence of the two movements could have the most profound consequences.

CONVERGENCE IN JAPAN

It is not surprising that the Japanese are said to have a 'nuclear allergy'. Since the bombing of Hiroshima government policy has at least publicly been restricted by the 'three nuclear principles' of never manufacturing nuclear weapons, never possessing them, and never permitting them to be brought into Japan. But, at least initially, this concern did not extend to nuclear power. Indeed there was a wide acceptance of the belief that 'Japanese who suffered from atomic bombs must take the lead in utilising nuclear power peacefully'.[52] By mid 1979 the country possessed an installed capacity of some 12 GWe of nuclear power, ranking second in the world as a producer of nuclear electricity.[53] Nevertheless, its programme has had mixed fortunes. This was demonstrated by the dramatic drop in estimates for the country's expected nuclear

STRATEGY AND STRUCTURE

capacity in the year 1985 – a fall from an anticipated 60 GWe in 1972 to around 21–26 GWe in 1980. Although a substantial capacity has been constructed, the rate of expansion therefore has fallen well below the expectations of the Japanese nuclear industry. The reasons for this decline are as mixed as the programme's performance.

On the one hand the programme has been cursed by an extraordinary level of technical problems and overall unreliability. On the average, between 1971 and 1977, Japanese reactors produced only half as much electricity as they were designed to produce.[54] This compared with a rate of 60 per cent for the world,[55] and with the confident expectation in the 1960s that nuclear reactors would produce electricity at 80 per cent of their maximum design capability.[56] However, equally important has been the growth of local citizens opposition which has caused plant interruptions, order cancellations, and long delays in reactor construction. Considerable achievements have been made by a relatively scattered set of citizens groups. But until recently, although steps had been taken towards greater national coordination, there was still insufficient coherence between the movement against nuclear power and the enormously strong Japanese movement against nuclear weapons.

Opposition was first localized at nuclear reactor sites where affected communities, usually fishermen and farmers, began to work against proposed nuclear reactors. Initial actions were against the proposed Onagawa reactor in Miyagi whose construction was planned to begin in 1970, and the proposed Kashiwazaki reactor, planned for construction between 1972–77.[57] It is a sign of this local opposition's strength that construction of the Onagawa reactor had still not commenced ten years later,[58] and that the Kashiwazaki reactor was not expected to be completed earlier than 1984.[59] While this strength varies markedly across Japan, the possibilities for the growth of national opposition to nuclear power received a substantial boost in 1974 with the launching of the nuclear ship *Mutsu*.

This 8 346 tonne surface ship was built by the Japanese Nuclear Ship Development Agency at a cost of $133 million.[60] Built with local technology, it was intended as a showpiece of Japanese nuclear skill. However, at Mutsu Bay where the ship was being built, doubts were growing. The local fishermen began to realize that radioactivity from the ship might destroy the edibility of the scallops and fish which provide their livelihood. Their fears were

not simply abstractions, since fishermen at Tsuruga in 1971 had already found radioactive Cobalt-60 contaminating abnormally large trepangs, mussels, and oysters. The contamination had come from the nuclear reactor erected there in the previous year.[61]

By 1972 there was enough opposition to achieve a court ruling keeping the *Mutsu* firmly anchored in port. In August 1974 the ship tried to sail. It was no easy task. All day it was imprisoned by an encircling ring of 300 fishing boats lashed together. These boats dispersed only when the evening brought an impending typhoon. Once into the sea, the ship set its reactor into operation only to discover that it was seriously leaking radiation. The crew attempted to improvise shielding by packing the face of the reactor with socks filled with polythene and 34 kilograms of borated rice. Eventually they were forced to shut down the reactor. It was less easy to shut down the resultant storm of opposition. The ship drifted for fifty-one days before being allowed into port, and then only on condition that a new home be found within six months. Later efforts to move it raised protests and increased opposition wherever this was attempted.[62]

Opposition spread to all the communities where nuclear reactors were sited, residents using a variety of tactics to block new proposals. Legal suits highlighted the problems of many plants. One of the first such suits, launched in 1973 against the Ikata plant in Ehime Prefecture, revealed generic problems in the steam pipes of the pressurized water reactors, as well as the perfunctory nature of the government's safety inspection procedures.[63] Another technique, the planting of the Spiderwort flower, was used to reveal the effects of radiation leakages. A Japanese scientist, Dr Sadao Ichikawa, had discovered a variant of this flower whose stamens consist of long easily-visible purple cells. The variant, however, is a mutation and on exposure to radiation some of the new cells that grow revert to pink or colourless varieties.[64] The 'Spiderwort strategy' of planting these near nuclear reactors to visibly demonstrate the mutating effect of the emitted radiation, had a profound effect on local farmers.

In addition, prefectural administrations were approached to refuse permission for reactor construction, and fishermen and farmers formed themselves into cooperatives, refusing to sell their land or their fishing rights to the nuclear companies. This proved a particularly successful strategy as the central government has insufficient powers to force the surrender of these rights to the

STRATEGY AND STRUCTURE 305

nuclear companies. By 1980 these strategies had made it almost impossible for the nuclear companies to find suitable new land for their proposed projects. Equally importantly, the movement was by then showing real signs of national coordination.

August 1975 had seen the first national meeting against nuclear power, with representatives gathered together from most of the localities in which reactors were to be sited.[65] Trends in this direction continued over the following years, including the holding of a National Anti-Nuclear Power Week in 1977 which, among other things, raised 100 000 signatures against nuclear power.[66] In April 1979, with the events at Three Mile Island still making headlines, this momentum increased substantially. Already widespread local opposition was causing delays to the six plants under construction and to the eight firm proposals for future reactors.[67]

In response to the accident, representatives met in Tokyo on 5 April 1979 to form a National Anti-Nuclear Power Struggle Committee.[68] They included representatives from power sites in fifteen prefectures. They were joined by the Chugoku Electric Power Company Union which had called a strike in support of Anti-Nuclear Power Week, 1977. Residents groups from Osaka and Tokyo also attended. What the government feared most was that this new mood of coordinated opposition would cause further delays to the already delayed programme. Their fear was rapidly justified.

All pressurized water reactors were temporarily forced to suspend operation while they were checked, leaving only seven of the existing nineteen reactors operating. More importantly, the construction of several reactors was delayed or deferred. The scheduled ground-breaking ceremony for the Onagawa reactor was indefinitely postponed on request of the prefecture's governor. In Kagoshima prefecture the mayor asked the Kyushu Company to cancel an order for the proposed Sendai reactor. In Hidaka the township refused permission for a site survey for a reactor proposed by Kansai Power Company, and decided to hold a referendum on the issue. In Kyoto the governor said he would cancel permission for the construction of the Kumihama plant. And in other places the proposed reconstruction of the Noto, Shimane 2, and Hohoku plants were all deferred. On 22 April 1979 local elections at Hohoku resulted in an overwhelming victory for anti-nuclear candidates.

In addition, eighty citizens groups in Tokyo concerned with a

GLOBAL FISSION

wide range of issues joined together in a committee to stop nuclear power. On Whitsunday, 1979 there were demonstrations and rallies in Osaka, Tokyo, and Kyoto, as well as at ten reactor sites. Below the surface of this perhaps temporary surge of activity was a less visible, but possibly even more significant groundswell: the beginnings of convergence on the issues of nuclear power and nuclear weapons.

As early as 1975 the Association of Atom Bomb Victims had held a series of seminars entitled 'From Atomic Bombs to Atomic Power'. There was, however, a series of obstacles to the entire anti-nuclear movement becoming involved in the struggle against nuclear power. It lay in the historic split that had occurred within the peace movement. The original movement, the Japan Council Against Atomic and Hydrogen Bombs (Gensuikyo) was formed in the aftermath of the bombing of Hiroshima and Nagasaki.[69] It exerted great influence on conservative and radical politicians alike, and developed an enormous following among the people. However, during 1962–63 it split over the crucial question of whether to oppose all nuclear weapons in all countries or, as the Communist Party wanted it, whether to restrict its attention to the capitalist world. Finally, those who opposed nuclear weapons in all countries, including the USSR, decided to form a new organization, Gensuikin.

Gensuikin was not to be taken lightly. It had the support not only of many citizens but also the powerful Japanese Socialist Party, together with the unions affiliated with it. As early as 1969 it began to question the wisdom of the 'peaceful' nuclear energy programme. Gensuikyo, however, although moving to a position of unconditional opposition to all nuclear weapons, continued to believe in the 'peaceful atom'. By 1977 Gensuikin's by now adamant opposition to nuclear power could be seen in a rally against nuclear power in Mito city. It was sponsored jointly by Gensuikin, the Socialist Party and, perhaps most significantly of all, by SOHYO, the largest council of trade unions in Japan. The combined strength of the peace and anti-nuclear power movements can open avenues to participation by the labour movement where none was previously available. In this case, SOHYO covers key unions involved in the nuclear industry, and workers covered by it have struck several times over the hazards of nuclear power. Undoubtedly, Gensuikin's large and powerful organization helped

STRATEGY AND STRUCTURE

link together the many citizens groups around the country. However, this strengthening was only a shadow of what could be should Gensuikyo too come to oppose nuclear power.

From outside Gensuikyo there seemed excellent pressing reasons for a move in this direction. Not only was the international nuclear reactor industry providing an increased threat of nuclear war, but the nuclear fuel cycle in Japan seemed to be taking the country ominously closer to the construction of nuclear weapons. Despite US objections, an experimental waste reprocessing plant had already been constructed at Tokai and, despite serious technical problems, was beginning to operate. Another much larger plant is planned if a site can be found. Already, right-wing forces within the ruling Liberal Democratic Party have openly advocated the development of a Japanese nuclear weapons capability. There are formal obstacles in the way of such a course, such as Japan's signature of the Nuclear Non-Proliferation Treaty. However, only three months notice is needed to withdraw from it. As Prime Minister Ohira pointed out to the Diet in 1978, Japan can lawfully possess nuclear weapons under its present constitution, provided they are to be used for 'self-defence'.[70]

The leadership of Gensuikyo nevertheless still seemed unconvinced by these arguments. However in the aftermath of the accident at Three Mile Island, with Gensuikin demanding an end to the operation of all nuclear plants, Gensuikyo seemed to be feeling the concern within its own ranks. Soon after the accident it demanded that all reactors be shut down until their 'complete safety is guaranteed'.[71] This shift in its position was not a temporary aberration. Nine months later, in January 1980, an important Gensuikyo meeting concluded with a demand that the nuclear administration should be 'rechecked' and made democratic. 'Where necessary' nuclear reactors should be closed and the construction of new reactors should cease. Despite this official position, many of the 200 people who attended expressed the view that a more clear-cut stance against nuclear power should be taken. Accordingly, it was agreed to carry out a further review of the policy.[72]

The pressure for a policy of opposition to nuclear reactors increased considerably in the first half of 1981 when it became public that a major leak (of more than 40 tonnes of radioactive coolant) had occurred at the Tsuruga power station, contaminating some fifty-six workers.[73] The news became a major scandal when it

308   *GLOBAL FISSION*

was discovered that not only had this accident been covered up by
nuclear authorities, but that it was only one of a series of major
accidents, four of which occurred in 1981.[74] The political setback
for the nuclear industry was just one in a long series of events
which was contributing, seemingly inexorably, to the growth in
Japan of a much stronger and more integrated citizens opposition
to both nuclear weapons and nuclear power.

As links develop between the opposition to nuclear weapons and
to nuclear power in Europe and the rich countries of the Pacific rim
the importance and potential of further developments based on
these links is also growing. Nowhere is this better demonstrated
than within the Pacific itself. There, nuclear power, nuclear
weapons, and the land rights of indigenous people blend into one
current of controversy, rich in implications not only for the people
of the Pacific, but for the whole world. In the battle over the human
use of nuclear fission significant developments are occurring in the
interplay between the countries of the Pacific and its rim. It is here
that the nuclear-free zone strategy may first come of age.

TOWARDS A NUCLEAR-FREE PACIFIC

With its strategic and economic importance, its sparse populations
and huge area, the Pacific has become a major sphere of US
influence, a nuclear proving ground, a target for nuclear vendors,
an area dotted with nuclear ships, submarines, and military bases, a
dumping ground for nuclear wastes, and a potential theatre for
World War III. The USSR has nuclear armed ships in the Pacific,
but no bases. In strategic terms, the Pacific is a US stronghold. It is
also a crucial staging base for any all-out conflict with the USSR,
and a likely theatre for a 'limited' nuclear exchange. Its complement
of nuclear weapons is enormous.

Hawaii is the headquarters of the US CINPAC (Commander-in
Chief Pacific), whose field of operations covers 70 per cent of the
world's oceans, two-thirds of its population, and half of its surface
area. Hawaii has some 100 US military installations, including
Pearl Harbour, a home port for nuclear submarines.[75] Under
CINPAC command, most, if not all, of the ultimately ten Trident
submarines are expected to operate in the Pacific. To its 4 000
nuclear warheads must be added the 8 000 to 12 000 warheads
already estimated to be stored and deployed around the Pacific
Basin.[76] Taking the form of missiles, shells, and anti-submarine

STRATEGY AND STRUCTURE 309

rockets, they can be found on board more than 120 ships and submarines of the US Pacific fleet, on aircraft, and at US bases, especially those located in Guam, Subic Bay in the Philippines, Japan, and the islands of Micronesia, a United Nations trusteeship under US administration.

The Marshall Islands is one group of atolls within Micronesia which has already suffered heavily from nuclear activity. The area is a US Inter-Continental Ballistic Missiles (ICBM) proving ground and has been subjected to test explosions for several decades. Four of the atolls, Bikini, Eniwetok, Rongelap, and Utirik, are admitted by the USA to have been severely contaminated. However, representatives of the thirty atolls comprising the Marshalls say that on several others many people are developing growths 'the size of limes'. A high rate of cancers, miscarriages, mutations, and early loss of hair indicate the prevalence of radiation sickness.[77] Of Rongelap's eighty-two inhabitants, thirty-three have developed thyroid problems. On Bikini, whose inhabitants were evicted in 1946 to make way for nuclear explosions, an ERDA study has found well-water exceeding federal limits for Strontium-90 and a dangerously high radiation background.[78] The inhabitants were allowed to return in 1968 but ten years later, as the health risk became increasingly obvious, the US government announced plans to again move the islanders.[79]

For those forced to leave their homeland twice in their lifetime the implications go beyond any health risks they have been subjected to. In a report on the relocation to an international conference in 1978, Mike Malone, editor of the *Micronesia Independent*, quotes Tomae Juda, a Bikini Island magistrate, as asking 'How do you repay the people that have already died without ever seeing Bikini again? And how do you repay the young people who may never, never see Bikini?'. As Malone makes clear, there is no answer to these poignant questions, for 'In a small island society, such as Bikini, land is scarce. It is sacred. It is part of the identity of the people'.[80]

In neighbouring French Polynesia, the people have been subjected to equal damage to their lives, this time from French tests.[81] In 1964 the French foreign legionnaires occupied Moruroa and Fangataufa atolls, despite the protestations of the Tahitian Territorial Assembly, and construction of a nuclear testing site began in earnest. Since then the region has been subjected to the fallout from forty-one atmospheric nuclear explosions. In 1973 the largest

310 GLOBAL FISSION

political protest ever held in Tahiti, a rally of 5 000 people, was organized by the autonomist parties in opposition to the tests. The pressure mounted as the Australian and New Zealand labour governments obtained an injunction against the French government at the International Court of Justice. As has already been mentioned, boycotts by trade unions and public protest in the two countries finally persuaded the French government to move to underground testing. However, while this reduced the threat to health posed by the radioactive fallout, there was still the dislocation to the local culture caused by the disruptive presence of some 20 000 foreign troops.

The background of increasingly forceful international opposition to nuclear weapons and nuclear power reinforces two compatible trends in the Pacific. First, the experience of those Pacific islanders who have been victims of radioactive contamination has created a deep sensitivity to nuclear issues. Second, in a number of islands the success of independence movements in the late 1960s and 1970s, has heightened the confidence, ability, and determination of people in these islands to resist nuclear activities in their region. Where nuclear activities are so entrenched that outright obstruction seems impossible, the demands tend to be restricted to increased compensation. For example, in Kwajolein atoll, one of the Marshall Islands which the USA uses as a missile tracking base, the Kwajolein landowners occupied the island for over a month during August 1979. They demanded that the USA renegotiate the ninety-nine year lease that had been agreed to by the Trust Territories government in 1964. Under that agreement the USA had paid a lump sum of $750 000, amounting to only $10 per 0.4 hectares a year, for the use of the island. At the same time, the landowners of nearby Roi Namur atoll, comprising some sixty men, women, and children, occupied that atoll also seeking compensation. The occupiers only left after the Pentagon agreed to meet them and negotiate on their demands.[82]

On the other hand, in the Marianas islands, where the people face a Japanese proposal to dump low-level nuclear wastes in the Northern Pacific Ocean, the North Marianas Commonwealth Legislature has responded by declaring its islands and waters a nuclear-free area.[83]

In Hawaii also, there is an increasing sensitivity to nuclear issues, with the governor, major political parties, and some key unions supporting a ban on the transport of nuclear wastes through

STRATEGY AND STRUCTURE
311

civilian ports. In French Polynesia on 26 July 1979 a tidal wave which seems likely to have been caused by a nuclear bomb which exploded at the wrong depth in its test shaft, overturned cars and injured people at Moruroa. It also strengthened voices in the Polynesian Assembly now urging that the area be declared nuclear-free.[84] Important though these developments are, they only mark a trend. In Palau that trend has reached a markedly more advanced stage.

Palau, a chain of islands rich in tropical vegetation, covers a 167 kilometre stretch within Micronesia close to its most western border. With a population of 14 000 people, it had been left relatively untouched until the early days of President John F. Kennedy's administration. Then the USA annexed it as a commonwealth (with internal self-government but no control over foreign affairs or defence), and began to channel funds into the islands. The result was the creation of a fairly prosperous business community dependent on the USA for further development, and perhaps even the continued maintenance of the economy. The economic inputs were not purely philanthropic in nature, as the USA sees the islands as having significant strategic value. Indeed, as early as 1971, the US government was talking of using 16 hectares of submerged land in Palau, together with adjacent shore land, to establish 'a very small naval support facility'. Since then evidence has mounted that what is envisaged is a forward base for Trident submarines.[85]

In 1974 processes were begun to establish a consortium of oil, banking, industrial, and military interests to develop the islands into multi-million dollar centres for the storage of oil, and transshipment by supertanker.[86] The proposal attracted Japanese support because Japan has only two deep-water harbours. Palau has deep water harbours that would allow oil to be trans-shipped to smaller vessels and then taken to other harbours in Japan. The plan involved the creation of heavy industry, refineries, and petrochemical plants on the largest island, Babeldaob. Elsewhere, some 12 000 to 14 000 foreign workers would have to be housed, and the reefs dredged and channels blasted. Later stages of the plan call for the construction of nuclear plants. While some businessmen and politicians eagerly embraced the plan, others, including many younger people, the chiefs, and the owners of land, feared deeply for their environmental and cultural heritage.

Opposition to the superport proposal grew rapidly, adding to the growing movement for Palauan independence. At the same time

312   GLOBAL FISSION

the consortium's interest began to wane, and in 1978, with the withdrawal of any active interest from Iran (one of the plan's principal backers), the plan was dropped. While this removed any immediate threat of nuclear power in Palau, it did not end the islands' role in nuclear developments in the Pacific.

The USA's trusteeship of Micronesia was to expire in 1981, and at that time the region's self-sufficiency was to be negotiated. However, when a constitution for the whole region was drafted in 1976, Palau decided to negotiate separately for its own constitution. Those who helped draft its new constitution had already benefited from discussions with others in the Pacific about their nuclear experiences, and in particular had been deeply concerned by the contamination of the Marshall Islanders.[87] The result was that in April 1979, when the government adopted its proposed constitution, the draft contained a clause stating:

> Harmful substances such as nuclear, chemical, gas or biological weapons intended for use in warfare, nuclear power plants and waste materials therefrom, shall not be used, stored or disposed of within the territorial jurisdiction of Palau without the express approval of not less than three-fourths (3/4) of the votes cast in a referendum submitted on this specific question.[88]

In July 1979 this constitution received the overwhelming endorsement of 92 per cent of the population. By any reasonable assessment that should have finalized the matter. But any such expectations were soon confounded.

The USA was not at all happy with the constitution. Not only did it place a major obstruction in the way of its hopes for a nuclear submarine base, but it also rendered illegitimate the movement of US nuclear weapons anywhere within the territory of Palau (extending 322 kilometres out from the 241 islands in the Palauan chain). Accordingly, the USA began to exert considerable pressure to have the constitution reconsidered and the clause removed. The legislature succumbed and withdrew the proposed constitution. However, at the following elections for the legislature in September, with one exception only candidates who supported the constitution were elected. When the amended constitution, with the nuclear-free clause deleted, was put before the people at another referendum in October it was decisively rejected by 70 per cent of the vote. In July 1980 the people of Palau once again were forced to

vote on a referendum, this time on the original nuclear-free constitution. They supported it with an overwhelming 78 per cent of the vote.[89]

Six months later, on 1 January 1981, Palau became an independent nation, the Republic of Belau, and the first country in the world to possess a constitution explicitly forbidding nuclear weapons and nuclear power within its borders. But even now, its nuclear-free clause was not secure. Despite the clearly expressed wishes of the people on three separate occasions, the US government had not relented in its quest to thwart them. Even as the country began its first year of independence, the US government was negotiating with Belau, seeking the adoption of a treaty to be known as the Free Association Compact. Under this treaty, the constitution's nuclear-free clause would be suspended in exchange for the provision of financial aid over the next fifteen years. However, as such a treaty would require a 75 per cent majority of the vote at a referendum, there seemed little prospect of this latest US manoeuvre succeeding. Instead, it seemed likely that the Palauan people would keep the door firmly closed on US nuclear ambitions in their area. In doing so they would set a precedent which would add considerably to the mounting opposition to nuclear weapons and nuclear power in many other parts of the Pacific.

While the Palauan people may well be able to hold their ground, others in the Pacific and its rim may find the battle for similar objectives considerably more difficult. This is especially so in countries where nuclear activities are already well established, and in which the governments are particularly tractable to outside pressure. For people in this predicament, valuable support might be available from the anti-nuclear movements in the rich countries of the Pacific rim. Already, exploration of this possibility has begun to produce positive results.

SUPPORT FROM THE RIM

The Pacific island nations generally do not have sufficient electricity requirements to even remotely justify the use of generating systems as big as nuclear reactors. By contrast, some of the underdeveloped countries within the Pacific rim have in the past provided a tempting target for the nuclear reactor vendors. In particular, some South-East Asian countries have been considered as major potential reactor purchasers.

314   *GLOBAL FISSION*

More recently, nuclear industry estimates have shown a growing realization that there is little potential for sales in most of these countries. Even Thailand, which in the early 1970s expressed intentions for a sizeable nuclear programme, has, with the help of a pamphlet about radiation leaks at the small research reactor at Kasetsart University (released by graduate students),[90] now firmly cancelled its nuclear plans.[91]

Only South Korea and Taiwan still maintain substantial nuclear programmes with targets of as many as forty-four and twenty-four nuclear reactors respectively operating by the end of the century.[92] These targets should be treated with some scepticism even though there is little visible evidence of opposition within these countries. Iran also had a target of some twenty-three nuclear reactors in 1978, but the subsequent political turmoil reduced this programme, at least for the foreseeable future, to zero.[93] Similarly, South Korea is not the most internally stable country. In 1978 its official plan was for over 20 per cent of its electricity to be provided by nuclear power by 1986.[94] However, its 1980 plan had reduced that to 8 per cent or 8 GWe.[95] Industry estimates suggest it is unlikely to reach this figure before 1990.[96]

As well as potential internal problems, the international links within the nuclear opposition are now making it possible for additional obstacles to be placed in front of these programmes, even though overt opposition in these countries is often dangerous. In the USA a group has already mounted a campaign against exports of reactors to Taiwan on the grounds that the first six nuclear reactors are destined for seismically active sites with serious geological faults.[97] Under US legislation, federal agencies must assess the environmental impact of US activities overseas.

Such legislation has already been invoked in relation to a planned nuclear reactor at Bataan in the Philippines. That plant was the first of an ambitious programme of eleven reactors whose output of more than 6 GWe would add to a massive hydroelectric project to give some 16 GWe by the year 2000.[98] There were strong reasons to oppose the plant.[99] First, the programme of which it is part is destined to be of little value to the majority of the population, 70 per cent of which is rural. About 90 per cent of the country's electricity is used in the urban-industrial enclaves.[1] Clearly the increased supplies were intended to power the factories of large multinational corporations which had taken advantage of lower than subsistence wage rates in the Bataan Free Export Zone to

STRATEGY AND STRUCTURE

employ a small part of the many impoverished unemployed. With 8 million children malnourished, massive unemployment, and 1.5 million additional mouths to feed each year, the ploughing of three times the government's entire agricultural investment in 1975 into one nuclear plant, rather than into the labour-intensive agricultural sector, seemed outrageous to many people.[2]

Second, a 1977 study had reported that the plant is of an inferior and defective design.[3] Its site regularly experiences earth tremors, and a major fault line runs through the region. It is also periodically struck by tidal waves, the last in 1971. There are therefore good reasons for opposition. It began as early as 1977 from local Catholic priests and students, and from farmers and fishermen.[4] Soon the fishermen were to find their livelihood threatened as erosion from the site works drastically decreased their catches of the local 'milkfish fingerlings'. A petition opposing the plant attracted the signatures of more than half the local people. However the opposition was operating under the constant threat of President Marcos's martial law administration. One leading activist, Ernesto Nazareno, was tortured and later 'disappeared' after reporting to the police as ordered.[5] Another was shot dead.[6]

The high level of repression characteristic of the Marcos regime is also a sign of weakness. Although not operating near Bataan, there are officially estimated to be 10 000 Muslim guerillas, and 3 000 to 4 000 armed men and women of the rapidly-growing New Peoples Army in action in the Philippines. With the additional increasing undercurrent of opposition, both among Catholic parish priests and more generally, Marcos too has his legitimation problems.[7] After the accident at Three Mile Island, Marcos thus found it necessary to suspend construction of the plant at Bataan while he held a (closed) inquiry into its safety.[8] Nevertheless, six months later construction of the non-nuclear parts was resumed.[9]

The dangers and difficulties of fighting the plant within the Philippines have been considerably offset by aid from the nuclear opposition in the USA. In 1978 Congressman Clarence Long, incensed partly by scandals associated with the ExIm Bank loan for the project, held hearings on the reactor. These revealed not only the hazards, but also the uneconomic nature of the project. He later proposed that the bank be banned from financing the sale. Later, after much lobbying and vacillation, the US State Department finally approved Westinghouse's export of the stainless steel reactor vessel for the plant.[10] However, the battle to stop the plant

was far from over. As late as August 1980 Westinghouse was girding itself to fight a court battle with the Natural Resource Defence Council and the Centre for Development Policy (CDP) over a charge that it was illegal for the reactor to be exported without taking account of its dangerous location. As the director of the CDP, said 'We've come to the conclusion that what we're doing is exporting the public interest movement, just as the technology has been exported'.[11]

Whether or not the Bataan reactor would ever be commissioned, the pressure that had been created around it appeared to have put an end to the more ambitious plans for a nuclear programme. As one industry analyst was reported to have put it, it seems likely to be the only plant that will be built in the Philippines 'for many years'.[12] Meanwhile, the events that surrounded it showed the potential strength of a combination of local opposition in an underdeveloped country, and support from the opposition in an industrialized nation on the rim. It is not the only example. Elsewhere, the strength of that combination has been demonstrated with even greater force.

In June 1978, President Carter announced that the USA, in conjunction with Japan, was considering using a Pacific island as a radioactive-waste storage dump. It was to be a regional 'fuel centre' complete with fuel reprocessing plant, an enrichment plant, and facilities to store plutonium and other high level waste. This ran counter to Japanese desire to have their own facilities under autonomous control, and the idea failed.[13] However, in February 1979 the idea reappeared in more modest form. Now the USA proposed to buy a Pacific island on which they would store the highly radioactive spent fuel rods. Likely candidates were Palmyra, Wake, or Midway.[14] There was immediately considerable opposition from the Pacific island nations. But the effectiveness of that opposition was enhanced when it met with the support of two industrialized countries on the rim.

In Australia and New Zealand there was sensitivity to the reaction the proposal could create in their own countries as well as to the reaction already developing in the Pacific island nations. New Zealand's conservative prime minister, Robert Muldoon, condemned the proposal outright, noting that the material could be stolen and turned into weapons. Australia, too, at the South Pacific Forum in June, endorsed a resolution condemning the proposal.[15] This reaction by two staunchly pro-USA national leaders, coupled

STRATEGY AND STRUCTURE 317

with the opposition of others at the forum, including the premier of the Cook Islands and the governor of Hawaii, represented a major setback for the proposal. By March 1980 a bill had become law which prohibited the US administration from initiating such a waste storage centre without explicit authorization from congress. According to sponsors of the bill, although the administration could continue with its feasibility study, the bill had effectively prevented the administration from implementing a Pacific island waste dump.[16]

GLOBAL COORDINATION

The growth of mutually supporting interaction between movements opposing the spread of nuclear weapons and reactors in the Pacific and its rim, constitutes just one important example of a similar process occurring world-wide. This development has the potential to greatly strengthen the effectiveness of the opposition. But although links have developed between different groups, forming an informal spiderweb of communication channels across the world, there is still little facility within the network to either strategically plan or coordinate action at the global level. Nevertheless, over the last few years of the 1970s, experimentation with developing such a facility began to take place.

In the Pacific, the first important development was the Conference for a Nuclear-Free Pacific. Invited to convene in Suva in 1975 by the ATOM (Against Testing on Mururoa) Committee, a Fijian group which had been organizing against the French tests, it proved remarkably successful. Despite the problems involved in gathering delegates together from such a vast area, eighty-eight people from movements in twenty Pacific and two European countries attended. The result was a Peoples' Treaty for a Nuclear-Free Pacific Zone, drafted by the conference and finalized by a Pacific People's Action Front set up by the conference to help implement its decisions.

The idea of a nuclear-free zone in the Pacific did not seem unrealizable. Already, a zone banning the entry of nuclear weapons had been set up in Latin America under the Treaty of Tlatelolco. Others had been recommended by the United Nations for Africa, the Middle East, and South Asia. Additionally, both the New Zealand and Australian labour governments had spoken approvingly of the concept for their region, and there were even sugges-

tions that the US State Department might support the idea.[17] However, when the New Zealand government circulated a draft treaty for such a zone in 1975, the USA fiercely attacked the concept. Under pressure from the USA, the Australian government withdrew,[18] and although a resolution calling for such a zone was subsequently put before the United Nations by the governments of New Zealand, Fiji, and Papua New Guinea, and was carried without open opposition, the USA's counter-pressure ensured that no subsequent action resulted.

While the USA remains unenthusiastic about the creation of such a zone it can only be regarded as a distant goal. But it is one still worth fighting for. That that fight would continue was made clear in October 1978 in Ponape (an island of Micronesia), when a second Nuclear-Free Pacific Conference convened and adopted the draft Peoples' Treaty. Its joint clauses, prohibiting nuclear weapons, nuclear reactors (apart from small research units), and nuclear fuels and wastes, seemed certain to gain little sympathy, at least initially, from the governments of the rich countries on the Pacific rim. But it was already gaining attention and sympathy from the governments of several Pacific island nations.

By the time the third Nuclear-Free Pacific Conference was called, in Hawaii in May 1980, this time with the backing of Japan's powerful Gensuikin, Palau had already adopted its nuclear-free constitution. To assist the growth of the movement, the conference now set up an ongoing communication centre – the Pacific Concerns Resource Centre. Signifying the growth of support for the idea among people in the region, the proposed zone was extended to include Japan, the Philippines, and Hawaii. Two months later, the government of the newly independent country of Vanuatu (formerly the New Hebrides) took office with a policy firmly endorsing the principles underlying the Peoples' Treaty.[19]

In September 1980 a Nuclear-Free Pacific Forum was convened in Sydney, Australia, to publicize the developments to date. It too endorsed the treaty's principles and set in motion a series of supporting activities. This time, not only was the Japanese group Gensuikin present in force, but so too was Gensuikyo, the other major Japanese peace organization which had previously expressed its opposition only to nuclear weapons. Now, for the first time, it strongly and unambiguously opposed an aspect of the nuclear fuel cycle: the proposed dumping of nuclear wastes in the Pacific by the Japanese government.[20] If that progress towards a position of

STRATEGY AND STRUCTURE 319

opposition to nuclear power could be maintained, then the effect could be a dramatic strengthening of the Japanese movement.

Accompanying this chain of important developments has been the holding of another international gathering of perhaps even greater significance. From discussions between some trade unionists at the third Nuclear-Free Pacific Conference, came first a preparatory meeting in Fiji in November 1980, and then invitations to trade unions in Pacific and Pacific rim countries to attend an inaugural Pacific Trade Union Conference. It would discuss actions that unions might take to increase pressure to make the Pacific nuclear-free.[21]

The Pacific Trade Union Conference was held in Vanuatu in May 1981.[22] Not only was it attended by 128 delegates from unions in thirteen countries, but it achieved a backing that would have seemed almost inconceivable even a few years earlier. It was officially endorsed by the national trade union bodies in Australia (the ACTU) and New Zealand (the New Zealand Federation of Labour), and by one of the national federations of labour in Japan (SOHYO). Each of these sent senior officials of their organizations to participate. Other trade unions sent delegates from Fiji, Guam, Hawaii, Kiribati, New Caledonia, Belau, Papua New Guinea, the Solomon Islands, Tahiti, and Vanuatu. The result was a unanimous declaration of their 'determination to campaign for a Nuclear-Free Pacific' including opposition to the 'testing and storage of nuclear weapons, the dumping of nuclear waste, uranium mining, proliferation of nuclear reactors, the presence of nuclear vessels and nuclear military bases, and the transport and storage of uranium, uranium waste, and nuclear weapons'.[23]

The conference also initiated the setting up of a Pacific Trade Union Forum, 'to coordinate an intensive trade union campaign for a Nuclear Free Pacific', and to develop cooperation among unions and workers in the Pacific region. In addition, the conference adopted an initial programme of actions,[24] and agreed to meet again soon in a second conference to discuss progress and possible further action.

The success of these conferences within the Pacific region is both a symptom and a cause of the developing interconnection and strength of the international movement. But this is not restricted to the Pacific. As early as 1977, Europe also experienced its first international anti-nuclear gathering. It was the Salzburg Conference for a Non-Nuclear Future, which was attended by citizens from over

twenty countries in Europe, North America, and the Pacific. Salzburg ushered in a new stage of development for the movement. On the one hand it was the platform for a major announcement by US nuclear proliferation specialist Paul Leventhal. His revelation, subsequently confirmed by others, was that in November 1968 a West German ship carrying 200 tonnes of uranium ore had 'disappeared' at sea.[25] It had later reappeared with a new name and owner, but no uranium cargo. That had been landed at Haifa, Israel, for transformation into weapons-grade plutonium in the Demona 'experimental' reactor. It was already believed to have produced enough plutonium for more than fifteen nuclear bombs. Leventhal's announcement received media attention world-wide and helped crush the credibility of 'safeguards'.

The Salzburg conference was thus an international platform. On the other hand it was a major forum for examining the development of the international battle over nuclear power. By the time the conference was over it was clear that the diverse movements from many countries had not only developed many shared objectives but had moved a significant distance towards achieving them.

The Salzburg style of meeting was a creative experiment which could well be profitably repeated. But since then the international gatherings called in Europe have tended to focus more sharply on developing particular actions. One such, of considerable importance, occurred a year after Salzburg in June 1978. Convened by sixty anti-nuclear organizations in Switzerland, it issued a call for an international demonstration against nuclear power. It was held on Whitsunday 1979. The span and strength of that demonstration, and that which followed it a year later, made one thing clear. Despite its amorphous form, not only the transmission of up to date information, but the generation of common action on an international scale is now possible within the diverse network of groups, organizations, and individual citizens comprising the movement for a non-nuclear world.

Over the years leading up to the 1980s, the development of this movement has been staggering. It has spread across the world, blocking not only the individual plans of immense nuclear corporations, but also the entire nuclear programmes of national governments. Across the world millions of people have said 'no' to the major physical and social transformation embodied in the nuclear enterprise, and with great effect have raised the vision of an alternative future based on different technological and social priorities.

It is too early in the battle over nuclear power to say what future will ultimately emerge. What can be said is that the form of the future is not yet irrevocably committed. As the movement grows in strength, so too do the efforts of the proponents of nuclear power to contain it. The outcome of the battle lies in the future. But perhaps, even now, its significance may begin to be judged.

13

BEYOND NUCLEAR POWER

Each new power won *by* man is a power *over* man as well . . . Man's conquest of nature, if the dreams of some scientific planners are realised, means the rule of a few hundreds of men over billions upon billions of men.

C. S. Lewis, *The Abolition of Man*, 1947

THE NATURE OF THE BATTLE

From the dropping of the bomb on Hiroshima to the beginning of the 1980s, major technological developments have gone hand in hand with far-reaching social and political transformations. To a large extent these have been based on the widespread availability of cheap energy in the advanced industrialized countries, and in particular on extraordinarily cheap oil. As new physical and political constraints move us out of this era, the choice of the path to be followed into the future is ready to be made. For the sort of society and future that we will experience, that choice will be profoundly significant.

During the 1970s one immense technological programme representing one possible form of the future, began to be deployed. Global in scale and staggering in its proposed proportions, the nuclear industry began to develop. But for the communities in which nuclear facilities began to be constructed, the novel physical hazards inherent in their operation ignited a controversy, and then a conflict, which has flared ever more intensely.

In those communities which the nuclear controversy has touched, it has taken away the easy acceptance of the present and raised difficult questions about the future. The nuclear enterprise's own transnational development has created an international

BEYOND NUCLEAR POWER

citizens response. While only loosely integrated through the communication channels of a network, that response is unified by the questions raised by nuclear technology itself. Although different communities begin by being concerned over very different questions, the objectives of different groups steadily begin to merge. Out of this begins to arise common action.

The result is the international battle over nuclear power. Its background is the world's political and economic organization which since World War II has been characterized by the steady growth of the increasingly centralized power of a few massive corporations and the state apparatus of most nations. The capital intensive, technically sophisticated, and centralized form of nuclear technology gives it the ability to play an important role in the formation of a future in which these trends are continued. By opposing that technology the movement against nuclear power raises the possibility of a different future.

What will the future be? It is a question with an assortment of possible answers. But in the industrialized countries during the last three decades, probably the most intense and widespread international conflict over which of the answers will become reality has been the battle over nuclear power.

In this battle, the power of the side which supports the nuclear industry may at first sight appear overwhelming. On that side in each country stands not only the industry itself, but the segment of the population which supports it, including often prestigious scientists, professional associations, and media people, as well as major political parties, large corporations, and usually key sections of the state to which the industry is intimately linked. On the other side is the diffusely organized network of diverse interlinked groups and organizations which form the movement against nuclear power. Especially at the beginning, the opposition which confronts the powerful forces supporting the nuclear enterprise is often made up of isolated communities and characterized by a low level of strategic planning and organization, and a lack of understanding of the forces against which it is pitting its strength.

The vulnerability of such groups is revealed by several documented instances of anti-nuclear groups being infiltrated by police or industry undercover agents – for example the Abalone Alliance in California, various community groups in Maryland, and the Clamshell Alliance in New Hampshire.[1] Yet from the beginning, despite failures to take elementary precautions or to plan ahead, in

many engagements the opposition has proved the stronger. The explanation does not lie in some hidden tactical or organizational superiority. It relates to substantial weaknesses in what is superficially the invincible combination of forces which it confronts.

Of these weaknesses, the most fundamental is also the most simple. It is that the state and the nuclear corporations are social structures. Their operation relies ultimately not only on the consent, but also the active participation of large sections of the community. And that participation depends on the legitimacy which the community attaches to their operation.

Whether it is the state as in the communist countries, or the corporations and the state as in the capitalist countries, the legitimacy of these organizations relies on their ability to provide an optimum distribution of resources and protect the health of the community. Yet in both social systems the nuclear enterprise is seen by many to accomplish quite the opposite. In this sense it challenges the very foundations on which the legitimacy, and therefore the power, of the dominant authority is based.

The battle over nuclear power thus becomes a struggle for the greatest legitimacy. In this a variety of tactics may be attempted, but there is no guarantee that any one will prove successful. The generally disastrous attempts by numerous governments to increase the legitimacy attached to their nuclear programmes by means of information campaigns, attest to this.

Contestants may pay too little attention to the effect their strategies will have on their legitimacy. Thus in West Germany at Brokdorf, when demonstrators attempted to force their way through police lines with the apparent intention of confronting the power of the state, the state did not lose. Similarly, but at the other extreme, in the USA a seemingly unending debate over whether cutting the fence surrounding the Seabrook reactor would constitute 'violence' had a paralysing effect on the movement. Framed in this abstract form the question is probably unanswerable, and the debate over it is extremely difficult to resolve. This is because for the industry, the state, and the opposition, the implications of an action depend not only on its form (the cutting of the wire), but also on its history and the social context in which it is carried out.

What then determines whether in one context an action will seem legitimate while in another it will be greeted with hostility by the community? There is no easy formula. Nevertheless, it is reasonable to say that an action which lies outside the usually

BEYOND NUCLEAR POWER

accepted rules of conduct in a society is more likely to be accepted as legitimate if: (a) it is clearly seen to be directed at preserving something which the community strongly believes society should guard, for example, life or property; (b) any available normally accepted channels have been explored and clearly demonstrated to be valueless, either because they ultimately will not work or because there is insufficient time for them to be used, and; (c) the hazard being opposed is clearly seen as more serious than any side effects of the action.

If these conditions are not fulfilled and the demonstrators are able to be styled as insincere or motivated merely by external political considerations, then violent measures by the state to stop them may seem perfectly legitimate. However unfairly, that was the outcome at Brokdorf. On the other hand, if a nuclear reactor is widely seen as being a serious threat to the life and property of the community, and there seem to be no other available channels for stopping its construction, then the infringement of the nuclear company's property rights by cutting its fences and occupying its land may seem completely legitimate, and a repressive response by the state may endanger its authority. That was the case at Wyhl.

Because of the local character of much of the opposition, the broad cross-section of the community which has quickly become involved, and a natural tendency to first exhaust the officially available channels, the opposition has often been accorded considerable legitimacy. Nevertheless, it cannot be said that it has been quick to make use of all its opportunities. In many places it has been slow to counter the nuclear proponents' campaign that the opposition is 'anti-technology' and that their approach is a prescription for economic crisis and 'freezing in the dark'. The opposition has also often been slow to relate the problems with nuclear power to other burning issues of the day. For example, in many places it has not drawn much attention to the intensifying attack on the right of workers to strike, increasing disparities of wealth within the community, and the labour-displacing effects of energy intensive industry which are all another side of the trend of which the nuclear enterprise is part. By raising these, the legitimacy of the opposition might be greatly increased among the labour movement.

But the actions of both sides are limited by the need to enhance their legitimacy, and the nuclear industry and the state have often found themselves markedly on the defensive in the battle over nuclear power. At each turn the unfolding saga of new hazards and

unsolved problems continues to undermine their position. By the beginning of the 1980s the result of this was manifested in the internationally stagnated state of the nuclear industry.

In some countries, such as the USA, the reason for the staggering slump in the nuclear industry appears at first sight to be its economics. Closer examination reveals the sensitivity of those economics to decisions by the state. Thus even here, the fundamental cause of the industry's problems lies in the success of the opposition in limiting state initiatives which could restore the nuclear industry to a more favourable economic condition. In other countries, most notably those which have been forced to national referenda over the issue, or in which court cases have effectively paralysed the nuclear programmes, the political conflict over nuclear power, and the remarkable success of the anti-nuclear movement to date, is more overtly visible.

In the final analysis, major power centres analogous to those embodied in the nuclear enterprise often lose not when they are confronted by superior force, but when they collapse seemingly of their own accord through loss of confidence or will, and with whole sections of their alliance moving over to the other side. In the second half of the 1970s that was at times the demoralizing prospect faced by the nuclear industry in a number of countries. Whether the industry survives in the next two decades will depend on the degree to which that process can be further developed by the opposition. But the survival of the nuclear industry is by no means all that is at stake.

On the one hand the nuclear enterprise has given rise to a citizens response which has proved remarkably effective in challenging the corporate state's legitimacy. But on the other, by its vulnerable and centralized nature it legitimates greater centralization of state authority. In its most extreme form this leads to the 'nuclear state' resulting from the severe security measures required to protect a plutonium-based energy economy.

More generally, nuclear technology by its nature can act as a tool to intensify trends towards centralization of power in the world's social organization. It is not necessary to suppose that those who frame nuclear programmes are consciously motivated by this general characteristic. But specific aspects of it, such as the role that nuclear technology can play in advancing a country's terms of trade, or the advantage of being the producer of a product which few can duplicate and on which whole nations could come to

BEYOND NUCLEAR POWER 327

depend, undoubtedly play a role in the strategy of both the large corporations and the agencies of the state.

While the overall effect of the enterprise may thus be to aggravate centralization of power, the attractions perceived by the industry may merely derive from this general property. Similarly, the concern over the physical hazards of nuclear power held by its opponents may have beneath it a more general, not yet necessarily consciously recognized, concern. That is, a perception of the characteristic which underlies many of the attractions of nuclear power for the other side: the centralization of power inherent in the technology.

FROM PHYSICAL HAZARDS TO POLITICAL CONCERN

Many citizens groups or individuals who oppose nuclear power do not place much emphasis in their campaign literature, or in their discussions, on the political implications of the nuclear enterprise. Especially initially, but also often much later, the focus of concern is the range of physical hazards which accompany the technology. It may even be directed more finely upon just one or two of these hazards. But over time these concerns spread and the focus broadens. A group first concerned mainly with problems posed by radioactive waste becomes also concerned about the danger of a loss of cooling accident. This process of 'diffusion' of different concerns is stimulated by the exchange of information through the network. But it also comes from a process of exploration. People beginning with one concern start to examine its implications. With richer and more sophisticated understanding come new concerns. Eventually, almost inevitably, this leads to a concern over the political relations inherent in the nuclear industry. It is worth pausing to see how.

If all the different concerns over the physical hazards of nuclear power were distilled into one succinct statement, it might be this: that it is a technology whose safety people deeply distrust. But that distrust must also apply to something else. People must have come not only to distrust the safety of the technology but also the authority of those who have assured them so confidently that nuclear power is safe. In this sense people distrust the entire nuclear enterprise − not only its technology, but the public and private organizations, the political parties, and those often prestigious

scientists who advocate and assist in the development of nuclear power.

For some who may have been unflinchingly conservative in the past, this questioning of authority is a major adventurous step opening a whole new world of previously inaccessible thought and discussion. Inevitably it leads to an analysis of the alternatives. Here it is tempting, having rejected the authority of the nuclear proponents, to nevertheless accept the claimed authority of the market place. This can lead opponents of nuclear power into complex and seemingly unending debates with its proponents over the comparative economics of different energy paths.

Debates within this framework are, however, usually inadequate. Amory Lovins and others have demonstrated that a convincing case can be levelled against nuclear power based on its present costs compared with those of alternative technologies, including solar systems and energy conservation measures.[2] But, as was shown in chapter 4, these comparative economics are themselves fundamentally determined by decisions in the political arena. Changes in government policy over what constitutes a 'safe' nuclear reactor may increase the cost of electricity generated by a new reactor to twice its present level, or alternatively may halve it. Changes in government priorities on energy research and development may similarly dramatically change the costs of energy produced by the alternatives. No sooner may a set of calculations have been carried out than the economic 'facts' on which it is based change through alterations to government or industry policy. In this situation, the choice ceases to be a technical one based on absolute technical criteria, but a social one based on society's preferences. Those preferences depend in the final analysis not on the number of kilowatts produced for each dollar, but on the quality of life that a given energy path may help foster.

There is even less basis for seeking to formulate quantitative measures for 'quality of life' than there is in comparing energy paths within a strictly economic framework. Quality of life is not resources consumed, energy generated, nor Gross National Product per capita. It is a part conscious, part sub-conscious judgement made by each person on the basis of the inevitably incomplete information to which we have access. That judgement may differ markedly depending on how each of us sees the world. It will depend on the framework through which we reject what we consider unimportant and place emphasis on what is important for the

BEYOND NUCLEAR POWER 329

future of ourselves, our children, and others. For those who have come to distrust the nuclear enterprise there is at least one likely common feature in the quality they seek for their future. It is that of a future in which the technology and institutions must be what the nuclear enterprise is not: a future in which technology takes such a form that it, and those who operate it, *can be trusted*.

Just as the distrust of the nuclear enterprise summarizes a whole range of concerns about nuclear power, so the requirement that the future energy path should be one that can be trusted also implies certain conditions about the form and management of the energy technology involved. Only a little exploration is needed to see how quickly, in a search for a better energy path, the boundary between technical and social aspects becomes so blurred that the exercise becomes a broader search for understanding.

EXPLORING THE IMPLICATIONS

There are many starting points from which we might seek to isolate some of the properties that an energy programme would need to possess for people to feel able to trust it. Let us pick out one and take a few steps along the train of argument that it can generate.

RESILIENCE AND DIVERSITY An energy programme based on one or two technical possibilities that may or may not prove too hazardous to use, or to use economically, leads to the prospect that unexpected sacrifices in safety or resources may have to be made to provide the energy on which society has become dependent. A more secure future might be provided by placing emphasis on resilience through the use of a variety of energy technologies. However, this is impossible if large, complex, capital-intensive technologies, requiring the major share of the energy generation to pay for their development, are to be used.

The requirement for resilience dictates the use of more diverse decentralized technology in energy programmes. The use of such technology is no impossible pipedream. There are by now an array of reports which attest to this.[3] One of these is the US Congress's Office of Technology Assessment's 1977 favourable report on the potential economics of solar technology in the USA. It points out that the form of energy technology developed as 'economic' depends very sensitively on policy choice. For example, it concludes:

There are few economies of scale in the onsite solar energy equipment examined in this study ... Within 10–15 years it may be possible to develop onsite solar devices capable of producing electricity at rates equivalent to the electricity produced by new utility (centralised) generating plant.[4]

Programmes that emphasize the use of decentralized technology provide an additional bonus. Both catastrophic and routine risks associated with energy programmes are often shared unequally across a community. Workers operating the energy systems suffer the worst of the routine risk, whether it be from radiation or from gas explosions in coal mines. The surrounding communities also suffer the most immediate and severe consequences of major accidents, for example from the fallout from a nuclear reactor core meltdown accident. The knowledge that one is suffering a substantially greater threat than the rest of the community is unlikely to inspire confidence that the energy programme is in one's best interests. But energy programmes based on decentralized technology spread the risk as evenly as possible across the community. They also allow each locality to exercise local judgement over the level and form of energy generation deemed acceptable.

The desire for greater diversity and resilience raises another condition for a more acceptable energy programme. Sometimes advocates of nuclear-based programmes will argue that the opponents are demanding the impossible: that nuclear power be proved absolutely safe before it can be acceptable. Such proof is of course impossible. But the charge is not an accurate explanation for people's opposition to nuclear power. It is not because nuclear technology has not been proven absolutely safe that people distrust it. Rather it is because it seems so unsafe.

Nuclear technology is distrusted because it is based on critically toxic materials which cannot be rendered safe chemically and which are capable of being transformed into the most devastating weapons; because these must be contained to a degree of precision which seems impractical to any person familiar with the errors, mechanical failures, and even corruption, which are a feature of the work place everywhere; and above all, because the technology has been carried from the experimental to the commercial scale in less than two decades. During this time many new hazards have been identified, and old ones have been found to be much more serious than early confident assurances suggested. Energy systems seen to be based on critically dangerous materials, still in many ways

BEYOND NUCLEAR POWER   331

experimental but incorporated into energy programmes as if they
were adequately understood, are not compatible with trust by the
community.

The implication is that where there are *prima facie* grounds for
challenging the safety of a proposed new energy technology, a sufficient experimental period must be allowed, and the deployment of
the technology must be sufficiently slow, to allow the community
to develop trust in its safety. This is incompatible with a high
energy growth programme which, in order to be sustained, locks
the society into dependence on a single, still largely experimental,
technology.

The need for resilience also requires emphasis on energy and
resource conservation. By doing more with less, the lifetime of
energy technologies which have proved acceptable may be
extended. This removes the pressure for an ever more frantic
development of new technologies whose performance and safety
are still highly controversial.

If decentralized, diversified, resilient technology, and a prudent
pace of net energy consumption growth, are among the conditions
for more acceptable energy programmes, then so too are the social
changes necessary to achieve them. This train of thought, starting
from the single criterion of the need for an energy programme
which can be trusted, thus inevitably raises questions of ownership,
participation, and control.

OWNERSHIP, PARTICIPATION, AND CONTROL
The physical conditions that may be generated in the above manner
extend well beyond those listed. But whether these relate to the
rate, direction, or form of the energy programme, there is always
the need for social judgements and strategies to achieve them. This
leads to a more fundamental set of social conditions. One that
stands out most clearly is that if people are going to be confident
about the long term direction of an energy programme they must
feel confident they will agree with the social and technical judgements underlying it. There is no reason for such confidence when
the critical decisions are made by a technical, political, or corporate
elite behind a screen of semi-secrecy or abstruse and mystifying
claims and language. On the contrary, an energy programme which
inspires confidence is one which has been guided through channels
which, to a maximum degree, are responsive to community wishes
and judgement. In short, a high degree of participatory democracy

332   GLOBAL FISSION

is fundamental to ensuring that an energy programme will be acceptable.

If people are to participate in decision making about energy options they must be able to evaluate the possibilities with confidence. This requires that they have free access not only to all available information, but also to the varying viewpoints about it. Communication channels must be available and open to all contenders. Access to the media must be made more equally available to all sides. In other words, the requirement of a participatory process invokes the need for the characteristics of an open society. This line of thought can be taken much further.

In the capitalist world the ownership of technologies is a major factor in constraining the degree to which people can participate in the decisions made about them. Where large corporations (such as Exxon or General Electric) have a virtual monopoly over a technology the political force with which it is pushed forward becomes greater while access to details of its construction and operation is diminished. This situation is most obvious with the large-scale 'mega-buck' technologies such as nuclear power plants, coal liquefaction plants, fusion plants, and solar satellites. These are precisely the forms of energy generation technology compatible with the needs of the largest corporations. Thus, as has been documented in considerable detail by R. Reece in his book *The Sun Betrayed*, these corporations worked extremely effectively during the 1970s to force US energy and research funding policy towards the support of large-scale, centralized solar technology, despite the by now well-documented possibility of equally efficient smaller-scale alternatives.[5]

The search for an acceptable energy programme has the potential to lead opponents of nuclear power to confront a most difficult social problem: the need to restructure social and economic power relations so as to eliminate the power of these corporations, at least in the energy field. At the very minimum it suggests that where large-scale energy technologies are chosen by the community their development and production must not be left under the control of a handful of immense corporations to be developed according to their profit–investment strategies.

But even the replacement of the present private ownership of energy production technology with public ownership through the state is an inadequate solution. Whether capitalist or socialist, a

highly centralized society and highly centralized technology are mutually compatible. In the USA, as in the USSR, this compatibility is equally well illustrated by the enthusiasm with which corporations and the state have pressed for increasingly centralized energy technology, parallel with their own increasing centralism. Thus the search for an energy programme which does not share the undesirable characteristics of the nuclear enterprise must come to grips with the need, not only for safer technological forms, for decentralization of energy technology, nor simply for public ownership of energy production, but most fundamentally of all for the decentralization of authority and control over energy policy and energy production. In the final analysis, nuclear power represents a supreme example of the inhuman potential of both late twentieth-century capitalism, and bureaucratic socialism. Only a sharing, participatory democracy, a self-managed society, stands truly in opposition to the world-wide trend towards centralized authority and control represented by the emergence of the nuclear enterprise.

The chain of argument followed here is merely an example of the way in which concern over one or another aspect of the nuclear enterprise leads to a questioning of the society that produces it. However, it would be totally false to suggest that, at least up to the beginning of the 1980s, the analysis given in this example would be held by even a fraction of the groups of opponents to nuclear power, let alone the vast numbers of people who have reached a position of opposition to a particular nuclear project, or to nuclear power in general. Nevertheless, there is evidence that, even if by many different sequences of steps, from many different starting points, the search for a better energy future is steadily leading in the same general direction.

THE POINT REACHED

What then is the anti-nuclear movement's perception of its own goals as it enters the decade of the 1980s? By now it is clear that in this form the question is not answerable. The 'movement' is too diverse in culture, composition, and class origins, too scattered and fragmented, and too amorphous to attribute to it a concrete position on anything except an opposition, for a wide variety of reasons, to nuclear power. Nevertheless, although there is no official platform that can be consulted, no representative assembly

that can be interrogated, and indeed no leaders who could validly claim to represent the developing understanding of this diverse movement, it is still possible to perceive indicative trends.

One of the first indications that the movement is already opposing not only the physical but also the political implications of the nuclear enterprise is the way in which it has almost instinctively organized itself. Throughout many citizens action groups can be found a deep aversion to any form of hierarchical organization, or centralized authority. There is much experimentation with democratic forms, and often strong opposition to traditional political tactics within the organization, such as 'getting the numbers' for a vote on policy. To avoid the 'tyranny of the majority' that can result, the Clamshell Alliance on the eastern seaboard of the USA, like many similar citizens groups, has for many years made its decisions by consensus, although this can be an extremely slow and difficult process. In the OOA in Denmark, decisions are made by a process which they call 'presentative democracy'. Those present at the organizational meetings make the decisions, and any member is entitled to be present.

It should not be imagined that any of these processes constitute the ideal formula for smooth decision making. Consensus processes can suffer from the tyranny of the small minority which refuses to yield over some deep-seated conviction, and totally open meetings may be stacked by a minority which, although active, holds very different views from the majority of those in the movement. But irrespective of the possible drawbacks, active experimentation in the search to achieve a non-hierarchical organizational form is a strong tendency throughout the length and breadth of the movement against nuclear power.

The aversion to centralization is revealed also in the network structure of the movement, both internationally and within individual countries. The structure is the bane of the nuclear industry and governments, since leaders are generally not available to be persuaded to the government viewpoint. It is almost incomprehensible to officers of hierarchical organizations that such a network could operate, let alone with such success. On the one hand, this can produce ludicrous attempts by industry analysts (such as the book *The Struggle for Power* by Peko-Wallsend's John Grover) to place a centralized structure on the movement artificially.[6] On the other hand, it can produce, as Carl Goldstein of the US Atomic Industrial Forum expressed it, 'apprehension and despair at the thought that

BEYOND NUCLEAR POWER 335

this sort of amorphous opposition has kept us off balance for the better part of ten years'.[7]

But although it has limitations the network does work. By maximizing individual autonomy and decision making for the component groups, and maximizing the flow of information between groups, a form of self management is produced which, even though groups come and go, allows the network as a whole to become stronger and more effective. The growth is a process of crystallization often not well understood even within the movement. It comes not so much as a consequence of design but as a reaction to the hierarchical social relations within the nuclear industry and the trends towards centralism which it represents.

The effectiveness of this network comes not only as a surprise to the nuclear industry, but often also to the opponents of nuclear power themselves. But as the success of the movement in retarding nuclear programmes in the 1970s begins to become clear, the apparently invincible power of the pro-nuclear forces begins to be de-mystified and the battle over nuclear power becomes recognized as one in which the outcome is far from decided.

Beyond the structure of the movement there is other evidence that nuclear opposition is a reaction not only to the physical, but also the political implications of the nuclear enterprise. In particular this is suggested by a broad trend that may be observed running across the world: the mutually supportive relationship between the opposition to nuclear power and regionalism. This can be seen in France (in Brittany, Normandy, and Strasbourg), in the UK (in Wales, Scotland, and the Orkney Islands), in Spain (in the Basque country and in Catalonia), in the Netherlands (in Groningen), and in West Germany and Switzerland (in Dreyeckland). Even in Yugoslavia it is probably no coincidence that the first citizens opposition should happen in the historic Croatian region in which Zadar is situated. Nor is it surprising to find nuclear opposition in French Polynesia coupled with the autonomist movement, and Belau and Vanuatu's declarations of independence accompanied by a statement that they are now nuclear-free countries.

In part this is because of the function these regions often fulfil. Because they are usually poor, relative to the rest of the country, they become the target for multinational corporations seeking to relocate their more labour-intensive mass production in areas with lower wage rates.[8] At the same time, the state depends least on these regions for political support. The result is a tendency to

'export pollution' from the centre to the periphery. Whether for nuclear or other similar projects, as this trend becomes more evident, alarm and then conflict develops around it.

Additionally, however, these regions retain both a strong sense of community, and their traditional values. They are thus most likely rapidly and clearly to perceive the threat that the nuclear enterprise poses to their quality of life. At the same time their sense of community makes possible a quick and often effective stand against it.

In those industrialized parts of the 'new world' which have attained self-government, such as Australia, the USA, and Canada, it is (with exceptions such as French Canada) generally more difficult to identify regions of historic regional identity. There are, however, two clearly identifiable groups which possess their own cultural identities, and who are rapidly developing their own movements for self-determination. These are the original native inhabitants of Australia, the Aboriginals, and the native North Americans, who now find themselves often desperately struggling to maintain some vestige of their traditions and land. It is often these very land areas which are the sites of major uranium deposits or other nuclear projects. In either case, in the struggle for self-determination by these groups, they once again find themselves pitted against the requirements of an industrialized, increasingly centralized economy. That the native peoples should become opponents of nuclear power is perhaps natural. But equally significant is the effect on the anti-nuclear movement which, by recognizing and taking up their demands for self-determination, further raises its own understanding of the power relations structured into the nuclear enterprise.

The structure of the movement as a network, the experimentation with novel participatory democratic forms, and the strong mutual compatibility of the movement against nuclear power and movements for regional autonomy or self-determination, are all strong indications that the movement against nuclear power, in its struggle to halt the spread of physical hazards posed by nuclear technology, has also begun to move against the trends towards centralized power which the nuclear enterprise represents.

While these characteristics help chart the overall direction of the seemingly random pattern of actions and counter actions which make up the battle over nuclear power, the same direction is more overtly evident in the goals and ideas expressed in the literature and

BEYOND NUCLEAR POWER

statements emanating from the movement. Over the last decade this literature has reflected a growing awareness of the political implications of nuclear power. Most of the early publications, including the ground-breaking works from FOE in the UK and elsewhere, and the pamphlets and newsletters, centred on the physical hazards presented by the technology. Later, about 1976, the argument began to include the analysis of the different possible energy paths. Undoubtedly the seminal work was a paper by Amory Lovins, 'Energy Strategy: The Road Not Taken', which was first published in *Foreign Affairs* in 1976.[9] Almost instantly it was widely reprinted and became the centre of a burgeoning controversy. Later published as a book, and subsequently backed by several other independent studies, it spelled out the physical and economic possibilities of a 'soft' energy path based on decentralized energy technology and energy conservation techniques which could be phased in without the use of nuclear technology in the USA.[10] Similar studies were carried out for many other countries from the UK, to Papua New Guinea, to Sweden, Denmark, and France.[11]

At the same time, other publications about the implications of nuclear power for social control began to appear. The study of the implications for the UK by M. Flood and R. Grove-White, which first appeared in October 1976, was undoubtedly extremely influential, and it was soon followed by a variety of other similar studies for countries ranging from Australia to the USA.[12] Over the last few years of the 1970s, the concept of an energy path based on decentralized technology became increasingly emphasized as the alternative necessary not only to counter the physical hazards posed by nuclear power, but also the sort of centralized power that it would help entrench. Thus, at the 1977 Salzburg Conference for a Non-Nuclear Future the declaration which was jointly agreed on not only stressed the physical hazards of the nuclear fuel cycle and the accompanying increased risk of nuclear weapons proliferation. It also noted:

Such a highly centralized and capital-intensive technology as nuclear power leads to a further concentration of economic and political power in all nations.

Compared to nuclear power, the technologies we propose are – for all nations – cheaper, more reliable, easier to deploy, less disruptive of existing social and cultural patterns, more

equitable, less damaging to the environment and more amenable to democratic control.[13]

As understanding has grown and spread within the movement it has begun to be expressed in some places in more recognizable political forms. Most obvious of these is the phenomenon of green parties. The Ecologists of France, the Green Party of West Germany, and the Citizen's Party in the USA, are all direct reflections of an explicit equating of the political and physical implications of the nuclear enterprise. It should be stressed that, at least at the national level, such parties face enormous obstacles in attaining success at the polls precisely because they are parties which support the decentralization of power. They are forced to fight within a highly centralized contest with policies on how to run a social structure to which they are, in essence, opposed. Nevertheless, as has been shown, at the local level of state or municipal elections, they have already proved they can be extremely effective.

While these parties are still confined to a few countries, and even in those cannot be said to represent a majority of those who oppose nuclear power, their growth and policies are indicative of the ideas developing in the rest of the movement. Dr Petra Kelly, elected chairperson of the Green Party of West Germany, which in March 1980 claimed to have over 1 million members, expressed those ideas this way:

> because plutonium is 'thalidomide for ever' and will make states into totalitarian systems, we rise up . . . and demand that the creativity and imagination of human beings be diverted from self-destruction and destruction of others to soft, renewable, decentralised energy systems and to an ecological society where one *is* and does not only *have*.[14]

In the global arena in the decade of the 1980s, the programmes of the nuclear enterprise represent one particularly forceful technical and social proposal which if implemented could powerfully affect the future of the world. But other available technologies, from the manipulation of genetic material to the construction of integrated circuits, may form the basis for industrial projects also pregnant with physical, social, and political implications. As these projects begin to be thrust forward and to spark a political response, the opposition movement to nuclear power is finding that it is not alone. It finds itself tentatively beginning to link, as a network, with

other movements starting to question the social and technological form of the future.

One movement with which it has much in common is the broader environmental movement which, although more diverse in its focus, has also grown to span the world. Another is the workers' movement for self-management of the technology of production, typified on the one hand by the green bans of the Builders' Labourers' Federation in Australia and on the other by the work-ins of the Strathclyde shipbuilders in Scotland, the watch-makers of the Lip factory in France, and the demands by the workers at Lucas Aerospace and Vickers, that their workers' plan for the production of more socially useful technology be imple-mented.[15] That these are no longer separate movements was symbolized neatly in September 1980, when the Green Party organized an Alternative Production Congress in Essen, West Germany. Invited participants included representatives of the Lucas Aerospace Combine Shop Stewards Committee, the Irish Transport and General Workers Union, shop stewards from the German armaments industry, as well as members of the Green Party.[16]

The focus of this workshop, ranging from alternative energy technology to the conversion of weapons manufacture to socially productive purposes, emphasizes the difficult and dangerous cross-roads that the world is approaching in the 1980s. The hazards of nuclear power, the threat of nuclear war, and the implications of increasing centralization of social control throw a shadow before us. The nuclear enterprise throws that shadow into stark relief. But is has also begun to produce its opposite. An opposition has sprung up across the world. Through its concern over the implications of nuclear power it has begun to demand access to the obscured world of science, to create a new sense of community, and to raise the possibility of a different path to an alternative future.

We have come to a point where it is possible to ask, what is the significance of the phenomenal growth of opposition to nuclear power? For the nuclear industry the answer is clear enough, encoded in the cancellations of orders and stagnated nuclear programmes which now beset it. Outside the industry there is an additional significance which extends beyond nuclear power. For the movement there is the slow realization of its achievements to date, the de-mystification of the apparently insuperable power of the state and the corporations, and the understanding that the

340   GLOBAL FISSION

future is not pre-determined, but ultimately lies in the hands of all of us. But even beyond these vital implications, and beyond the present realization of many who have come to oppose nuclear power, there is something else.

In 1932 Aldous Huxley wrote *Brave New World*, his classic vision of the future towards which he believed society was heading.[17] Seventeen years later George Orwell wrote his corresponding classic, *Nineteen Eighty-Four*.[18] While these books described two apparently very different futures, they possessed one common key feature. Each foresaw a future for both the capitalist and socialist countries dominated by overwhelming centralized power. In *Nineteen Eighty-Four* that power was expressed through the totalitarian control measures available to a technologically advanced police state. In *Brave New World* equally authoritarian control was exercised, but by a benevolent administrative technocracy which, through sophisticated technological means such as genetic engineering, maintained a society which although based on an extreme form of consumerism rather than overt repression, also gave most people a life which was equally meaningless and stultified.

As we move through the decade which includes 1984 there is evidence that important social mechanisms in both the capitalist and non-capitalist world are moving their societies towards a period of greatly increased centralized social control. The attempt to develop a major international nuclear industry forms part of that evidence. In 1946, in a foreword to his book, Aldous Huxley wrote:

> It is probable that all the world's governments will be more or less completely totalitarian even before the harnessing of atomic energy; that they will be totalitarian during and after the harnessing seems almost certain. Only a large-scale popular movement toward decentralisation and self-help can arrest the present tendency toward statism.[19]

Now it is clear that such a movement has begun. It is most discernible where the tendency towards centralism is most pronounced. Its presence is most evident in the world-wide popular movement against nuclear power. The success of that movement to date can give all who view with concern not only the gross physical hazards presented by that technology, but also the broader global trend towards centralized social control, considerable grounds for believing that a better future is not only possible, but achievable.

NOTES

1 FROM BOOM TO CRASH

1 K. Sakuma on behalf of the Japanese Central Organizing Committee for sending a National Delegation to the UNO demanding an international treaty completely banning nuclear weapons and immediate measures to ban their use, *An Introductory Report on the damage and after-effects of the atomic bombing of Hiroshima and Nagasaki Addressed to the Secretary General of the United Nations*, Hiroshima, 16 July 1976, p. 7.

2 A Citizen's Group to Convey Testimonies of Hiroshima and Nagasaki, *Give Me Water: Testimonies of Hiroshima and Nagasaki*, Tokyo, Japan, October 1976, pp. 8, 9. (Obtainable from Fujiko Tochiki, Metropolitan Apt. 603, 49–19, Horinouchi 3-Chome, Suginami Tokyo 166, Japan.)

3 A. Booth, 'Atomic bombs and human beings', *International Social Science Journal*, vol. XXX, no. 2, 1978, p. 383.

4 *News Chronicle*, 7 August 1945, p. 1.

5 A very readable description of these early developments is given in L. Bickel, *The Deadly Element*, Macmillan, London, 1979.

6 R. Jungk, *Brighter than a Thousand Suns*, Penguin, Harmondsworth, 1958, p. 110.

7 D. Dietz, *Atomic Energy in the Coming Era*, Dodd, Mead & Co., New York, 1945, p. 155.

8 Ibid.

9 Jungk, *Brighter than a Thousand Suns*, p. 182.

10 *News Chronicle*, 7 August 1945, p. 4.

11 Ibid., 8 August 1945, p. 4.

12 Dietz, *Atomic Energy*, p. 178.

13 *News Chronicle*, 7 August 1945, p. 1.

14 Ibid., 8 August 1945, p. 4.

15 Ibid., p. 1.

342   *GLOBAL FISSION*

16 Ibid., p. 4.

17 Ibid.

18 Dietz, *Atomic Energy*, pp. 11–41.

19 M. Schulman, 'The Impact of Three Mile Island', *Public Opinion*, June–July 1979, p. 7.

20 S. Novick, *The Electric War: The fight over nuclear power*, Sierra Club Books, San Francisco, 1976, p. 40.

21 G. A. Modelski, *Atomic Energy in the Communist Bloc*, Melbourne University Press, 1959, p. 109.

22 I. C. Bupp and J. C. Derian, *Light Water: How the Nuclear Dream Dissolved*, Basic Books, New York, 1978, p. 63.

23 Ibid., p. 60.

24 D. Burn, *Nuclear Power and the Energy Crisis: Politics and the Atomic Industry*, Macmillan, London, 1978, p. 98.

25 H. P. Metzger, *The Atomic Establishment*, Simon & Schuster, New York, 1972, p. 85.

26 R. S. Lewis and J. Wilson with E. Rubinowitz, *Alamogordo Plus Twenty-Five Years*, Viking Press, New York, 1971, p. 144.

27 W. C. Patterson, *Nuclear Power*, Penguin, Harmondsworth, 1976, p. 72.

28 Ibid., pp. 253–5.

29 'Nuclear Power Costs', *Twenty-Third Report by the Committee on Government Operations, House of Representatives Ninety-Fifth Congress*, House Report No. 95–1090, US Government Printing Office, Washington, 1978, p. 39.

30 Burn, *Nuclear Power*, p. 26, fig. 3.1.

31 Bupp and Derian, *Light Water*, p. 49.

32 Ibid., p. 73, fig. 4.1.

33 Burn, *Nuclear Power*, p. 27, fig. 3.2.

34 T. R. Malthus, *An Essay in the Principle of Population*, Dent, London, 1973, p. 311.

35 'Alternative Long-Range Energy Strategies', *Joint Hearing before the Select Committee on Small Business and the Committee on Interior and Insular Affairs, U.S. Senate 94th Congress, Interior Committee Serial No. (94–17) (92–137)*, US Government Printing Office, Washington, December 1977, p. 1569.

36 'Buyers Guide: Reports on nuclear programs around the world', *Nuclear News*, vol. 18, no. 3, mid Feb. 1975, p. 26.

37 'Nuclear Power Costs', *Report*, p. 39.

38 'DOE Units Differ on the Outlook for U.S. Nuclear Capacity', *Nucleonics Week*, 7 August 1980, p. 4.

39 'Buyers Guide', *Nuclear News*, p. 43.

40 *General Review of the Long Term Program*, Japan Atomic Energy Commission, 1972, p. 67.

41 *AMPO-Japan-Asia Quarterly*, vol. 17, no. 1, Winter 1975, p. 25.

NOTES

343

42 'The Energy Crisis and Japan's Response to it', *Energy in Japan*, no. 29, June 1975, p. 8.

43 'Policy Choices in the Age of Diversified Energy Sources', *Energy in Japan*, Institute of Energy Economics, Tokyo, no. 36–1, March 1977, p. 5.

44 'Japan's Target of 49 GW Nuclear Capacity by 1985 is "Virtually Impossible" ', *Nucleonics Week*, 30 June 1977, p. 11.

45 See, for example, N. Barrett, *The Nuclear Power Experience in Japan: Exposing the Myth*, FOE Australia, June 1977, p. 4, and 'Japan's Nuclear Situation and Anti-Nuclear Movement', *Gensuikin News*, 1 August 1980, p. 1.

46 'Buyers Guide', *Nuclear News*, p. 42.

47 'West German Nuclear Power Capacity in 1985 Will Amount to Only 34 000 Mw', *Nucleonics Week*, 30 December 1976, p. 5; OECD (NEA)/IAEA, *Uranium: Resources, Production and Demand*, December 1977, table 8, p. 28, footnote.

48 'Buyers Guide', *Nuclear News*, p. 44.

49 Ibid., p. 32.

50 Ranger Uranium Environmental Inquiry, *First Report*, Australian Government Publishing Service, Canberra, 1976, p. 45, fig. 6.

51 Mid-range estimate calculated from OECD Nuclear Energy Agency, *Eighth Activity Report 1979*, OECD, Paris, 1980, table III, p. 69.

52 A. B. Lovins and L. H. Lovins, *Energy/War: Breaking the Nuclear Link*, FOE, San Francisco, 1980, note 142, p. 64.

53 'INFCE Moves Toward Lower Nuclear Power Target for 2000', *Nucleonics Week*, 15 February 1979, p. 4.

54 'Nuclear Power: The Crisis in Europe and Japan', *Business Week*, 25 December 1978, p. 44.

55 Ibid., p. 48.

56 Australian Atomic Energy Commission, *Annual Report*, Australian Government Publishing Service, Canberra, 1974–75, p. 19; 1978–79, p. 15; 1979–80, p. 12.

57 'Nuclear Power Costs', *Report*, p. 40.

58 'Nuclear Dilemma: The atom's fizzle in an energy-short world', *Business Week*, 25 December 1978, p. 54.

59 'Nuclear Power Costs', *Report*, p. 39.

60 Australian Atomic Energy Commission, *Annual Report*, 1974–75, p. 19; 1978–79; p. 15; 1979–80, p. 12.

61 R. A. McCormack (president, Power Systems Group, General Atomic Company), in an address to the Atomic Industrial Forum annual conference, San Francisco, 17 November 1975, quoted in 'The Price of Power' (editorial), *Energy Policy*, vol. 4, no. 1, March 1976, p. 2.

62 'Nuclear Power Costs', *Report*, p. 39.

63 'Nuclear Dilemma', *Business Week*, p. 54.

344 GLOBAL FISSION

64 Quoted in 'Nuclear Power: Its future in the U.S. looks grim', *Los Angeles Times*, 28 March 1980, p. 2.

---

2 MESSAGE FROM PENNSYLVANIA

1 'Technicians testify they left pump valves open at 3-mile Island', *Boston Globe*, 31 May 1979.

2 G. Shaw and R. Gillette, 'At Three Mile Plant', *Boston Globe*, 9 April 1979.

3 J. G. Kemeny (chairman), *Report of The President's Commission on The Accident at Three Mile Island*, US Government Printing Office, Washington, October 1979, p. 86.

4 A. Roberts, 'Middletown U.S.A. – in the nuclear shadow', *Arena 52*, 1979, p. 8.

5 Union of Concerned Scientists, *The Nuclear Fuel Cycle*, MIT, Cambridge, 1975, p. 74.

6 Ibid., p. 94.

7 American Physical Society, *Reviews of Modern Physics*, vol. 47, Supplement no. 1, Summer 1975.

8 AEC Internal Files on 1964–65 Update of Wash–740. Released June 1973. Quoted in Union of Concerned Scientists, *The Risks of Nuclear Power Reactors: A Review of the NRC Reactor Safety Study WASH-1400 (NUREG-75/014)*, UCS, Cambridge, Mass., August 1977, p. 146.

9 'Three Mile Island: An Accident Just Waiting to Happen', *Nucleonics Week*, 5 April 1979, p. 3.

10 Shaw and Gillette, 'At Three Mile Plant'.

11 Office of Inspection and Enforcement: US Nuclear Regulatory Commission, *'Investigation into the March 28, 1979 Three Mile Island Accident by Office of Inspection and Enforcement NUREG-0600'*, Investigative Report No. 50–320/79–10, USNRC: Washington D.C., August 1979, pp. IA–1, IA–120. Unless stated otherwise, times are taken from the 'Operational Sequence of Events' contained in this report.

12 'Three Mile Island: "An Accident Just Waiting to Happen"', *Nucleonics Week*, 5 April 1979, p. 3.

13 W. J. Lanouette, 'No Longer can the NRC say . . .', *Bulletin of the Atomic Scientists*, June 1979, p. 6.

14 'Nuclear Power Plant: Oversight Hearings', *Before a Task Force of the Subcommittee on Energy and the Environment of the Committee on Interior and Insular Affairs*, U.S. House of Representatives, serial no. 96–8, US Government Printing Office, Washington, 9, 10, 11, 15 May 1979, p. 129.

15 *Supplemental Views by Members of the President's Commission on the Accident at Three Mile Island*, Washington, October 1979, p. 93.

NOTES   345

16 Kemeny, *Report of The President's Commission*, p. 94.

17 J. R. Hallam, *Three Mile Island*, Research Paper No. 2, FOE Australia, p. 21; see also 'Nuclear Power Plant: Oversight Hearings', *Task Force*, pp. 52–4.

18 L. Torrey, 'The Week they almost lost Pennsylvania', *New Scientist*, 19 April 1979, p. 174.

19 See 'Latest Analysis of the Course of The Three Mile Island Nuclear Plant Accident', *Nucleonics Week*, 5 April 1979, p. 5; and Office of Inspection and Enforcement, *Investigation, NUREG-0600*, p. 4.

20 'Reactor Safety, Instrumentation, Coolant Natural Circulation Worry ACRS', *Nucleonics Week*, 19 April 1979, p. 5.

21 'Secrecy, frisbee and silly-putty', *Not Man Apart*, 17 June 1979, p. 13; based on an article by S. Lovelady and twenty-seven other reporters in the *Philadelphia Enquirer*, 8 April 1979.

22 Torrey, 'The Week they almost lost Pennsylvania', p. 175.

23 'Three Mile Island: A Chronicle of the Nation's Worst Nuclear Power Accident', *New York Times*, 14 April 1979.

24 Torrey, 'The Week they almost lost Pennsylvania', p. 175.

25 'Three Mile Island: A Chronicle', *New York Times*.

26 Ibid.

27 Torrey, 'The Week they almost lost Pennsylvania', p. 176.

28 'NRC Commissioner's TMI Meetings Transcript', *Nucleonics Week*, Special Transcript Issue, 26 April 1979, p. 10; R. Smith, 'Three Mile Island seen as Laboratory Holding Data Unavailable so Far', *Nucleonics Week*, 26 April 1979, p. 2.

29 E. Diamond, 'Three Mile Island: How Clear was the TV's Picture?', *TV Guide*, USA, 4 August 1979.

30 Sir B. Flowers (chairman), 'Nuclear Power and the Environment', *Royal Commission on Environmental Pollution*, 6th Report, Her Majesty's Stationery Office, London, 1976, p. 23.

31 'Three Mile Island: Accident Appears to be Well Under Control', *Nucleonics Week*, 5 April 1979, p. 2.

32 'Latest Analysis of the Cause of the Three Mile Island Nuclear Plant Accident', *Nucleonics Week*, 5 April 1979, p. 5.

33 'President's Commission on TMI Sums up New Information', *Nucleonics Week*, 7 June 1979, p. 6.

34 Ibid.

35 'NRC Commissioner's TMI Meetings Transcript', *Nucleonics Week*.

36 'Three Mile Island: A Chronicle', *New York Times*, p. 4.

37 'NRC Commissioner Differed on Accident Evacuation Order', *Nucleonics Week*, 12 April 1979, p. 7.

38 Kemeny, *Report of the President's Commission*, p. 134.

39 'Three Mile Island: Accident', *Nucleonics Week*, p. 1.

3 THE US RESPONSE

1 See W. C. Patterson, *Nuclear Power*, Penguin, Harmondsworth, 1976, pp. 214–15 for a very readable description; for more detail see D. Ford, H. Kendall, and L. Tye, *Browns Ferry: The Regulatory Failure*, Union of Concerned Scientists, Cambridge, Mass., 10 June 1976.

2 'Nuclear Nightmare', *Time Magazine*, 9 April 1979, p. 15.

3 Ibid.

4 'Nuclear Power Plant: Oversight Hearings', *Before a Task Force of the Subcommittee on Energy and the Environment of the Committee on Interior and Insular Affairs*, US House of Representatives, serial no. 96–8, US Government Printing Office, Washington, 9, 10, 11, 15 May 1979, pp. 5–8.

5 'Nuclear Industry Downhearted by Three Mile Island Accident? Never!', *Nucleonics Week*, 26 April 1979, p. 5.

6 'Nuclear Nightmare', *Time Magazine*, p. 16.

7 'Nuclear Industry Downhearted', *Nucleonics Week*, p. 5.

8 'Officials quick to alter answers about radiation', *Journal News*, Rockland City, New York, 30 March 1979.

9 'Nuclear Industry Downhearted', *Nucleonics Week*, p. 5.

10 C. Allan, 'Crisis town at end of its tether', *Australian*, 4 April 1979.

11 *The New York Times Index*, 1–15 April 1979, p. 10.

12 The figures were 57 per cent 'for', 31 per cent 'against' (1978), and 47 per cent 'for', 45 per cent 'against' (1979); see M. Schulman, 'The Impact of Three Mile Island', *Public Opinion*, June–July 1979, p. 8.

13 'Briefs', *Nucleonics Week*, 10 May 1979, p. 8.

14 G. Shaw and R. Gillette, 'At Three Mile Plant', *Boston Globe*, 9 April 1979.

15 C. Cookson, 'Three Mile Island in a Sea of Inquiries', *New Scientist*, 17 May 1979, p. 525.

16 See, for example, W. C. Patterson, *Nuclear Power*, Penguin, Harmondsworth, 1976; R. Nader and J. Abbotts, *The Menace of Atomic Energy*, Outback Press, Melbourne, 1977; E. W. Titterton and F. P. Robotham, *Uranium – Energy Source of the Future?*, Abacus, London, 1979; J. McPhee, *The Curve of Binding Energy*, Farrar, Straus and Giroux, New York, 1974; Ranger Uranium Environmental Enquiry, *First Report*, Australian Government Publishing Service, Canberra, 1976; A. B. Lovins and L. H. Lovins, *Energy/War: Breaking the Nuclear Link*, FOE, San Francisco, 1980; and references contained therein.

17 W. Overend, 'A Nuclear Pitch at the Airport', *Los Angeles Times*, 10 April 1980.

NOTES

18 H. J. Evans, K. E. Buckton, G. E. Hamilton, and A. Carothers, 'Radiation-induced chromosome aberrations in nuclear-dockyard workers', *Nature*, vol. 277, 15 February 1979, pp. 531–4.

19 T. Najarian and T. Colton, 'Mortality from Leukaemia and Cancer in Shipyard Nuclear Workers', *Lancet*, 13 May 1978, pp. 1018–20.

20 L. Torrey, 'Row about nuclear submarine cancers', *New Scientist*, 1 January 1981, p. 6.

21 T. Mancuso, A. Stewart, and G. Kneale, 'Radiation exposures of Hanford workers dying from cancer and other causes', *Health Physics*, 33, 1977, p. 369.

22 See, for example, J. A. Reissland, 'An assessment of the Mancuso Study', *National Radiological Protection Board*, NRPB-R79, Harwell, Oxon., 1978, and references contained therein.

23 See 'Level best on radiation', *New Scientist*, 11 January 1979, p. 75.

24 Ibid.; see also J. W. Gofman, 'The Question of Radiation Causation of Cancer in Hanford Workers', *Health Physics*, vol. 37, November 1979, pp. 617–39, for a detailed re-work of Mancuso et al.'s data, substantially confirming their principal conclusion of an association of cancer and radiation doses below the permissible dose rate.

25 See, for example, I. Bross and N. Natarajan, 'Risk of Leukemia in Susceptible Children Exposed to Preconception, In Utero and Postnatal Radiation', *Preventative Medicine*, vol. 3, 1974, pp. 361–4.

26 L. A. Sagan, 'The Mancuso Study: A Comment', *Atom*, 262, August 1978, p. 207.

27 See, for example, Subcommittee on Energy and the Environment, United States House of Representatives, *Proceedings of a Congressional Seminar on Low-Level Ionizing Radiation*, Environmental Policy Institute, Washington D.C., 1977; and Congressional Environmental Study Conference, Environmental Policy Institute, and Atomic Industrial Forum, *Radiation Standards and Public Health: Proceedings of a Second Congressional Seminar on Low-Level Ionizing Radiation*, Congressional Environmental Study Conference, Environmental Policy Institute, and Atomic Industrial Forum, Washington D.C., 1978.

28 US cancer specialist Dr Joseph Wagoner, quoted in Elmer Lammi, 'Uranium Miner Cancer Epidemic Change to U.S.', *The Tenessean*, 20 June 1979.

29 J. Raloff, 'Radiation Hazards: The Military Experience', *Science News*, vol. 113, no. 6, 1 April 1978, p. 92; 'Fear on Effects of Atom Tests Grew Gradually', *New York Times*, 13 May 1979; 'Rediscovering the Past', *Time Magazine*, 2 July 1979, p. 21.

30 B. Curry, 'U.S. ignored study linking A-tests, cancer: '65 report cited deaths in Utah', *Boston Globe*, 8 January 1979.

31 'High-level activity on low-level radiation', *New Scientist*, 8 March 1979, p. 747.

348 GLOBAL FISSION

32 'Officials quick', *Journal News*.

33 C. Mohr, 'Califano Now Says a Cancer Death from Nuclear Accident is Possible', *New York Times*, 4 May 1979.

34 'NRC, Others Alleged to Be Misleading Public on TMI Low-Level Radiation', *Nucleonics Week*, 19 April 1979, p. 1.

35 Ibid.

36 Mohr, 'Califano Now Says', *New York Times*.

37 Union of Concerned Scientists, *The Risks of Nuclear Power Reactors*, UCS, Cambridge, Mass., 1977, p. 4.

38 For an excellent dissection of the Rasmussen Reactor Safety Study, see Union of Concerned Scientists, *Risks*, 1977. For a review of relevant critical literature, see also 'Why Don't People Believe Rasmussen?', *Eco*, vol. VIII, no. 1, 1977, p. 7.

39 J. O. Membrino, 'Scientists ask closing of 16 nuclear plants', *Boston Globe*, 27 January 1979.

40 'U.S. Nuclear Utilities Not Deterred by Three Mile Island Accident', *Nucleonics Week*, 3 May 1979, p. 9.

41 'Rasmussen Sees Gaps in Safety Research, Says "We Have A Lot to Learn" ', *Nucleonics Week*, 26 April 1979, p. 3.

42 'Three Mile Island Seen As Laboratory, Holding Data Unavailable So Far', *Nucleonics Week*, 26 April 1979, p. 2.

43 Ibid.

44 J. Schwartz, 'Harrisburg: counting the cost', *Nature*, vol. 278, 12 April 1979, p. 589.

45 Union of Concerned Scientists, *The Nuclear Fuel Cycle*, MIT, Cambridge, 1975.

46 D. Pisello, 'The Zirconium Connection', *Ecologist*, vol. 9, nos 4–5, August 1979, p. 116.

47 'Reactor Safety, Instrumentation, Coolant Natural Circulation Worry ACRS', *Nucleonics Week*, 19 April 1979, p. 5.

48 L. Torrey, 'Engineering solutions to nuclear accidents', *New Scientist*, 15 November 1979, p. 503.

49 'TVA Initiating Major Changes for its Nuclear Plants as Result of TMI', *Nucleonics Week*, 7 June 1979, p. 9.

50 D. Martin, 'House Probe Contradicts B & W on Impact of Operator Error', *Nucleonics Week*, 24 May 1979, p. 4.

51 'TMI Operators Trapped in Web of System Complexity', *Nucleonics Week*, 24 May 1979, p. 9.

52 'Control Room Technology Termed Obsolete', *Nucleonics Week*, 7 June 1979, p. 8.

53 'The First NRC Inspection and Enforcement Report on the TMI Sequence', *Nucleonics Week*, 31 May 1979, p. 8.

54 'President's Commission on TMI turns up New Information', *Nucleonics Week*, 7 June 1979, p. 6.

NOTES 349

55 'Nuclear Industry Downhearted', *Nucleonics Week*, p. 5.

56 'Task Force head says N-accident repeat is likely', *Boston Globe*, 22 May 1979.

57 D. Burnham, 'Fundamental Changes for Nuclear Safety, says U.S. Commission', *Australian Financial Review*, 1 November 1979, p. 9.

58 *Report of the National No Nukes Strategy Conference*, Louisville, KY, 16–20 August 1978, mentions some 330 citizens groups opposed to nuclear power but, as the report indicates, this is only the tip of an iceberg.

59 R. S. Lewis, *The Nuclear-Power Rebellion: Citizens Vs. The Atomic Industrial Establishment*, Viking Press, New York, 1972; S. Novick, *The Electric War: The fight over nuclear power*, Sierra Club Books, San Francisco, 1976; A. Gyorgy and Friends, *No Nukes: everyone's guide to nuclear power*, South End Press, Boston, 1979; H. Wasserman, *Energy War: Reports from the Front*, Lawrence Hill, Connecticut, 1979.

60 S. Novick, *The Electric War: The fight over nuclear power*, Sierra Club Books, San Francisco, 1976, p. 184.

61 P. E. Steiger, 'Officials Push Diablo Plants despite Major Quake Fault', *Los Angeles Times*, 29 June 1977.

62 For further detail on the history of the Clamshell Alliance see A. Gyorgy and Friends, *No Nukes: everyone's guide to nuclear power*, South End Press, Boston, 1979, pp. 396–402.

63 Quoted in Gyorgy and Friends, *No Nukes*, p. 396.

64 For further detail see Gyorgy and Friends, *No Nukes*, pp. 393–5.

65 See 'State Legislative Action on Nuclear Power Dropped Dramatically in 1978', *Nucleonics Week*, 4 January 1979, p.10; and 'Nuclear Power Costs', *Twenty-Third Report by the Committee on Government Operations, House of Representatives Ninety-Fifth Congress*, House Report No. 95–1090, US Government Printing Office, Washington, 1978, p. 14.

66 'New Wave of Moratorium Bills and Initiatives Spawned by TMI', *Nucleonics Week*, 14 June 1979, p. 10.

67 J. Byrne, 'California asks for shutdown of plant', the *Age*, Melbourne, 3 April 1979, p. 8.

68 'Anti-nuke protests spread across U.S.', *Workers World*, 6 April 1979; 'Nuclear Protests held across U.S.', *Los Angeles Times*, 2 April 1979.

69 M. Berkowitz, 'Thousands Rally in Largest U.S. No-Nukes Protest', *WIN Magazine*, 3 May 1979, p. 14.

70 '1,000 stage protest', *Boston Globe*, 9 April 1979.

71 'Protesters Rally Coast to Coast', *Hartford Courant*, 8 April 1979.

72 'Ellsberg among 284 held in Nuke protest', *Philadelphia Bulletin*, 30 April 1979; 'N-Plant Mishap Still Fuels Protests', Daily Press, *Newport News*, VA, 29 April 1979; 'Nuclear dangers cited at Mt. Taylor protest', *New Mexico Daily Lobo*, 2 May 1979.

350   GLOBAL FISSION

73 'Officials quick', *Journal News*.
74 'N-Plant Mishap Still Fuels Protests', Daily Press, *Newport News*, VA,
    29 April 1979.
75 'Washington: 10,000 protest nuclear power', *Nuclear Newsletter*,
    Saskatoon Environmental Society, Saskatchewan, Canada, vol. 3, no.
    7, 20 June 1979.
76 T. Oliphant, 'Thousands form N-Protest', *Boston Globe*, 7 May 1979.
77 Ibid.

4   POLITICAL ECONOMICS

1 'Nuclear Power Costs', *Hearings before a Subcommittee of the Committee on
   Government Operations, House of Representatives Ninety-Fifth Congress*,
   12–19 September 1977, US Government Printing Office, Washington, 1977, p. 1181.
2 C. Komanoff, *Nuclear Plant Performance Update 2*, Komanoff Energy
   Associates, 475 Park Ave South, New York City 10016, USA, 1978.
3 For example, the cancellation of the two Haven plants for coal
   stations, 'Three Wisconsin Utilities have decided to cancel the Haven
   Nuclear Plant', *Nucleonics Week*, 6 March 1980, p. 14; the decision to
   build a coal station instead of the two nuclear reactors proposed for
   Long Island Lighting Company in New York, 'Second Thoughts on
   Nuclear Plants', *New York Times*, 3 February 1980; and the unprecedented announcement by Virginia Electric and Power Co. (VEPCO)
   in October 1979 that it may convert two partially completed reactors
   to coal. 'Vepco to Switch to Coal?', *Guardian*, 24 October 1979, p. 22.
   In December 1980 VEPCO confirmed that it would cancel the North
   Anna No. 4 plant already 10 per cent complete with $165 million
   already spent on it, and would build no more nuclear reactors before
   the year 2000. A key reason given was the additional capital costs due
   to regulational changes since Three Mile Island; see B. Toohey, 'U.S.
   utility abandoned work on new N-Plant', *Australian Financial Review*, 4
   December 1980, p. 35.
4 Figures given by A. J. Parisi (former energy correspondent for the
   New York Times, and now an editor of *Petroleum Intelligence Weekly*), in
   A. J. Parisi, 'Nuclear Power Falls on Hard Times', *National Times*,
   7–13 June 1981, p. 23.
5 M. Gravel, *Mike Gravel Newsletter*, 29 September 1973.
6 'Now Comes the Fallout', *Time Magazine*, 16 April 1979, p. 26.
7 Ibid., p. 25.
8 'Bankruptcy Feared', *Boston Globe*, 18 April 1979.
9 'Pennsylvania freezes power utilities' rates', *Boston Globe*, 20 April
   1979; 'Pennsylvania Consumer Advocate Starts Battle to Protect TMI
   Ratepayers', *Nucleonics Week*, 12 April 1979, p. 2.; 'U.S. Nuclear Utili-

NOTES   351

ties Not Deterred by Three Mile Island Accident', *Nucleonics Week*, 3 May 1979, p. 9.

10 'Three Mile Island firms get a rate increase', *Boston Globe*, 16 June 1979.

11 'TMI-1 Is Not Producing Electricity Because of Discriminatory Action', *Nucleonics Week*, 18 October 1979, p. 13.

12 'Thornburg Holds Forth on: TMI-1, TMI-2, NRC and the States Role', *Nucleonics Week*, 26 July 1979, p. 1.

13 'Hearings on Three Mile Island Begin', *Arizona Republic*, 16 October 1980.

14 'With the Events of Three Mile Island-2 Accident Indelibly Etched on Their Minds', *Nucleonics Week*, 26 April 1979, p. 15.

15 '$400m plan to fix Three Mile Island', *Australian Financial Review*, 14 August 1979.

16 J. G. Kemeny (chairman), *Report of The President's Commission on The Accident at Three Mile Island*, US Government Printing Office, Washington, October 1979.

17 'Bernard Rusche to Play Major Role In Three Mile Island Cleanup', *Nucleonics Week*, 19 April 1979, p. 6.

18 'Met. Ed. Is Likely To Be Set Back In Plan To Purge Krypton From The TMI-2 Containment', *Nucleonics Week*, 27 September 1979, p. 11.

19 'Three Mile Island cleanup costs zoom astronomically', *Australian Financial Review*, 12 November 1980, p. 33.

20 'Cost of cleaning impasse over Three Mile Island', *Australian Financial Review*, 22 May 1981, p. 61.

21 'Utility doubles cost of Three Mile repair', *Boston Globe*, 9 August 1980.

22 'Company Eliminates Jobs', *Albuquerque Journal*, 1 May 1979.

23 'Federal Aid Asked for N-Plant Owners', *Times Herald Newport News*, VA, 9 May 1979.

24 'GPU suit blames U.S. for accident', *The Reading Times*, PA, 9 December 1980.

25 'Cost of cleaning impasse', *Australian Financial Review*.

26 D. D. Comey, 'The Uneconomics of Nuclear Power', *Skeptic Magazine*, Issue no. 14, 1976.

27 Komanoff, *Nuclear Plant*, p. 7.

28 C. Komanoff, quoted in 'Capacity factors take a dive', *Not Man Apart*, October 1979, p. 11.

29 'Nuclear Electricity Generation for May 1979', *Nucleonics Week*, 28 June 1979, p. 15.

30 'Licensing Stability Now a Total Loss, Says Industry Man', *Nucleonics Week*, 17 May 1979, p. 3.

31 E. Marshall, 'NAS Study on Radiation takes the Middle Road', *Science*, 204, 1979, pp. 711–14.

32 'Nuclear Accident', *Newsweek*, 9 April 1969, p. 19.

352   GLOBAL FISSION

33 'Atom Workers in Danger', *Herald*, Melbourne, 29 June 1978, p. 13.
34 'Radiation Exposure Increasing', *World News*, Albuquerque, New Mexico, 1–7 April 1979.
35 'Science Watch', *New York Times*, 27 February 1979.
36 'Radiation Exposure Increasing', *World News*.
37 A. Martin, 'Occupational radiation exposure in LWRs increasing', *Nuclear Engineering International*, January 1977, p. 32.
38 Ibid., p. 33.
39 'Nuclear Blowdown', *Not Man Apart*, December 1976.
40 C. Komanoff, 'Three Mile Island: The Economics Fallout', draft of article to be published in the *Critical Mass Journal*.
41 J. Vinocur, 'Outlook Held Bleak for Manufacturers of Atomic Reactors', *New York Times*, 21 September 1977, pp. A1, D6.
42 C. Komanoff, 'The End of Nuclear Power', *The New York Review of Books*, vol. LXXVI, no. 8, 17 May 1979.
43 I. C. Bupp, J. C. Derian, et al., 'The Economics of Nuclear Power', *Technology Review*, February 1975, p. 15.
44 Ibid., p. 25.
45 R. K. Lester, *Nuclear Power Plant Lead-Times*, The Rockefeller Foundation/The Royal Institute of International Affairs, New York/London, November 1978, fig. 5.
46 Ibid., p. 24.
47 A. W. Murphy, et al., *The Licensing of power plants in the United States*, Seven Springs Center, Mt Kisco, New York, January 1978.
48 Lester, *Nuclear Power Lead-Times*, p. 26.
49 'Nuclear Costs will be Twice Those of Coal by 1986–87, Says Komanoff', *Nucleonics Week*, 19 April 1979, p. 2.
50 'Commonwealth Edison Says It Already Has Begun to Comply', *Nucleonics Week*, 2 August 1979, p. 6.
51 Lester, *Nuclear Power Lead-Times*, p. 20.
52 Ibid., table IV; and K. W. Sieving and E. A. Saltarelli, *Lead times in the deployment of commercial nuclear power for United States LWR plants*, report prepared for the International Consultative Group on Nuclear Energy, NUS Corporation, NUS-3138, Rockville, Md., April 1978.
53 D. Fishlock, 'The World Nuclear Industry After Harrisburg', *Financial Times*, 26 April 1979, p. 2.
54 Calculated from A. M. Lovins, *Soft Energy Paths: Toward a Durable Peace*, Penguin, Harmondsworth, 1977, pp. 105–6.
55 Kemeny, *Report of the President's Commission*, p. 64, Recommendation 6.
56 'NRC Chairman Joseph Hendrie is "Rabidly Opposed" to a Provision', *Nucleonics Week*, 26 July 1979, p. 10.
57 'Reactors in Jeopardy', *Australian Financial Review*, 7 November 1979, p. 2.

NOTES 353

58 'Goodwill in Congress for Nuclear is Now Destroyed, says Industry Man', *Nucleonics Week*, 19 April 1979, p. 10.

59 'Licensing Stability', *Nucleonics Week*, p. 3.

60 'Higher Capital Costs for Nuclear Could Be Accident Result', *Nucleonics Week*, 26 April 1979, p. 1.

61 Ibid.

62 'Bank America stops new nuclear loans', *Australian Financial Review*, 17 May 1979, p. 7.

63 J. Hoare, 'U.S. downturn hits market', *Australian Financial Review*, 18 September 1980, p. 21.

64 The initial draft report version of M. Lönroth and W. Walker *The Viability of the Civil Nuclear Industry* was circulated confidentially. From the report given in 'Western Countries said to Have Five Years to Avert Nuclear Vendor Fallout', *Nucleonics Week*, 27 September 1979, p. 1, its comments were rather more forthright than those contained in the later publicly released document.

65 From the publicly released report: M. Lonroth and W. Walker, *The Viability of the Civil Nuclear Industry*', The Rockefeller Foundation/The Royal Institute of International Affairs, New York/London, 1979, p. 96.

66 Quoted in *Powerline*, vol. 4, no. 11, June 1979.

67 R. Lenzer, 'Nuclear-plant investor big loser', *Boston Globe*, 1 April 1979.

68 ' "The Atom Has Fizzled." This Remark Made By A Senior DOE Official Has Upset', *Nucleonics Week*, 5 July 1979, p. 13.

69 'The Nuclear Carcass Seems To Be More Attractive To The Anti-Nuclear Movement', *Nucleonics Week*, 28 June 1979, p. 13.

70 'Nuclear Power Costs', *Hearings*, p. 1064.

71 Ibid., p. 1065.

72 'Energy: Receiving the Lions Share', *Science News*, vol. 109, no. 52, 24 January 1976.

73 'Total Cost of TMI for NRC In FY-79 Reckoned at $13–14 Million', *Nucleonics Week*, 14 June 1979, p. 5.

74 'CENTEC/URENCO'S New Enrichment Price is $100/SWU Plus Inflation', *Nucleonics Week*, 13 March 1975, p. 1; this accords well with the earlier estimate by the chairman of the UK's Nuclear Power Company that commercially enriched uranium would cost twice its present price if it were produced commercially – see 'Fundamental Questions', *New Scientist*, 26 June 1975, p. 710.

75 Ranger Uranium Environmental Inquiry, *First Report*, Australian Government Publishing Service, Canberra, 1976, p. 45, fig. 6.

76 Quoted in G. D. Friedlander, 'Decommissioning Commercial Reactors', *Electrical World*, 15 February 1978.

77 E. Kahn, et al., *Investment Planning in the Energy Sector*, Energy and

354 GLOBAL FISSION

Environment Division, Lawrence Berkeley Laboratory, Berkeley, 1 March 1976, p. 89.

78 'Nuclear Power Costs', *Hearings*, p. 571.

79 Ibid., p. 595.

80 Congressional Budget Office, *The Export Import Bank: Implications for the Federal Budget and Credit Market*, Staff Working Paper, 27 October 1976, p. 17.

81 G. C. Duffy and G. Adams, *Power Politics: The Nuclear Industry and Nuclear Exports*, Council on Economic Priorities, New York, 1978, p. 57, table XI.

82 Ibid., p. 56.

83 Ibid., p. 58.

84 R. Nader and J. Abbotts, *The Menace of Atomic Energy*, Outback Press, Melbourne, 1977, p. 285.

85 US Nuclear Regulatory Commission, *Reactor Safety Study – An Assessment of Accident Risks in US Commercial Nuclear Power Plants*, Report No. WASH-1400, NUREG-75/014, US Government Printing Office, Washington, 1975.

86 Nader and Abbotts, *The Menace of Atomic Energy*, p. 227.

87 H. Brown (ed.), 'Nuclear Fishing: A Preliminary Exploration into the Hidden Costs of Nuclear Power', *College of Science in Society*, Wesleyan University, USA, 1978.

88 Unpublished study by Joseph Bowring prepared for the US Energy Information Administration's Office of Economic Analysis at the request of the House Subcommittee on Energy and Power, and reported in 'Government funds prop up the nukes', *Australian Financial Review*, 9 January 1981, p. 17, and 'Nuclear power highly subsidized', *Milwaukee Journal*, 28 December 1980. The $40 billion subsidy was reckoned as: research and development $23.6 billion; foreign reactor sales promotion $237.4 million; uranium market promotion $2.5 billion; fuel enrichment aid $7.1 billion; and waste management, spills cleanup, and unpaid decommissioning costs $6.5 billion.

89 'Nearly 400 Nuclear Industry Leaders were In Washington This Week to Lobby Congress', *Nucleonics Week*, 10 May 1970, p. 12.

90 'Nuclear Power Costs', *Twenty-Third Report by the Committee on Government Operations, House of Representatives Ninety-Fifth Congress*, House Report No. 95–1090, US Government Printing Office, Washington, 1978, p. 50.

91 C. Mohr, 'President Contends It Is Impossible to "Abandon" U.S. Nuclear Energy', *New York Times*, 11 April 1979.

92 'Senate is Next Big Test as House Rebuffs Moratorium on Licensing', *Nucleonics Week*, 21 June 1979, p. 7.

93 L. Staltzfus, 'Legislative Alert', *The Mobilizer*, September 1979, p. 2; see also W. Weaver, 'Energy Bill Advances in Senate', *New York Times*, 21 September 1979, p. D12.

NOTES

355

94 'Things Needed to Make Possible Acceptance of More Nuclear Power Plants', *Nucleonics Week*, 24 May 1979, p. 11.

95 See, for example, A. B. Lovins, *Soft Energy Paths: Toward a Durable Peace*, Penguin, Harmondsworth, 1977; R. Stobaugh and D. Yergin (eds), *Energy Future: Report of the Energy Project at the Harvard Business School*, Random House, New York, 1979; A. B. Lovins and L. H. Lovins, *Energy/War: Breaking the Nuclear Link*, FOE, San Francisco, 1980; and references contained therein.

96 B. Greenman, 'The Impact of Alternative Policy Reactions to Three Mile Island', (National Impact), *NERA*, vol. 1, 80 Broad Street, New York, NY10004, quoted in 'Nuclear Moratoriums', *Goundswell*, June 1979.

97 V. Taylor, *The Easy Path Energy Plan*, Union of Concerned Scientists, Massachusetts, September 1979.

98 Lovins, *Soft Energy Paths*.

99 'Nuclear Power Costs', *Report*, pp. 51–2.

1 Ibid., p. 50.

2 Office of Technology Assessment, Congress of the United States, *Application of Solar Technology to Today's Energy Needs*, US Government Printing Office, Stock No. 052–003–00539–5, June 1978, pp. 18–24.

3 'Nuclear Power Costs', *Report*, p. 57.

4 Office of Technology Assessment, *Application of Solar*, p. 59.

5 'Met-Ed: selling you what you've already paid for', *Harrisburg Magazine*, July 1979.

6 'Top Nuclear Industry Execs Reportedly Planning PR Drive', *Tennessean*, 9 June 1979, p. 15.

7 'The OECD's Nuclear Energy Agency Will Soon Launch A Campaign', *Nucleonics Week*, 18 January 1979, p. 11.

8 D. Burnham, 'Pro-nuclear lobbies in U.S. step up their campaign', *Australian Financial Review*, 28 December 1979, p. 10.

9 Ibid.

10 C. Wolff, 'The "Energy Advocates" ', *New Age*, June 1979, p. 33.

11 Ibid.

12 Jan Carlin, first woman nuclear engineer at Westinghouse, quoted in Susan Jaffe, 'The Pro-Nuke Lobby Targets Women', *Ms Magazine*, June 1980, p. 28.

13 Jaffe, 'The Pro-Nuke Lobby Targets Women', *Ms Magazine*, June 1980, p. 30.

14 T. Lange, 'Pro-nuke rally draws big crowd', *In These Times*, 12–18 September 1979, p. 6.

15 D. Schaal, 'NYC Antinuclear action the largest in U.S. history: 250,000 demand end to nuclear energy', *Guardian*, 3 October 1979.

16 C. Joyce, 'Reagan looks fine to pro-nuclear lobby', *New Scientist*, 13 November 1980, p. 414.

17 '$3 billion security bill OK'd', *Arizona Republic*, 10 November 1979.

18 P. Gwynne, 'Reagan rides against the regulators', *New Scientist*, 17 July 1980, p. 183.

19 R. J. MacDonald, 'Reagan plans to beat the environmentalists and start big project', *Business Review*, 9–15 November 1980, p. 21.

20 'As Much as 400 GW of Nuclear Capacity May Be Needed By 2000 In the U.S., D.O.E.', *Nucleonics Week*, 1 May 1980, p. 5.

5 ORIGINS OF THE OPPOSITION

1 C. Goldstein, 'The Opposition Movement and its Leaders', presented at the AIF-SVA *International Workshop, Nuclear Power and the Public*, Geneva, 26–9 September 1977.

2 California Energy Resources Conservation and Development Commission, *Nuclear Fuel Reprocessing and High Level Waste Disposal: An Interim Report*, NAO-6A:25, July 1977, pp. 54–8.

3 Cited in 'Proposals to postpone Atomic Power', *Bulletin of the Atomic Scientists*, vol. 3, no. 10, January 1946, p. 281.

4 'Plans for International Control Take Shape', *Bulletin of the Atomic Scientists*, vol. 1, no. 8, April 1946, pp. 1–15.

5 H. Lisco, 'Radiation Hazards and Radiation Sickness', *Bulletin of the Atomic Scientists*, vol. 2, nos 9–10, November 1946, p. 26.

6 'The American Proposal for International Control', presented by Bernard M. Baruch to the UN Atomic Energy Commission on 13 June 1946, *Bulletin of the Atomic Scientists*, vol. 3, no. 10, 1946, p. 5.

7 'The Carnegie Draft Convention – Legal Sub-Committee, Carnegie Endorsement Committee for International Peace', *Bulletin of the Atomic Scientists*, vol. 2, nos 1 and 2, July 1946, pp. 15–19.

8 W. De Laguna, 'What is Safe Waste Disposal?', *Bulletin of the Atomic Scientists*, vol. 15, no. 1, 1959, p. 35.

9 N. N. Semenov, 'The Future of Man in the Atomic Age', *Bulletin of the Atomic Scientists*, vol. 15, no. 3, March 1959, pp. 123–6.

10 J. S. Lehac, 'Eve's Fight Against the Atom', *Poughkeepsie Journal*, 17 November 1974.

11 A. M. Butler, F. G. Keyes, and A. Szent-Gyorgy, 'Sea Disposal of Atomic Wastes', *Bulletin of the Atomic Scientists*, vol. 16, no. 4, April 1960, p. 141.

12 Les Amis de la Terre, *L'Escroquerie Nucléaire*, Éditions Stock, Paris, 1978, p. 276.

13 D. Nelkin, 'Nuclear Power and Its Critics: A Siting Dispute', in D. Nelkin (ed.), *Controversy: Politics of Technical Decisions*, Sage Publications, London, 1979, pp. 49–67.

14 See S. Novick, *The Electric War: The fight over nuclear power*, Sierra Club Books, San Francisco, 1976, chapter two, for further background to Commoner's remarkable role.

NOTES

15 Ibid., ch. 36.

16 R. S. Lewis, *The Nuclear-Power Rebellion: Citizens Vs. The Atomic Indus-trial Establishment*, Viking Press, New York, 1972, pp. 7–25.

17 Ibid., pp. 269–97.

18 M. Naoto, and H. Kazuko, 'Japanese Nuclear Power Industry and Its Problems', First Draft, Institute of Policy Studies Pacific-Asia Resources Center, Tokyo, July 1979, pp. 3–26.

19 Quoted in J. Berger, *Nuclear Power: The Unviable Option*, Dell, New York, 1977, p. 58.

20 *Testimony of Dale G. Bridenbaugh, Richard B. Hubberd, and Gregory C. Minor before the Joint Committee on Atomic Energy*, 18 February 1976, available from the Union of Concerned Scientists, Washington, D.C.

21 Testimony of Robert D. Pollard before the Joint Committee on Atomic Energy, 28 January 1974, quoted in P. Faulkner (ed.), *The Silent Bomb*, Vintage, New York, 1977, appendix B.

22 *Nuclear Power Focus of a Burgeoning Controversy: A declaration by members of the American technical community*, FOE USA, 6 August 1975.

23 Gensuikin, *A Summing up of the Movement and Organisation of the Japan Congress Against Atomic and Hydrogen Bombs*, Gensuikin Press, Tokyo, 3 April 1970, p. 2.

24 For a more detailed history of CND see P. Duff, *Left, Left, Left*, Alison and Busby, London, 1971, pp. 118–257.

25 Duff, *Left*, p. 132.

26 Ibid., pp. 226–44.

27 For some histories of the Vietnam War see 'The Pentagon Papers', *New York Times*, Bantam, New York, 1971; D. Halberstam, *The Best and the Brightest*, Pan, London, 1972; and F. Fitzgerald, *Fire in the Lake: the Vietnamese and Americans in Vietnam*, Macmillan, London, 1972; and references contained therein.

28 P. Jallée, *The Third World in World Imperialism*, Monthly Review Press, New York, 1971; S. George, *How the other half dies*, Penguin, Harmondsworth, 1976.

29 M. Weber, *The Theory of Social and Economic Organization*, The Free Press, Glencoe, Ill., 1964, pp. 124–5.

30 Quoted in D. Dickson, *Alternative Technology*, Fontana, London, 1977.

31 For a graphic account of the events see P. Seale and M. McConville, *French Revolution 1968*, Penguin, Harmondsworth, 1968.

32 E. R. Richardson, *The Politics of Conservation – Crusades and Controversies 1897–1913*, University of California, Berkeley, 1962.

33 L. K. Caldwell, 'Administrative Possibilities for Environmental Control', in F. F. Darling and J. P. Milton, *Future Environments of North America*, The Natural History Press, New York, 1966, pp. 648–71.

34 R. Carson, *Silent Spring*, Houghton Mifflen, Boston, 1962.

35 National Academy of Sciences-National Research Council, Commit-tee on Natural Resources, *Publication 1000A*, 1962, p. 33.

358   *GLOBAL FISSION*

36   P. R. Ehrlich, 'World Population: A Battle Lost?', *Stanford Today*, Series 1, no. 22, Winter 1968.

37   B. Commoner, 'Nuclear Pollution: The Myth of Omnipotence', *Environment*, March 1969.

38   The author is indebted to Nina Gladitz and Heidi Knott of Teldok films for providing a transcript of their excellent film describing the struggle over nuclear power at Wyhl: N. Gladitz (producer), 'Better Active today than radioactive tomorrow', Teldok Productions, Schillerstrasse 52 D–7800, Freiburg, West Germany, Transcript of Film, 1977. Except where otherwise noted, the detail of the events at Wyhl is from this transcript. Additional sources are A. Gyorgy and Friends, *No Nukes: everyone's guide to nuclear power*, South End Press, Boston, 1979; M. Lucas, 'We are no parties – we are the people!', MS of article written for *Peace News*, Nottingham, February 1975; P. Taylor, 'The Struggle Against Nuclear Power in Central Europe', *Ecologist*, vol. 7, no. 6, July 1977, p. 217; and interviews carried out by the author with the assistance of Heidi Knott in Freiburg and Wyhl in December 1977.

39   For further detail on the occupation of Markolsheim, see P. Taylor, 'The Struggle Against Nuclear Power in Central Europe', *Ecologist*, vol. 7, no. 6, p. 218.

40   N. Gladitz (producer), 'Better Active today than radioactive tomorrow', Teldok Productions, Freiburg, West Germany, Transcript of Film, 1977.

41   Cited in M. Lucas, 'We are no parties – we are the people!', MS of article written for *Peace News*, Nottingham, February 1975.

42   'Court demands PV rupture protection for Wyhl', *Nuclear Engineering International*, World Digest, April 1977.

---

6   THE STATE AND THE INDUSTRY

1   Calculated from M. Lönroth and W. Walker, *The Viability of the Civil Nuclear Industry*, The Rockefeller Foundation/The Royal Institute of International Affairs, New York/London, 1979, pp. 84–5, table 5.

2   Lönroth and Walker, *The Viability*, pp. 14–15, fig. 3.

3   G. C. Duffy and G. Adams, *Power Politics: The Nuclear Industry and Nuclear Exports*, Council on Economic Priorities, New York, 1978, p. 46, table VIII.

4   Lönroth and Walker, *The Viability*, p. 89.

5   Ibid., p. 28, table 2.

6   J. Hoare, 'U.S. downturn hits market', *Australian Financial Review*, 18 September 1980, p. 21.

7   Lönroth and Walker, *The Viability*, p. 70.

NOTES

8 Ibid., p. 63, table 4.

9 Ibid., p. 64.

10 'Threat to W. German nuclear industry seen', *Australian Financial Review*, 6 September 1979, p. 2.

11 Lönroth and Walker, *The Viability*, p. 90.

12 The inspiration and much of the detail for the analysis of the historical role of the state in 'pushing' nuclear power forward has been drawn from A. P. Roberts, 'The Energy Cuckoo', in N. Barrett, P. Hayes, and J. Nicholls, *Atoms for the Poor?*, Community Aid Abroad, Melbourne, November 1977, p. 6.2.1. The author is also indebted to Dr Roberts for several interesting discussions about the theme of this article.

13 J. Baner, 'Public Organisation of Electric Power', quoted in S. Novick, *The Electric War: The fight over nuclear power*, Sierra Club Books, San Francisco, 1976, p. 131.

14 S. Kuhn, *Scientific and Managerial Manpower in the Nuclear Industry*, Columbia University Press, 1966, p. 90.

15 Ibid., p. 96.

16 C. Allardice and E. Trapnell, *The Atomic Energy Commission (U.S.)*, Praeger, New York, 1974, p. 93; and A. Roberts, 'The Energy Cuckoo', in N. Barrett, P. Hayes, and J. Nicholls, *Atoms for the Poor?*, Community Aid Abroad, Melbourne, November 1977, p. 6.2.5.

17 Kuhn, *Scientific*, p. 82.

18 J. Hogerton, 'The Arrival of Nuclear Power', *Scientific American*, 218, 2, February 1978, p. 23.

19 H. Nau, *National Politics and International Technology: Nuclear Development in Western Europe*, John Hopkins Press, Baltimore, 1974, p. 75.

20 Kuhn, *Scientific*, p. 98.

21 A. P. Roberts, 'The Energy Cuckoo', in N. Barrett, P. Hayes, and J. Nicholls, *Atoms for the Poor?*, Community Aid Abroad, Melbourne, November 1977, p. 6.2.14.

22 Kuhn, *Scientific*, p. 115.

23 For further detail of the moves towards public power see S. Novick, *The Electric War*, 1976; and R. Morgan, T. Riesenberg, and M. Troutman, *'Taking Charge: A New Look at Public Power'*, Environmental Action Foundation, Washington, 1976.

24 Hogerton, 'The Arrival', p. 23.

25 D. G. Arnott, 'Power: Technical Considerations', *Marxist Quarterly*, April 1956, p. 114.

26 Hogerton, 'The Arrival', p. 23.

27 Nau, *National Politics*, p. 75.

28 P. Jalée, 'The Third World in World Economy', *Monthly Review Press*, New York, 1969, p. 91.

29 Ibid., pp. 43–7, 54–9.

360   *GLOBAL FISSION*

30 M. Wallace, *Resources and International Conflict: The Onrushing Crisis*, Society for Systems Research, 1974, p. 173.

31 J. Sabato and J. Ramesh, 'Atoms for the Third World', *Bulletin of the Atomic Scientists*, March 1980, p. 43.

32 Ibid., p. 38.

33 Ibid., p. 41.

34 For further detail see R. Alvarez, *Statement of Robert Alvarez, before the House Science and Technology Subcommittee on Nuclear and Fossil Energy Research and Development 4 March 1977*, Reprinted by Environmental Policy Centre, 317 Pennsylvania Avenue, S.E. Washington D.C., 20003.

35 For a more detailed analysis of the Carter policy see J. E. Falk, 'Australia, the New U.S. Nuclear Policy and the International Contestation over Nuclear Power', *Arena*, no. 47–8, 1977, p. 23.

36 'Learning to Live with a Dangerous Gift!', *Time Magazine*, 2 May 1977, p. 11.

37 K. Lang, *Information on Nuclear Energy in the Federal Republic of Germany*, IAEA-CN-36/81, vol. 7, 1977, p. 129.

38 Ibid., p. 134.

39 Interview with Frank Ploenes, Sabine Breustedt, Gunter Hopfermuller (BUU Hamburg), 9 December 1977, and Roland Vogt (BBU Berlin), 7 December 1977.

40 BUU, *Case Study: The Struggle against the Atomic Power Plant Brokdorf*, Hamburg, 1977, p. 2.

41 Ibid., p. 4.

42 C. Simpson, 'Gorleben Soll Leben: A desire for Life', *WIN*, 26 October 1978, p. 7.

43 M. Lucas, 'The "Uses" of Terrorism', *Fifth Estate*, Detroit, May 1977.

44 M. Lucas, 'Kalkar', Unpublished MS, Berlin, November 1977.

45 Translated from *Frankfurter Allgemeine*, 14 May 1977.

46 V. Rich, 'Berufsverbote continues in spite of resolutions', *Nature*, vol. 282, 6 December 1979, p. 549.

47 From an answer to a question in the German Federal Parliament in 1979 reported in Rich, 'Berufsverbote', p. 549.

48 Further detail on the use of Berufsverbote is contained in C. Forsyth, 'Berufsverbote', *Nation Review*, vol. 9, no. 1, 20–26 October 1979, p. 1. The content is in agreement with interviews undertaken by the author in Berlin, Hamburg, and Freiburg in December 1977. Further background is given in S. Cobler, *Law, Order and Politics in West Germany*, Penguin, Harmondsworth, 1976.

49 R. Jungk, *The Nuclear State*, tr. Eric Mosbacher, John Calder, London, 1978, p. 81.

50 Ibid.

51 Ibid.

NOTES   361

52 D. Hayes, J. Falk, and N. Barrett, *Redlight for Yellowcake*, FOE Australia, 1977, p. 21.

53 P. Sieghart, 'Guarding nuclear materials and civil liberties', *Bulletin of the Atomic Scientists*, May 1980, p. 33.

54 R. W. Ayres, 'Policing Plutonium: the Civil Liberties Fallout', *Harvard Civil Rights – Civil Liberties Review Law*, 1975, p. 369.

55 Sir B. Flowers (chairman), 'Nuclear Power and the Environment', *Royal Commission on Environmental Pollution*, 6th Report, Her Majesty's Stationery Office, London, 1976, p. 23, chapter VII.

56 Ranger Uranium Environmental Inquiry, *First Report*, Australian Government Publishing Service, Canberra, 1976, chapter 14.

57 Nuclear Energy Policy Study Group, *Nuclear Power Issues and Choices*, Ballinger, Cambridge, Mass., 1977, chapter 10.

58 International Commission of Jurists, 'Plutonium and Liberty', *Justice*, London, 1978.

59 M. Flood and R. Grove-White, *Nuclear Prospects*, FOE UK, 1976.

60 Ibid.; see in particular, pp. 24–8.

61 R. Grove-White, 'Nuclear power: the threat to personal freedom', *Nature*, vol. 282, 20–7, December 1979, p. 775.

62 'Problems in the Accounting for and Safe-guarding of Special Nuclear Materials', *Committee on Small Business, U.S. House of Representatives, Subcommittee on Energy and Environment*, Washington D.C., 26 April, 7, 27 May 1976.

63 'Intelligence Activities and the Rights of Americans', *U.S. Senate, Select Committee to study Government Operations With Respect to Intelligence Activities*, 1975.

64 D. Warnock and K. Bossong (eds), *Nuclear Power and Civil Liberties: Can We Have Both?*, Citizens Energy Project, Washington, 1978.

65 Quoted in Warnock and Bossong, *Nuclear Power*, p. 52.

66 Ibid., p. 53.

67 'South Africa to ban news on uranium', *Australian Financial Review*, 9 March 1978, p. 33.

68 Ranger Uranium Environmental Inquiry, *Second Report*, Australian Government Publishing Service, Canberra, 1977.

69 *Approved Defence Projects Protection Act 1947–1973*, section 4.

70 *The Environmental Protection (Nuclear Codes) Bill 1978*, sections 49, 48, 13.

71 'Uranium laws are repugnant', editorial, *Age*, Melbourne, 26 May 1978.

72 Quoted in Sieghart, 'Guarding nuclear materials', p. 34.

73 Sieghart, 'Guarding nuclear materials', p. 35.

74 See, for example, R. Miliband, *The State in Capitalist Society*, Quartet Books, London, 1973; and R. J. Barnet and R. E. Muller, *Global Reach: The Power of the Multinational Corporations*, Simon and Schuster, New

362   *GLOBAL FISSION*

York, 1974; and references contained therein.

75 Miliband, *The State*, p. 3.

76 H. Perlmutter, 'Super-Giant Firms in the Future', *Wharten Quarterly*, Winter, 1978, cited in R. J. Barnet and R. E. Muller, *Global Reach: The Power of the Multinational Corporations*, Simon and Schuster, New York, 1974, p. 27.

77 *International Investment Production*, U.S. Council, International Chamber of Commerce, New York, 1969, mimeo, cited in R. J. Barnet and R. E. Muller, *Global Reach: The Power of the Multinational Corporations*, Simon and Schuster, New York, 1974, p. 27.

78 R. J. Barnet and R. E. Muller, *Global Reach: The Power of the Multinational Corporations*, Simon and Schuster, New York, 1974, p. 230.

79 By 1970, petroleum companies accounted for 72 per cent of all natural gas production and reserves, 20 per cent of coal production and 30 per cent of coal reserves, 25 per cent of uranium milling, and 50 per cent of uranium reserves. *Small Business Committee, House of Representatives*, 1971, cited in Barnet and Muller, *Global Reach*, p. 428.

80 Calculated from 'The Fortune Directory of the 500 Largest U.S. Industrial Corporations', *Fortune*, 5 May 1980, p. 276.

81 'The Fortune Directory', *Fortune*, p. 276.

82 A. J. Large, 'Bees get buzz and the sun shines bright', *Australian Financial Review*, 15 October 1980, p. 36; evidence that birds or humans passing through the microwave beam would be quite likely to cook, and a growing scientific controversy over the safety of existing permitted levels of microwave exposure, has not dampened interest in the proposal. A combined NASA and US Department of Energy Symposium, held in Nebraska in mid 1980 to evaluate progress on the assessment of its feasibility, concluded that there were no technical 'show stoppers' to prevent construction of some sixty power satellites to be launched between 2010 and 2040. See L. Torrey, 'A trap to harness the sun', *New Scientist*, 10 July 1980, p. 124.

83 R. Munsen, 'Ripping off the sun', *The Progressive*, September 1979, p. 12.

84 'U.S. Government "aiding" takeover of solar power industry', *Australian Financial Review*, 19 November 1980, p. 21.

85 Munsen, *Ripping off the sun*, p. 14.

86 J. O'Connor, *The Fiscal Crisis of the State*, St Martins Press, New York, 1973; J. Habermas, *Legitimation Crisis*, Beacon Press, Boston, 1975; C. Offe, 'Crises and Crisis Management: Elements of a Political Crisis Theory', *International Journal of Politics*, Fall, 1976, p. 29. For an excellent analysis of the various issues at stake in developing this analysis of the state see B. Frankel, *Marxian Theories of the State: A Critique of Orthodoxy*, Arena Publications, Monograph no. 3, Melbourne, 1978.

NOTES 363

87 'Another blow to Brokdorf', *Nuclear Engineering International*, World Digest, March 1977.

88 'Nuclear Rubbish Tip Row', *Herald*, Melbourne, 17 November 1976, p. 13.

7 REFERRED TO THE PEOPLE

1 For an analysis of the three stages of development of the nuclear controversy in Switzerland see C. Zangger, 'Interaction between Control and Acceptance of Nuclear Energy in Switzerland: Aims and Implementation', *IAEA Bulletin*, vol. 20, no. 3, p. 50.

2 For a more detailed description of the development of the power industry see I. C. Bupp and J. C. Derian, *Light Water: How the Nuclear Dream Dissolved*, Basic Books, New York, 1978, pp. 137–44; L. Daleus, 'A Moratorium in Name Only', *Bulletin of the Atomic Scientists*, October 1975, p. 27; A. Jamison, 'On the politicization of energy in Denmark and Sweden', *Nordisk Forum*, 15, vol. 12, no. 3, 1977, pp. 23–9; N. E. Abrams, 'Nuclear Politics in Sweden', *Environment*, vol. 21, no. 4, May 1979, p. 6.

3 L. Daleus, 'A Moratorium in Name Only', *Bulletin of the Atomic Scientists*, October 1975, p. 28. This reactor was based on the principle that natural uranium can be used as a fuel if the core is immersed in a 'moderator' designed to produce many neutrons with slow speeds when bombarded by the products of the fissions that occur in natural uranium. A suitable moderator is heavy water. This is water in which the abundance of deuterium, an isotope of hydrogen, has been increased. Heavy water is produced by an electrolysis process.

4 Daleus, 'A Moratorium', p. 28.

5 I. C. Bupp and J. C. Derian, *Light Water: How the Nuclear Dream Dissolved*, Basic Books, New York, 1978, p. 138.

6 P. Weish, *The Nuclear Debate in Austria*, Ludwig Boltzman Institut für Umweltwissenschaften und Naturschutz, April 1977, p. 1.

7 For more detail of the relevant acts see Zangger, 'Interaction', p. 48.

8 For a description of the occupation of Kaisergaust see 'Nuclear Power Stop!', *Agenor*, no. 65, 1977, p. 10.

9 'World List of Nuclear Power Plants', *Nuclear News*, August, 1980, p. 101.

10 'Nuclear Power Stop', *Agenor*, no. 65, p. 76.

11 Weish, *The Nuclear Debate*, p. 1.

12 S. Thunell and L. Litjegren, 'Energy Policy for Greater Prosperity', *Social Democratic Party*, Sweden, undated, p. 3.

13 Daleus, 'A Moratorium', p. 32.

14 I am indebted for much of the detail in this account of the Finnish

364 GLOBAL FISSION

debate to Bruno Bärs' excellent paper, B. Bärs, 'The Nuclear Conflicts in Finland', unpublished MS, 1978; as well as M. Roos, 'Nuclear Power in Finland', unpublished MS, Department of Nuclear Physics, University of Helsinki, 1975; and Energiapolüttinen yhdistys – Vaihtoehto Ydinvoimalle (EVY), *Nuclear Energy Situation in Finland*, 11 December 1978.

15 The Energy Commission and Siting reports are cited in Bärs, 'The Nuclear Conflicts', p. 20 as *1974:112*, and *1974:57* respectively.

16 EVY, *Nuclear Energy Situation in Finland*, 11 December 1978, p. 2.

17 Bärs, 'The Nuclear Conflicts', p. 20.

18 M. Roos, 'Nuclear Power in Finland', unpublished MS, Department of Nuclear Physics, University of Helsinki, 1975, p. 2.

19 Bärs, 'The Nuclear Conflicts', pp. 17, 19.

20 Ibid., pp. 15–18.

21 Finnish Communist Party Congress: 1966 and 1968, cited in Bärs, 'The Nuclear Conflicts', p. 15.

22 Zangger, 'Interaction', p. 52.

23 W. C. Patterson, 'Austria's Nuclear Referendum', *Bulletin of the Atomic Scientists*, January 1979, p. 1.

24 Bupp and Derian, *Light Water*, p. 143.

25 A. Jamison, 'On the politicization of energy in Denmark and Sweden', *Nordisk Forum*, 15, vol. 12, no. 3, 1977, p. 31.

26 S. Frigren, 'Public Education for Energy Policy Decisions', *Conference on Nuclear Power and Its Fuel Cycle, Salzburg*, IAEA-CN-36/287, vol. 7, 1977, p. 139.

27 Daleus, 'A Moratorium', p. 32; Bupp and Derian, *Light Water*, p. 141.

28 This is certainly the view of Professor Dean Abrahamson who has played a central role in the Swedish debate. See D. Abrahamson, T. B. Johanson, P. Steen, and W. Barnaby, 'Sweden Debates its Peaceful Atom', *Bulletin of the Atomic Scientists*, November 1979, p. 29.

29 Bupp and Derian, *Light Water*, p. 141.

30 Daleus, 'A Moratorium', p. 33.

31 For more details of Miljöcentrum's style of work see D. Shapely, 'Sweden: Naderism Blooms in the North Country', *Science*, vol. 182, 12 October 1973, p. 145; M. Burros, 'The Unlikely Crusader', *Washington Post*, 29 September 1977, p. F1.

32 I. Wood, 'Interview with Bjorn Gillberg' (private communication), 1979.

33 L. Daleus, 'Sweden and Nuclear Power', *Current Sweden*, The Swedish Institute, no. 177, November 1977, p. 4.

34 D. Hinricksen, 'Sweden's "reservoir" of uranium', *Sweden–Now*, vol. 4. 1978.

35 'The Largest Antinuclear Demonstration Ever Held in Sweden', *Nucleonics Week*, 3 June 1976, p. 10.

36 I. Wood, 'Interview with Ingbar Carlsson, Energy Spokesman, Social Democrats' (private communication), 15 November 1979.

NOTES 365

37 Frigren, 'Public Education', p. 141.

38 Some suggestive additional evidence for the working of this process may be found in a study of attitudes to nuclear power in Boston. Of the respondents, those with 'low knowledge' about nuclear power supported it by four to one, while those with 'high knowledge' were evenly divided on the issue. Similar, although less polarized trends, were found at three other places (all nuclear sites) in the USA. See R. E. Kasperson, G. Berk, et al., 'Public Opposition to Nuclear Energy: Retrospect and Prospect', in C. T. Unseld, D. E. Morrison, et al., *Sociopolitical Effects of Energy Use and Policy*, Study of Nuclear and Alternative Energy Systems, Supporting Paper 5, Report to the Committee on Nuclear and Alternative Energy Systems, National Academy of Sciences, Washington, 1979, pp. 280–3.

39 Zangger, 'Interaction', p. 53.

40 Patterson, 'Austria's Nuclear Referendum', p. 1.

41 P. Weish, 'The Nuclear Referendum in Austria: Its Antecedents and Consequences', *FOE International Conference-Workshop*, Rome, 17–20 May 1979, p. 2.

42 S. Powell, 'Austria Wary on N-Power', *Herald*, Melbourne, 24 January 1977, p. 11.

43 I am indebted to the OOA for their two detailed accounts of the development of the nuclear issue in Denmark, as presented in OOA, 'OOA – a short presentation', Ryesgade 19, 2200 Copenhagen N, Denmark, 1977; S. Christiansen, 'The Danish Information Campaign on Atomic Power 1974–1977', OOA, Denmark, June 1977. Material from Jamison, 'On the politicization', pp. 23–9, has also proved valuable.

44 OOA, 'OOA – a short presentation', p. 1.

45 D. Hinricksen and P. Cawood, 'Fresh Breeze for Denmark's Windmills', *New Scientist*, 10 June 1976, p. 567.

46 Ibid.

47 'Tvindmill Turning', *WISE*, no. 1, May 1978, p. 7.

48 S. Blegaa, et al., 'Skitze til Alternative Energiplan for Danmark', OVE & OOA, Denmark, 1976.

49 S. Christiansen, 'The Danish Information Campaign on Atomic Power 1974 – 1977', OOA, Denmark, June 1977.

50 OOA, 'OOA – a short presentation', p. 1.

51 For a list of actions in Switzerland to 1978 see A. Gyorgy and Friends, *No Nukes: everyone's guide to nuclear power*, South End Press, Boston, 1979, pp. 323–5.

52 Zangger, 'Interaction', p. 52.

53 'Swiss Vote for Nuclear Power', *Age*, Melbourne, 20 February 1979, p. 8.

54 From 'World List of Nuclear Power Plants', *Nuclear News*, August 1979, p. 76.

55 W. Barnaby, 'Sweden', *Nature*, vol. 264, 4 November 1976, p. 9.

366   *GLOBAL FISSION*

56 Gyorgy and Friends, *No Nukes*, p. 333.

57 Christiansen, 'The Danish Information Campaign', p. 10.

58 'Scandinavia's anti-nuclear march', *Peace News*, 21 August 1977.

59 S. Bergstrom, Information Officer, Ministry of Industry, Sweden, *Memorandum*, September 1978.

60 Abrams, 'Nuclear Politics in Sweden', *Environment*, vol. 21, no. 4, May 1979, p. 7.

61 Ibid., p. 11.

62 W. Barnaby, 'Sweden's new government wrestles with nuclear power', *Nature*, 30 November 1978.

63 Ibid.

64 'The Swedish Nuclear Inspectorate Has Recommended Startup', *Nucleonics Week*, 5 April 1979, p. 10.

65 'News Releases' from *WISE*, 11 April 1979.

66 H. Rehnvall, 'Political Chaos in Sweden', *New Scientist*, 19 April 1979.

67 'Chancellor may quit over N Poll defeat', *Age*, Melbourne, 7 November 1978, p. 7.

68 'The Swiss Nuclear Initiative was Defeated By a Narrow Margin', *Nucleonics Week*, 22 February 1979, p. 6.

69 'An Intensive Campaign to Inform and Influence Swiss Voters', *Nucleonics Week*, 1 February 1979, p. 11.

70 Bergstrom, *Memorandum*, p. 6; W. Barnaby, 'Swedish Experts Recommend Nuclear Energy-without reprocessing', *Nature*, vol. 272, 23 March 1978.

71 T. B. Johansson and P. Steen, *Solar Sweden*, Secretariat for Future Studies, Stockholm, 1977. For a good summary of this report see also G. Taylor, 'Sweden strides towards a solar society', *New Scientist*, 24 August 1978, p. 550.

72 *The Swedish referendum on nuclear power*, Swedish Embassy, Press Department, 23 March 1980.

73 Weish, 'The Nuclear Referendum', p. 3.

74 Ibid.

75 D. Dangelmayer, 'Swiss prepare for nuclear referendum', *New Scientist*, 8 February 1979, p. 366.

76 'The Swiss Are Not Making Any Official Prognosis On The Outcome of the National Vote', *Nucleonics Week*, 15 February 1979, p. 9.

77 Dangelmayer, 'Swiss prepare'.

78 'Switzerland Referendum Lost', *WISE*, no. 4, March 1979, p. 5.

79 T. Kalvemark, Department Head, Swedish Board of Universities and Colleges, *Swedish Public Discussion on Nuclear Power*, Preprint, no. 245, (6), February 1980, pp. 8, 9.

80 Weish, 'The Nuclear Referendum', p. 4.

81 'Austria's GRT Has Solved The Problem of Its Reprocessing Contract', *Nucleonics Week*, 11 January 1979.

NOTES

82 'Swiss Vote', *Age*, Melbourne, p. 8.

83 W. Dullforce, 'Nuclear power vote in Sweden fails to clear political air', *Australian Financial Review*, 28 March 1980, p. 32.

84 'Sweden's Nuclear Referendum Wound Up With Everyone Claiming a Victory', *Nucleonics Week*, 27 March 1980, p. 13.

85 Dullforce, 'Nuclear power vote', p. 32.

86 W. Barnaby, 'Swedes vote on their nuclear future', *Nature*, vol. 284, 13 March 1980, p. 117.

87 'Sweden to Press U.S. for "Generic" Approval of Reprocessing', *Nucleonics Week*, 1 May 1980, p. 1.

88 S. Drake and A. Collings, 'A Nod to Nuclear Power', *Newsweek*, 7 April 1980, p. 45.

89 'Activists Back on the Streets of Sweden', *WISE*, 61, 16 September 1980.

90 W. Barnaby, 'Limited nuclear programme favoured', *Nature*, vol. 284, 3 April 1980, p. 390.

91 'Activists', *WISE*.

92 Weish, 'The Nuclear Referendum', p. 4.

93 'The Austrian Parliament Will Probably Vote By The End of This Year', *Nucleonics Week*, 17 July 1980, p. 9.

94 'Swiss Blockade', *WISE*, no. 1, May 1978, p. 11.

95 'Switzerland faces two nuclear power station shocks', *Guardian*, 6 November 1979.

96 Ibid.

97 'Local Authorities Around Basel Are Again Opposing the Kaiseraugst Nuclear', *Nucleonics Week*, 25 October 1979, p. 5.

98 'A Committee Backing on Antinuclear Statute in the Swiss Canton', *Nucleonics Week*, 6 March 1980, p. 14.

99 'Swiss A-plans blocked', *Guardian*, 21 March 1980.

1 'Switzerland's NOK has given up plans for the Ruethi Nuclear Plant', *Nucleonics Week*, 6 March 1980, p. 18.

2 'Swiss Voters Decisively Approved a Partial Revision', *Nucleonics Week*, 24 May 1979, p. 10.

3 R. Blystone, 'Iceland's Energy Clean, Hot, Free for those living above volcano', *Journal Herald*, 25 May 1979.

4 'Norway Has No Thought of Nuclear Power Development Before 2000', *Nucleonics Week*, 6 March 1980, p. 15.

5 EVY, 'Finland Anti Nuke Newspaper', *WISE Bulletin*, 35–308, 1980.

6 EVY, 'Home Review (Finnish News)', reprinted, *International Nuclear News Service*, no. 13, June 1980, p. 9.

7 'First large scale demonstration in Finland', *WISE*, vol. 2, no. 4, May–June 1980, p. 2.

8 J. Karlsen, 'Uranium Mining Postponed in Greenland', *Keep it in the Ground, International Newsletter, WISE*, no. 4, March 1980, p. 4.

368   *GLOBAL FISSION*

9 Ibid.

10 'Denmark and Sweden are Jointly Studying Safety Considerations', *Nucleonics Week*, 19 April 1979, p. 12.

11 'Denmark: Campaigning to Close Barsebäck', *WISE*, no. 5, May–June, 1979, p. 15.

12 'Calendar of Events: May', *WISE*, no. 5, May–June, 1979, p. 19.

13 'Denmark: Campaigning', *WISE*, p. 15.

14 Christiansen, 'The Danish Information Campaign', p. 8.

15 T. Ogg, 'Europe nears crossroads in uranium power debate', *Australian Financial Review*, 11 September 1979, p. 17.

---

8 CENTRAL CONTROL, REGIONAL REVOLT

1 R. Miliband, *The State in Capitalist Society*, Quartet Books, London, 1973, p. 49.

2 'World List of Nuclear Power Plants', *Nuclear News*, August 1979, p. 81.

3 Ibid.

4 For a more detailed account see W. C. Patterson, *Nuclear Power*, Penguin, Harmondsworth, 1976, p. 162.

5 See, for example, Patterson, *Nuclear Power*; A. B. Lovins and J. Price, *Non Nuclear Futures: The Case for an Ethical Energy Strategy*, FOE and Ballingers Publishing Co., London, 1975, and A. B. Lovins, *Soft Energy Paths: Toward a Durable Peace*, Penguin, Harmondsworth, 1977.

6 Sir B. Flowers (chairman), 'Nuclear Power and the Environment', *Royal Commission on Environmental Pollution*, Her Majesty's Stationery Office, London, 1976, p. 205, conclusion 50.

7 For a journalistic account of the evidence see *Windscale: A summary of The Evidence and Argument*, Guardian, 1978.

8 I. Breach, 'Case Dismissed', *New Scientist*, 9 March 1978, p. 634.

9 J. Bugler, 'The Windscale Verdict', *New Statesman*, 10 March 1978.

10 'N-protests hit Britain, U.S.: Thousands rally over Windscale', *Age*, Melbourne, 1 May 1978, p. 6.

11 N. Lucas, 'Energy in the French Plans', *International Relations*, May 1978, p. 225.

12 For more detail of the history of the French nuclear industry see I. C. Bupp and J. C. Derian, *Light Water: How the Nuclear Dream Dissolved*, Basic Books, New York, 1978, and the references contained therein.

13 I am particularly indebted to Les Amis de la Terre for their detailed description of the development of the French opposition to nuclear power 1960–77 contained in Les Amis de la Terre, *L'Escroquerie Nucléaire*, Editions Stock, Paris, 1978, pp. 299–330. Except where indicated otherwise, the detail given for the early development of the movement (1960–74) is drawn from this source.

14 Bupp and Derian, *Light Water*, p. 305.

NOTES 369

15 A. Chausseborg, 'Le Débat sur l'Énergie a l'Assemblée: Les députés ne refusent pas le fait nucléaire, mais demandent qu'il soit mieux contrôlé', *Le Monde*, 16 May 1975, p. 10.

16 A. Gyorgy and Friends, *No Nukes: everyone's guide to nuclear power*, South End Press, Boston, 1979, p. 341.

17 R. Howard, 'The French Nucleocrates Battle On', *The Elements*, November 1978, no. 44, p. 5.

18 'Nuclear Power Stop', *Agenor*, no. 65, 1977.

19 Ibid., p. 24.

20 Howard, 'The French Nucleocrates', p. 6.

21 This point is discussed in more detail by Michael Lucas in 'Malville', unpublished MS, Berlin, November, 1977.

22 Howard, 'The French Nucleocrates', p. 6.

23 Les Amis de la Terre, *L'Escroquerie Nucléaire*, Éditions Stock, Paris, 1978, p. 310.

24 Ministère de l'Industrie et de la Recherche: pamphlet dated 15 November 1975.

25 Les Amis de la Terre, *L'Escroquerie Nucléaire*, p. 315.

26 The following account of the opposition to nuclear power in Flamanville is taken from the author's interviews with members of CRILAN and CCPAH during a visit in October 1977.

27 Comité Contre la Pollution Atomique dans la Hague, *La Hague: Impact Ecologique de l'Usine de Retractement*, CCPAH, Cherbourg, August 1977.

28 'L'Usine de la Hague', SNPEA-CFDT, la Hague, July 1976.

29 Syndicat CFDT de l'Énergie atomique, *L'Électronucléaire en France*, Éditions de Seuil, Paris, 1975.

30 Howard, 'The French Nucleocrates', p. 5.

31 CFDT, 'Energy Policy: the CFDT positions', Syndicat National du Personnel de l'Énergie Atomique CFDT, Bureau National BP No. 2, Paris, May 1978.

32 'Les Amis de la Terre et la CRILAN souhaitent enlargir les convergences avec la gauche politique et syndicale contre l'extension de la Hague', *La Manche*, 13 November 1978.

33 'Despite a Massive Demonstration, the French received the First Japanese Spent Fuel', *Nucleonics Week*, 25 January 1979, p. 3.

34 J. Spivak, 'Nuclear-Power Plans Unchanged in Europe Despite Rising Protests', *Wall Street Journal*, 3 April 1979.

35 These included 3 000 demonstrating at Nogent-sur-Seine on 2 June 1979, 'Gentle Sabotage at French Nuclear Site', *WISE*, Communique, 17 June 1979; sabotage of equipment at the site a week later, 'International Profile II', *Financial Times*, 8 May 1979, and the collection of 36 000 signatures; and a high tension pylon blow up in Malville in May, 'Gentle Sabotage at French Nuke Site', *WISE*, 17 June 1979.

36 F. J. Prial, 'Breton villagers in long struggle against N-plant', *Boston Herald American*, 18 March 1980.

37 'Police out in force at trial of nuclear protesters', *Guardian*, 18 March

1980; 'Plogoff opposition continues despite violent police reprisals', *WISE*, vol. 2, no. 4, May–June 1980, p. 3; 'Plogoff? Round One for Survival', *Agenor*, no. 80, October 1980.

38 'Mounting Tension at New Nuke Site in Brittany, *WISE*, News Communique no. 35, 12 February 1980.

39 'Larzac: Time to help', *Agenor*, no. 72, December 1978.

40 R. Clavaud, 'Uranium fever: farmers start to resist', *Guardian*, 27 July 1980.

41 Prial, 'Breton villagers'.

42 'Plogoff opposition', *WISE*, p. 3; 'Police out', *Guardian*.

43 'Plogoff opposition', *WISE*, p. 3.

44 Prial, 'Breton villagers'.

45 Ibid.

46 'Massive Antinuke Demonstration at Plogoff, France!', *WISE*, News Communique, 27 May 1980.

47 J. Bugler, 'Irish reactor bogged down in debate', *New Scientist*, 16 March 1978, p. 712.

48 This and following detail is drawn from B. Trench, 'The Anti-Nuclear Movement in Ireland', unpublished MS, Dublin, 1979, unless otherwise noted.

49 'Munster Mashed', *Peace News*, 18 April 1980.

50 'Locals Force Withdrawal of Uranium Explorers', *Keep it in the Ground Newsletter*, no. 4, March 1980, p. 7.

51 For details of withdrawal from Fintown see *Locals Force Withdrawal*, p. 7; for details of withdrawal from Glenleighan see 'Munster Mashed', *Peace News*.

52 'Irish Transport and General Workers' Union', *Annual Delegate Conference*, 29 May–1 June 1979.

53 J. Caroll, 'Electricity Supply Board Union against nuclear power', *WISE*, vol. 2, no. 1, November–December 1979, p. 9.

54 J. Caroll, 'Trade unions to call for referendum, *WISE*, vol. 2, no. 3, March–April 1980, p. 6.

55 D. Nolan, 'Carnsore reactors in cold storage', *WISE*, vol. 2, no. 4, May–June 1980, p. 14.

56 I am indebted to Dave Smith for his detailed account of the movement in the UK published in D. Smith, 'The Anti-Nuclear Movement in Scotland', *Science for People*, no. 40–41, Autumn 1978, p. 44. Except where otherwise noted, much of the detail that follows on Scotland is drawn from that article.

57 Except where otherwise noted the following detail on Wales is drawn from *Portskewett no thanks*, pamphlet, FOE, London; J. Marjoram, 'South Wales Groups Successfully Thwart Industry Plans', *WISE*, no. 5, May–June 1979, p. 9.

58 'The Central Electricity Board has Shelved Indefinitely', *Nucleonics Week*, 24 April 1980, p. 8.

NOTES

59 I am indebted to Michael Flood for the detail in his excellent booklet on the Torness conflict: M. Flood, *Torness, keep it green*, FOE, London, 1979. Except where otherwise noted much of the detail on Torness is drawn from this booklet; from Smith, *The Anti-Nuclear Movement*, p. 44; and from L. Bradburn and J. Leach (former activists with the Torness Alliance), (private communication), 1980.

60 'Antinuclear Campers Say it in Flowers', *Guardian*, 7 May 1979.

61 'Totally Opposed', *SCRAM Energy Bulletin*, no. 12, June–July 1979, p. 1.

62 *SCRAM Energy Bulletin*, no. 13, August–September 1979, p. 2.

63 'Totally Opposed', *SCRAM Energy Bulletin*.

64 'Our M.P. Working for You', *SCRAM Energy Bulletin*, no. 13, August–September 1979, p. 2.

65 'People against nukes . . .', *SCRAM Energy Bulletin*, October–November 1979, p. 1.

66 'High Cost Nuclear Go-ahead', *Guardian*, 15 April 1980.

67 M. Morris, 'Orkney any drilling opposed', *Guardian*, 21 March 1979.

68 For further detail see Morris, 'Orkney any drilling'; and R. Moody, 'Orcadia versus urania', *Peace News*, 20 April 1979.

69 'Scotland Orkney Islands Drilling Opposed', *WISE*, no. 5, May–June, 1979, p. 11.

70 R. Moody, 'Orcadia versus urania', *Peace News*, 20 April 1979.

71 B. Wilson, 'Orkney sinks uranium plan', *Guardian*, 30 December 1979.

72 See 'A Promise to Move Mountains', FOE, Aberdeen, Scotland, 1977, and Smith, 'The Anti-Nuclear Movement', p. 44.

73 For a description of the HARVEST plan see Sir J. Hill, 'Nuclear Waste Disposal', *Electronics and Power*, May 1978, p. 347.

74 'Radioactive Waste', *Telegraph Sunday Magazine*, 1979, pp. 17, 20.

75 'Scots Nuclear Dustbin Fears', *Sydney Morning Herald*, 11 April 1977.

76 P. Hetherington, 'Anger at Limit on Nuclear Dumping Inquiry', *Guardian*, 31 March 1980.

77 Smith, 'The Anti-Nuclear Movement', p. 46.

78 'Radioactive Waste', *Telegraph*, p. 17.

79 Scottish Campaign to Resist the Atomic Menace, *Nuclear Waste – the plain facts*, SCRAM, 2A Ainslie Place, Edinburgh, 1979, p. 2.

80 'Scots Nuclear Dustbin', *Sydney Morning Herald*.

81 P. J. Roche, 'Britain's First Nuclear Waste Inquiry?', *Ecologist*, vol. 10, no. 4, April–May 1980.

82 P. Lesley, 'Anti Waste Dumping Tactics', *WISE*, 59, 2 September 1980.

83 A. Raphael, 'Maggie pushes plan for 20 U.K. Nuclear reactors', *Observer*, 14 October 1979.

84 'Britain's Central Electricity Generating Board is to Study Five Sites', *Nucleonics Week*, 6 March 1970, p. 16.

372   GLOBAL FISSION

85 'Time to heed the Experts', *New Scientist*, 20 November 1979, p. 674.

86 Ibid.

87 British Nuclear Fuels Limited, *The Leakage of Radioactive Liquid into the Ground*, Windscale, UK, 18 March 1980.

88 M. Morris, 'Union joins complaints on leak at Windscale', *Guardian*, 2 August 1980.

89 'Sweden Appears to React Strongest to the Three Mile Island Accident', *Nucleonics Week*, 12 April 1979, p. 9.

90 Spivak, 'Nuclear-Power Plans'.

91 'French National Petition Campaign on Energy', *Translated Text of the Petition*, (private communication), 1979.

92 D. Ediger, 'French Socialists Prepare to War Against Government Nuclear Program', *Nucleonics Week*, 26 July 1979, p. 5.

93 'International Profile II', *European Energy Report, Financial Times*, no. 33, 8 May 1979.

94 'French Workers Reveal Reactor Risks', *Nuclear Newsletter*, 24 October 1979.

95 'Despite Continuing French Concern Over Cracks', *Nucleonics Week*, 18 October 1979, p. 9.

96 'Anti-Nuclear Strike', *WISE*, vol. 2, no. 1, November–December 1979, p. 9.

97 J. Ritter, 'France faces a cold winter', *Nature*, vol. 282, 29 November 1979, p. 436.

98 'La Hague Workers Demanding Several Months Stoppage for Overhaul of Plant', *Nucleonics Week*, 24 April 1980, p. 2.

99 Ritter, 'France faces a cold winter', p. 436.

 1 S. Milson, 'France's Left celebrates win', *Sydney Morning Herald*, 12 May 1981, p. 1; 'Confusion after euphoria', *Sydney Morning Herald*, 13 May 1981, p. 5.

 2 Milson, 'France's Left celebrates', p. 1; P. Smark, 'How the candidates see the issues', *Sydney Morning Herald*, 9 May 1981, p. 10.

 3 J. Columbani, 'Socialists Woo their Ally', *Guardian*, 7 June 1981, p. 11; 'Socialists move fast on radical ideas', *Sydney Morning Herald*, 11 August 1981, p. 5. The suspended reactors were Chooz, Golfech, Le Pellerin, Nogent-sur-Seine, and Penly; 'Cinq centrales nucléaires gelées aujourd'hui', *Liberation*, 30 July 1981, p. 6.

---

9 THE INTERNATIONAL ARENA

1 Except where indicated otherwise, the details of the Whitsunday demonstrations are drawn from 'The Anti-Nuclear Demonstrations on Which the Sun Never Set: Round-up on world-wide anti-nuclear

NOTES

demonstrations, 2–3–4 June', *WISE*, News Communique, 6 June 1979.

2 Swiss Association of Anti-Nuclear-Power-Organisations, International Coordination Meeting of Antinuclear organisations, *Minutes of Proceedings 24–25 June*, Basel, 22 July 1978.

3 For more detail see 'May 24–26 Whitsunday International days for anti-nuclear action', *WISE*, vol. 2, no. 5, July–September 1980, p. 3.

4 *Keesings Contemporary Archives*, 16 January 1976, p. A27517.

5 'Nuclear Rubbish Tip Row', *Herald*, Melbourne, 17 November 1976, p. 13.

6 D. Gardiner, 'Mounting Violence in Spanish Anti-Nuclear Protests', *Financial Times European Energy Report*, Issue no. 37, 3 July 1979, p. 9.

7 'Spain's Energy Plan Sweeps Through Parliament after Marathon Debate', *Nucleonics Week*, 2 August 1979, p. 7.

8 Trillo 1 and 2 (Madrid); Vandellos 1, 2, and 3, and Asco 1 and 2 (near Barcelona in Catalonia); Valdecaballeros 1, 2, Almaraz 1, 2, Sayago and Garonya (Extremadura); Regodola, Lemoniz 1 and 2 (Eskuadi).

9 Gardiner, 'Mounting Violence', p. 10.

10 *Not Man Apart*, 15 August 1976.

11 '150 000 AKW-Gegner demonstrieren in Bilbao!', *Die Internationale*, no. 28, August 1977, p. 23.

12 S. Drake and M. Acoca, 'The Power Struggle', *Newsweek*, 24 April 1978.

13 Ibid.

14 See 'Large Basque Crowd Protests Plans for Nuclear Plant in Navarre', *New York Times*, 16 July 1977; J. Roca, 'Nuclear in Spain', *The Elements*, February 1978. For a particularly good account of the detail of demonstration see '200 000 in Bilbao Protest Nuclear Plant', *Rouge*, 17 July 1977, tr. Intercontinental Press.

15 '150 000 AKW-Gegner', *Die Internationale*, p. 23.

16 Commission for a Non-Nuclear Basque Coast, 'Basques: big demo, little bomb', *WISE*, no. 1, May 1978, p. 12; Gardiner, 'Mounting Violence', p. 9.

17 'Basques: big demo, little bomb', *WISE*, p. 12.

18 Ibid.

19 There is disagreement over whether the warning was given ten minutes or half an hour before the blast took place; see *Peace News*, 7 April 1978, and 'Basque Nationalists', *Peace News*, 19 May 1978.

20 Drake and Acoca, 'The Power Struggle'.

21 CANC (Catalonia), 'News Release', in *No Nuke News*, April–May 1979, p. 38.

22 'Two Town Councils Demand an Immediate Stop on the Lemoniz Nuclear Plant', *Nucleonics Week*, 3 May 1979, p. 10.

23 'An IAEA Team is Basically Satisfied with the Lemoniz Nuclear Plant', *Nucleonics Week*, 17 May 1979, p. 14.

374 GLOBAL FISSION

24 J. M. Arteta, 'The Spanish Police let loose against anti-nuclear demonstration – a Young Basque demonstrator killed by a bullet', *Liberation*, 5 June 1979.

25 Gardiner, 'Mounting Violence', p. 9.

26 'Basque Terrorists Attacked The Equipos Nucleares Factory', *Nucleonics Week*, 15 November 1979, p. 8.

27 B. Trench, 'Massive protest may stall reactor', *New Statesman*, 9 November 1979, p. 716.

28 'Spain Delays Nuclear Plant; Protest by 100 Mayors Ends', *Los Angeles Times*, 9 August 1979; 'Spanish Mayors protest against nuclear plant', *Newport News Times-Herald*, 30 August 1979.

29 'Spain Delays', *Los Angeles Times*.

30 'The Valdecaballeros Town Council voted 6–2 to Approve', *Nucleonics Week*, 6 March 1979, p. 13.

31 Trench, 'Massive protest', p. 716.

32 Ibid.; 'Spain's Nuclear Power Plan Got Tacit Approval and Even Reinforcement', *Nucleonics Week*, 5 July 1979, p. 10.

33 Gardiner, 'Mounting Violence', p. 9.

34 'Madrid worried over autonomy momentum', *Guardian*, 3 March 1980.

35 Except where otherwise noted, the detail of the early development of the opposition to nuclear power in the Netherlands is drawn from E. van der Hoeven, 'Fact Sheet on the Nuclear Situation in the Netherlands', Mileudefensie, 20 April 1977.

36 Interview with Lisbeth van Driel (research officer for the NVV) and Hans Van Poelje (Research Department, FNV), Amsterdam, 2 December 1977.

37 Interview with Erick van der Hoeven, Milieudefensie, Amsterdam, 30 November 1977.

38 Interview with Odillia Boele, Milieudefensie, Amsterdam, 1 December 1977.

39 IKV, 'Information on the IKV-Campaign Against Nuclear Weapons', The Hague, 1977, p. 1.

40 J. de Onis, 'Vance Asks Brazilians to Curb Nuclear Program', *New York Times*, 23 November 1977.

41 A particularly good account was given in F. Fullgraf, 'The Peaceful Bomb', transcript of programme on radio WDR/SFB, West Germany, 24 May 1977.

42 'Briefs', *Nucleonics Week*, 19 April 1979, p. 9.

43 'Demonstration of Thousands at the Salt Dunes', tr. from *NRC Handelsblad*, 5 June 1979, p. 3.

44 N. Hildyard, 'European Elections', *Ecologist*, vol. 9, nos. 4–5 August 1979, p. 167.

45 'Demonstrations of Thousands', *NRC Handelsblad*, p. 3.

46 'Briefs', *Nucleonics Week*, p. 8.

NOTES 375

47 '20 000 protest NATO missiles', *Boston Globe*, November 1979, reproduced in *No Nuclear News*, Round 3, no. 3, November 1979, p. 19.

48 'Dutch Parliament rejects new missiles', *Nature*, vol. 282, 13 December 1979.

49 'Dutch Government Unveils Plan to Build 3 Nuclear Power Plants', *New York Times*, 8 July 1980.

50 Interview with Erik van der Hoeven, Milieudefensie, Amsterdam, 30 November 1977.

51 'Dutch Opinion 60% Against More Nuclear Power Plants', tr. from *NRC Handelsblad*, 31 October 1979.

52 D. Nelkin and M. Pollak, 'French and German Courts on Nuclear Power', *Bulletin of the Atomic Scientists*, May 1980, p. 37.

53 'Energy Debate in Parliament', *NEI World Digest*, June, 1977.

54 See P. Taylor, 'The Struggle Against Nuclear Power in Central Europe', *Ecologist*, vol. 7, no. 6, July 1977, p. 221.

55 UKAEA, 'Nuclear Power in Germany', *Atom*, 271, May 1979, pp. 124–5.

56 'Added Nuclear Power Facilities in Germany Delayed Up to 6 Years', *Wall Street Journal*, 6 August 1979.

57 M. Getler, 'West German Nuclear Business Dries Up', *Washington Post*, 1978.

58 A. Johansen, 'Environmentalists stopper Germany's nuclear energy', *New Scientist*, 22 March 1979, p. 934.

59 S. McQueen, 'Gorleben Hearings Start with Germany's Nuclear Future Riding on Them', *Nucleonics Week*, 29 March 1979, p. 11.

60 G. Stewart, 'Revolution casts long shadow', *Canberra Times*, 20 March 1979.

61 'Demonstration of Thousands', *NRC Handelsblad*.

62 M. Bazin, The politics of power in Brazil', *Nature*, vol. 284, 24 April 1980, p. 655.

63 P. Eisner, 'Brazil's $16m nuclear plant program faces fresh delays', *Australian Financial Review*, 9 February 1981, p. 23.

64 From a speech by Marcio Moreira Alves, former federal deputy of the opposition party Movimento Democratico Brasileiro, March 1978, quoted in A. Gyorgy and Friends, *No Nukes: everyone's guide to nuclear power*, South End Press, Boston, 1979, p. 376; Eisner, 'Brazil's $16m', p. 23.

65 The government's estimates for Brazil's total hydroelectric capacity have increased from 100 GWe (in 1975) to 213 GWe (in 1981). W. Hoge, 'Brazil's not-so-enthusiastic entry into the atomic era', *Australian Financial Review*, 21 May 1981, p. 16.

66 R. Smith, 'Nuclebras will Run Brazilian Program after Angra-2', *Nucleonics Week*, 31 May 1979, p. 6.

67 R. Smith, 'Antinuclear Forces in Brazil Gathering Momentum', *Nucleonics Week*, 7 June 1979, pp. 12, 13.

68 'Brazil curbs N-power plans', *Boston Globe*, 14 October 1979.

69 'Anti-nuclear demonstration in Brazil', *WISE*, vol. 2, no. 4, May–June 1980, p. 3.

70 'Appeal by Scientists', *Boston Globe*, 20 July 1980.

71 'Vote for referendum on nuclear power in Rio Grande do Sul in Brazil', *WISE*, vol. 2, no. 5, July–September 1980, p. 8.

72 'Brazil: fourth and fifth reactors placed', *WISE*, vol. 2, no. 5, July–September 1980, p. 20.

73 Eisner, 'Brazil's $16m', p. 23.

74 'Anti-nuclear election pressure', *NEI World Digest*, July 1976.

75 *Frankfurter Allgemeine*, 23 March 1976; a similar figure was given much later by P. Kelly (chairperson of the Green Party in West Germany), 'Speech to the Harrisburg Rally', Trafalgar Square, FOE, UK, March 1980.

76 ' "Schwarz oder rot, wir schlagen euch tot": Bürgerinitiativen – Stopp für den Staat?', *Der Spiegel*, no. 13, 21 March 1977, p. 35.

77 'Bonn is Facing Issue of Nuclear Waste Disposal', *New York Times*, 12 November 1978; Interview with Roland Vogt and Michael Schroerer, BBU, Berlin, 7 December 1977.

78 J. Vinocur, 'Party splits threat to future in West Germany', *Australian Financial Review*, 21 June 1978, p. 13.

79 W. Ellis, 'Germans finding few buyers for N-plants', *Boston Globe*, 23 November 1978.

80 'German breeder funds blocked', *NEI World Digest*, June 1977.

81 J. Car, 'West Germany's nuclear power headaches', *Australian Financial Review*, 20 February 1979.

82 'The Mood Was Glum Last Week At The European Nuclear Conference', *Nucleonics Week*, 17 May 1979, p. 12.

83 'Further Nuclear Plant Construction In West Germany Has Won Qualified Endorsement', *Nucleonics Week*, 21 June 1979, p. 9.

84 'Schism on N-power threatens Schmidt', *Australian*, 4 December 1979, p. 5.

85 'Nuclear Rubbish', *Herald*, Melbourne, p. 13.

86 For further descriptions of the fight over Gorleben see C. Simpson, 'Gorleben Soll Leben: A desire for Life', *WIN*, 26 October 1978, pp. 4–8.

87 Simpson, 'Gorleben Soll Leben', p. 6.

88 Translation cited in 'Nuclear Rubbish', *Herald*, Melbourne, p. 13.

89 'Nuclear Rubbish', *Herald*, Melbourne, p. 13; P. Wood, 'Gorleben – the part the Greens played', *Ecologist*, vol. 9, nos. 4–5, August 1979, pp. 165–6.

90 A. Dicks, 'The Crisis in German nuclear plans', *Australian Financial Review*, 28 May 1979, p. 2.

91 Johansen, 'Environmentalists', p. 934.

NOTES 377

92 Letter from the Bürgerinitiative Lüchow–Dannenburg to 'all friends', 10 July 1978; *Elbe–Jeetzel–Zeitung*, 17 August 1978.

93 'Echo in German Politics', *WISE*, no. 2, July 1978, p. 4.

94 Dicks, 'The Crisis', p. 2.

95 'Lower Saxony Will Get at Least $275-Million Compensation Payments', *Nucleonics Week*, 4 January 1979, p. 14.

96 Johansen, 'Environmentalists', p. 934.

97 Ibid.

98 J. Spivak, 'Nuclear Power Plans Unchanged in Europe Despite Rising Protests', *Wall Street Journal*, 3 April 1979.

99 'Massive Demonstration in Hannover', *WISE News Release*, 3 April 1979.

1 J. Vinocur, 'Bonn Puts Brake on Nuclear Race', *Tenneseean*, 17 May 1979.

2 'Gorleben: First Trees are Cut Down', *WISE*, 21 September 1979, p. 20.

3 K. Hopfer, 'Germany to decentralise nuclear waste treatment', *Nature*, vol. 281, 11 October 1979.

4 'Two More Reprocessing Plants as a Substitute for Gorleben (W. Germany)', *WISE*, January 1980.

5 For details of opposition to waste storage and reprocessing at Lingen, see 'Demonstration against nuclear Park', *WISE*, vol. 2, no. 5, July–September 1980, p. 4.; at Gorleben, see 'Gorleben: Thirty-three days in the Republic of Free Wendland', *Die Tageszeitung*, 21 June 1980; at Rabenau, see 'Looking for a Reprocessing Plant in Hessen', *WISE News Communique*, 56, 1980; at Wuergassen, see '14 000 objections against dry storage', *WISE*, vol. 2, no. 2, July–September 1980, p. 22; and at Gundremmingen, see '... and 15 000 against compact storage', *WISE*, vol. 2, no. 2, July–September 1980, p. 22.

6 *Umweltmagazin*, no. 11–12, 1979, p. 8.

7 Ibid., no. 5, 1980, p. 2.

8 R. Boyes, 'Green power forces its way into West German politics', *Australian Financial Review*, 21 March 1980.

9 G. Korporaal, 'Germany's Schmidt Runs Into Flak', *Australian Financial Review*, 23 February 1981, p. 10.

10 'Euromissiles: German "Greens" Launch Campaign', *Europe*, no. 3071 (new series), 5 February 1981, p. 4.

11 J. Geddes, 'West Germans in new dispute over energy "solution"', *Australian Financial Review*, 30 January 1981, p. 22; G. Korporaal, 'Germany's Nuclear Dilemma', *Australian Financial Review*, 27 February 1981.

12 A. Tucker, 'Bonn shuts four A-plants', *Guardian*, 1 March 1981, p. 7.

13 Ibid.

10 COUNTER-CURRENTS WITHIN THE LEFT

1 A. Kollontai, 'The Workers' Opposition (1921)', re-issued as *Solidarity Pamphlet*, no. 7, Reading, UK, undated.

2 A. Roberts, *The Self-Managing Environment*, Alison and Busby, London, 1979, p. 115.

3 For much more on this see the excellent analysis in C. Claudin-Urondo, *Lenin and the Cultural Revolution*, Harvester, Hassocks, Sussex, 1977.

4 V. I. Lenin, *Collected Works*, 4th edn, English version, vol. 26, p. 110, 'Can the Bolsheviks retain state power?' End of September – I (14) October 1917.

5 Lenin, *Collected Works*, vol. 27, pp. 349–50, 'Left-Wing Childishness and the Petty-Bourgeois Mentality', 5 May 1918.

6 For more on this see S. Pollard, *The Idea of Progress*, Pelican, Harmondsworth, 1971.

7 D. Dickson, *Alternative Technology*, Fontana, London, 1977, p. 56.

8 *Voprosy Kultury pri diktature proletariata* (Problems of Culture under the dictatorship of the proletariat), Moscow, 1925; cited in Claudin-Urondo, *Lenin*, p. 60. For detail of the debate over Proletkult, see I. Deutscher, *The Prophet Unarmed*, Oxford University Press, London, ch. III, pp. 164–200.

9 The actual quote is a little obscure. John Gunther, who may have obtained it first-hand while interviewing Lenin, gives it as 'Soviet Russia equals socialism plus electrification', in John Gunther, *Inside Russia Today*, Harper, London, 1958, p. 354. However, *Time Magazine*, 8 September 1947, gives it as 'Socialism is electrification plus Soviet Power'. For unannounced reasons, *Time Magazine*, 23 January 1956, changes this to 'Communism is Soviet Authority plus electrification'.

10 Lenin, *Collected Works*, vol. 30, p. 377, 'Speech at All-Russia Conference of Directors of Adult Education, Divisions of Gubernia Education Departments', 25 February 1920.

11 For details of the programme in the 1960s, see Roberts, *The Self-Managing Environment*, p. 112; for the 1970 programme, see A. M. Petrosyants, et al., 'Prospects of the Development of Nuclear Power in the USSR', *Soviet Atomic Energy*, 31 (4), October 1971, pp. 1067–8.

12 Australian Atomic Energy Commission, *Annual Report*, Australian Government Publishing Service, Canberra, 1974–75, table 2, p. 19.

13 Ibid., 1977–78, table 1, p. 12.

14 P. Stoler, 'Soviets Go Atomaya Energia', *Time Magazine*, 30 October 1978, p. 47.

15 Ibid.

NOTES 379

16 Calculated from 'World List of Nuclear Power Plants', *Nuclear News*, August 1980, pp. 106–7.

17 D. Salter, 'Soviet nuclear energy plans unshaken by safety debate', *Financial Times*, 28 December 1979.

18 Ibid.

19 S. White, 'Split leadership leaves Atommash in bits', *New Scientist*, 2 March 1978.

20 Stoler, 'Soviets Go Atomaya Energia', p. 48

21 Ibid.

22 J. Hallerbach, *Hammer, Sickel und Atom*, to be published, cited in A. Gyorgy and Friends, *No Nukes: everyone's guide to nuclear power*, South End Press, Boston, 1979, p. 356.

23 P. R. Pryde and L. T. Pryde, 'Soviet Nuclear Power', *Environment*, vol. 16, no. 3, April 1974, p. 26.

24 A. N. Komarovskiy, *Design of Nuclear Plants*, Atomizdat, Moscow, 1965, tr. Israel Program for Scientific Translations, 1968, p. 159.

25 J. M. Iacovino Jr, 'Soviet Nuclear Standards', *Environment*, July–August, 1974, p. 41.

26 Salter, 'Soviet nuclear energy'.

27 P. Hofmann, 'Russia Continuing Work on Plutonium Reactors', *International Herald Tribune*, 22 April 1978.

28 B. Belitzky, 'Removing radioactive rubbish in the USSR', *New Scientist*, 26 February 1976, pp. 437.

29 Z. Medvedev, 'Two decades of dissidence', *New Scientist*, vol. 72, November 1976, p. 64.

30 Quoted in Z. Medvedev, 'Facts behind the Soviet nuclear disaster', *New Scientist*, 30 June 1977, p. 761.

31 Z. Medvedev, *Nuclear disaster in the Urals*, Angus & Robertson, London, 1980; see also Medvedev, 'Facts behind the Soviet nuclear disaster', p. 761.

32 A. Cockburn, 'The Nuclear Disaster They Didn't Want to Tell You About', *Esquire*, 25 April 1978, p. 39.

33 R. Pollock, 'Soviets Experience Nuclear Accident', *Critical Mass Journal*, January 1978, vol. 3, no. 10, p. 9.

34 'The wasteful truth about the Soviet nuclear disaster', *New Scientist*, 10 January 1980.

35 'The Soviets Have Admitted That They Have Had At Least Two Serious Accidents', *Nucleonics Week*, 26 April 1979, p. 10.

36 V. Rich, 'Fire threatened fast reactor cooling system, says unofficial report', *Nature*, 31 January 1980.

37 Quoted in P. Gall, 'Top Soviet Scientist Drawn to Defend Nuclear Power as Debate Rises', *Nucleonics Week*, 13 May 1976, p. 12.

38 Gall, *Top Soviet Scientist Drawn*, p. 13.

39 Cited in Sozialistisches Osteuropakomitee, *Kernkraftwerke in Osteuropa*, Sonderinfo, May 1978.

380   GLOBAL FISSION

40 'Reactors now loom in U.K.', *Age*, Melbourne, 20 December 1979, p. 8.

41 A detailed report of this article is given in S. White, 'Soviets worried about nuclear power too', *New Scientist*, 8 November 1979, p. 419. A partial translation is given in N. Dollezhal and Y. Koryakin, 'The Soviet Nuclear Energy Programme', extracts tr. from an article of the same name in *Kommunist*, no. 14, published in *Marxism Today* (UK), December 1979, p. 23.

42 N. Dollezhal and Y. Koryakin, 'The Soviet Nuclear Energy Programme', extracts tr. from an article of the same name in *Kommunist*, no. 14, published in *Marxism Today* (UK), December 1979, p. 24.

43 S. White, 'Soviets worried about nuclear power too', *New Scientist*, 8 November 1979, p. 420.

44 Salter, 'Soviet nuclear energy'.

45 S. White, 'Nuclear Power and the Five-Year Plan', *New Scientist*, 21 April 1977, p. 129.

46 'Comecon's Projects Include Production of Atomic Energy', *Wall Street Journal*, 29 June 1979.

47 The analysis of the Eastern Bloc countries draws heavily on the detail in Sozialistisches Osteuropakomitee, *Kernkraftwerke*.

48 L. Albrecht, 'Genehmigungsverfahren für Kernkraftwerke', *Kernenergie*, January 1977.

49 Quoted in 'Kernkraftwerke in DDR', *Die Zeit*, 27 May 1977.

50 Cited in 'DDR unterstützt Atomgegner', *Frankfurter Rundschau*, 9 November 1977.

51 Cited in 'Discussion of Nuclear Power Demanded', *WISE*, 15 November 1979, quoted from *Tageszeitung*, 6 November 1979.

52 V. Rich, 'Czech Chartists claim two died in nuclear accident', *Nature*, vol. 276, 7 December 1978, p. 551.

53 *Polityka*, 17 September 1977.

54 'Geringste Gefahr. UdSSR/Kernenergie', *Der Spiegel*, 31 January 1977.

55 N. Stanic, 'Yugoslavs Swear Off Turnkey as they gear up for a bigger nuclear future', *Nucleonics Week*, 7 June 1979, p. 16.

56 Details of the opposition to the reactor proposal are drawn from G. Ronay, 'First victory for East Europe's anti-nuclear lobby', *New Statesman*, 9 November 1979; 'Three Mile Island fuels rebellion in Yugoslavia', *Guardian*, 24 April 1979; V. Rich, 'Yugoslavia hits nuclear snag', *Nature*, 8 March 1979; V. Rich, 'Yugoslavia changes nuclear site after protest', *Nature*, 13 December 1979; 'Yugoslavia Moves Site for Nuclear Station from a Tourist Area', *Wall Street Journal*, 29 June 1979.

57 G. Ronay, 'First victory for East Europe's anti-nuclear lobby', *New Statesman*, 9 November 1979.

58 See, for example, J. D. B. Miller and T. H. Rigby (eds), *The Disinte-*

NOTES 381

grating Monolith: Pluralist Trends in the Communist World, Australian National University, Canberra, 1965, pp. 46–66.

59 V. Rich, 'Yugoslavia hits nuclear snag', *Nature*, 8 March 1979, and references contained therein.

60 R. Breeze, 'China gears for the economic challenges of the eighties', *Australian Financial Review*, 5 February 1980, p. 28.

61 'China tells about its energy plans for the future', *The Elements*, June 1979, no. 50, p. 2, drawn from an interview in *China Reconstructs*, April 1979.

62 'Chinese French Nuclear Deal', *The Elements*, February 1978.

63 R. Burt, 'White House Endorses French Sale of a Nuclear Power Plant to China', *New York Times*, 25 November 1978.

64 'China Appears to Be Delaying Signing Contracts for Two Framatome Units', *Nucleonics Week*, 24 May 1979, p. 8.

65 'The Chinese Government has confirmed that it has cancelled letters of intent', *Nucleonics Week*, 19 July 1979, p. 7; 'China Cancels Order for 2 Nuclear Plants from French Concerns', *Wall Street Journal*, 12 July 1979.

66 'China planning to Design and Build Two Nuclear Plants in 1980's', *Nucleonics Week*, 20 March 1980, p. 1.

67 For more on this, see for example, P. Filo della Torre, et al., *Euro-Communism: Myth or Reality*, Penguin, Harmondsworth, 1979.

68 'Pressure on in Italy to revitalise shelved government nuclear plan', *Australian Financial Review*, 4 February 1981, p. 23. For further details of the local opposition which has brought Italy's programme to an effective halt see L. Holloway, 'Energia Nucleare? No Grazie!', *Chain Reaction*, vol. 5, no. 3, 1980, p. 28.

69 'Sweden Appears to React Strongest to the Three Mile Island Accident', *Nucleonics Week*, 12 April 1979, p. 9.

70 'A Sinister-Sophisticated Plot to Discredit the Socialist Party and to Divide the Labor Movement', Pamphlet authorized by R. Hearn, PO Box 303, Richmond, Vic., n.d., c. 1976.

71 In much of the analysis of the trade union movement and nuclear power I am indebted to L. K. Dalton, *Trade Unions and Nuclear Power: an international survey*, Spokesman Press, London, 1977, and to many valuable discussions with Les Dalton.

72 Dalton, *Trade Unions*, p. 2.

73 For much more on the U.S. labour movement and nuclear power see R. Logan and D. Nelkin, 'Labour and Nuclear Power', *Environment*, vol. 22, no. 2, March 1980.

74 *AFL-CIO News*, 27 March 1976, p. 1.

75 Logan and Nelkin, 'Labour'.

76 AFL-CIO and Construction Employers, *The Nuclear Power Construction Stabilisation Agreement: A Contribution to the National Energy Program*, Building and Construction Trades Department, 1978.

77 Roberts, *The Self-Managing Environment*, p. 170.

78 S. Jaffe, 'The Tyranny of the Working Class', *Village Voice*, 8 October 1979, p. 12.

79 'Shop Steward Wins Round in Atom Suit', *New York Times*, 18 March 1980.

80 Logan and Nelkin, 'Labour'.

81 United Auto Workers, 'Resolutions', *25th Constitutional Convention*, 15–19 May 1977, p. 19.

82 'Les Amis de la Terre et la CRILAN souhaitent enlargie les convergences avec la gauche politique et syndicale contre l'extension de la Hague', *La Manche*, 13 November 1978.

83 'Karen Silkwood: From Activist to Protest Symbol', *New York Times*, 19 May 1979; 'Silkwood Heirs Awarded $10.5 million in a Setback to Nuclear Industry', *New York Times*, 19 May 1979.

84 'Japan's Move After the TMI Mishap', *Price News*, August 1979.

85 'Belgian Union Shuts Down Nuclear Power Plant', *WISE* News Communique, reprinted in *No Nuke News*, Round 3, no. 7, March to mid April 1980, p. 15.

86 'Con Ed. Welders Fear A-Plant Job, But Company Says Work or Quit', *New York Times*, 24 March 1978.

87 'Windscale claim over cataracts', *Guardian*, 12 March 1980.

88 P. Charton, 'Another A-worker dies of cancer', *Guardian*, 12 March 1980.

89 M. Morris, 'Union joins complaints on leak at Windscale', *Guardian*, 2 August 1980.

90 Quoted in Dalton, *Trade Unions*, p. 16.

91 For more details see A. B. Lovins, *Soft Energy Paths: Toward a Durable Peace*, Penguin, Harmondsworth, 1977.

92 J. F. Caroll and P. K. Kelly (eds), *A Nuclear Ireland?*, Irish Transport and General Workers Union, Dublin, 1979, pp. 47–8.

93 Quoted in Dalton, *Trade Unions*, p. 11.

94 See Dalton, *Trade Unions*, pp. 8–12 for further details.

95 'The French Organizations Which Launched a National Petition', *Nucleonics Week*, 10 April 1980, p. 10.

96 Quoted in Dalton, *Trade Unions*, p. 9.

97 J. Roulston, 'Trade Unions, Uranium Mining and Civil Liberties', in *Uranium and Civil Liberties*, Movement Against Uranium Mining, Melbourne, 1978, p. 4.

98 See *Uranium and Civil Liberties*.

99 'The Right to Strike by French Nuclear Industry Workers', *Nucleonics Week*, 3 July 1980, p. 12.

1 'Strengthening of Struggle Against Nuclear Power Plant', resolution of the 1976 Congress of SOHYO, calls for 'strengthening the fight against nuclear power plants'. Translation of resolution was provided to the author by Jim Frazer, Secretary, Australian Railways Union,

NOTES 383

Victorian Branch (private communication), 26 September 1977. SOHYO's position of total opposition to all aspects of the nuclear fuel-cycle and to nuclear weapons was confirmed in its statements, and interviews with SOHYO delegates, at the Pacific Trade Union Conference, Vanuatu, 28–31 May 1981.

2 'Submission of the Confederation of Canadian Unions to the British Columbia Royal Commission of Inquiry', *Keep it in the Ground* Newsletter, August–September 1980.

3 TGWU motion reprinted in 'Union Moves', *SCRAM Energy Bulletin*, August–September 1979, p. 8.

4 'More Unions turn against nuclear', *WISE*, vol. 2, no. 5, July–September 1980, p. 9.

5 'Labour meeting big success', *WISE*, vol. 2, no. 5, July–September 1980, p. 9.

6 'Unionists say yes to "No Nukes" future', *Guardian*, 22 October 1980.

7 W. Golenson (ed.), *Comparative Labor Movements*, Russell and Russell, New York, 1968, p. 238.

8 R. J. Roddewig, *Green Bans*, Hale and Ironmonger, Washington, 1979, p. 65.

9 For a more detailed analysis of the Green Bans movement see Roddewig, *Green Bans*, p. 12.

10 F. K. Crowley (ed.), *A New History of Australia*, Heinemann, Melbourne, 1974, p. 455.

11 THE AUSTRALIAN EXPERIENCE

1 *Age*, Melbourne, 14 March, 1975.

2 For further analysis of the Labor government's policy and attitudes see 'Australia: A Quarry for Uranium Hunters?', *Bank of N.S.W. Review*, no. 14, April 1975, p. 5; G. Smith, 'Fuelling Up for Disaster', *Arena*, no. 42, 1976; J. Camilleri, 'Uranium Exports: Commercial Incentives Versus Nuclear Dangers', *Australian Outlook*, 20, 1 April 1976.

3 Ranger Uranium Environmental Inquiry, *Second Report*, Australian Government Publishing Service, Canberra, 1977, pp. 33–40.

4 *Aboriginal Land Rights (Northern Territory) Act 1976*, 3 (1).

5 See A. Moyal, 'The Australian Atomic Energy Commission: A Case Study in Australian Science and Government', *Search*, vol. 6, no. 9, September 1975, pp. 373–4; and references contained therein.

6 Ibid., pp. 375–7; opposition by the South Coast Trades and Labour Council is discussed in S. Wilson, 'History of the South Coast Trades and Labour Council', in South Coast Trades and Labour Council, *Trade Union Directory*, Wollongong, 1978–79, p. 17.

384 GLOBAL FISSION

7 Quoted from the submission by the Victorian Branch of the Amalgamated Metal Workers and Shipwrights Union to the Ranger Inquiry, in 'Metal Workers Oppose Uranium', *Arena*, no. 43, 1976, p. 14.

8 Ibid.

9 'ACTU Congress 1975 Resolution on Reserves, Environment and Conservation As Amended', adopted ACTU, September 1975.

10 'The Australian Uranium Industry', *Atomic Energy in Australia*, AAEC, Australia, April 1976, p. 20. More details of the history of the Australian uranium industry may be obtained from D. Hayes, J. Falk, and N. Barrett, *Redlight for Yellowcake*, FOE Australia, 1977; and M. Elliott (ed.), *Ground for Concern*, Penguin, Ringwood, Vic., 1977.

11 *Report of the Senate Select Committee on Water Pollution*, Australian Government Publishing Service, Canberra, 1970, p. 73.

12 Ranger Uranium Environmental Inquiry, *First Report*, Australian Government Publishing Service, Canberra, 1976, pp. 109, 110, 152, 159.

13 Ibid., p. 185, Principal Finding 3.

14 'Maralinga: The human toll', *Sydney Morning Herald*, 18 April 1980, p. 3.

15 Ranger Uranium, *First Report*, p. 185, Recommendation 7.

16 Ibid., p. 6

17 Ibid., p. 186, Final Recommendation.

18 For details of government statements accompanying the announcement see *Uranium – Australia's Decision*, Australian Government Publishing Service, Canberra, August 1977.

19 Announced on 24 May 1977 by the Prime Minister, *Australian Financial Review*, 25 May 1977, p. 16.

20 For a more detailed examination of the 'safeguards', see J. E. Falk, 'Australia, The New U.S. Nuclear Policy and the International Contestation over Nuclear Power', *Arena*, nos. 47–8, 1977, p. 29.

21 Ranger Uranium, *First Report*, p. 147.

22 J. Byrne, 'Nuclear Facts Disputed', *Age*, Melbourne, 31 August 1977, p. 4.

23 For further detail, as well as a political analysis of the Australian uranium controversy, see J. Camilleri, 'Nuclear controversy in Australia: the uranium campaign', *Bulletin of the Atomic Scientists*, April 1979, p. 40.

24 For further details see J. E. Falk, 'Movement Against Uranium', *Arena*, no. 46, 1977, pp. 31–3.

25 Morgan Gallup Polls cited in *The Bulletin*, 13 August 1977.

26 *Age* Poll in 'Majority back the mining and export of uranium', *Age*, Melbourne, 15 June 1977, p. 5. (The Morgan Gallup Poll figure was 33 per cent, Morgan Gallup Polls, *The Bulletin*.)

27 *Age* Poll in 'U Support drops', *Age*, Melbourne, 17 October 1977, p. 1.

NOTES

385

28 For further details see 'Dock Showdowns: Glebe Island Demo', *Uranium Deadline*, vol. 2, no. 4, 1977, pp. 3–4.

29 ALP Resolution National Conference 7 July 1977, printed in full in *Age*, Melbourne, 8 July 1977, p. 3.

30 ACTU Resolution quoted in full in 'Uranium', *ACTU Executive Recommendation to 1977 Congress*, ACTU, Melbourne, 1977.

31 'ACTU Executive-Recommendation Re Uranium to the Special Conference of Affiliated Unions held 10th February, 1978, in the Sydney Trades Hall', ACTU, Melbourne, 1978.

32 Parliamentary Library, *Parliamentary Handbook of the Commonwealth of Australia*, Australian Government Publishing Service, Canberra, 1978.

33 Ranger Uranium, *Second Report*, p. 9.

34 An *Age* Poll published in January 1979 showed that 58 per cent of Australians believe that Aboriginal communities should have the right to refuse to allow mining on their traditional Aboriginal land. 'Majority supports black land rights', *Age*, Melbourne, 2 January 1979, p. 4.

35 *Aboriginal Land Rights (Northern Territory) Act 1976*, 23 (3).

36 J. G. Yunupingu, 'Letter from Black to White', *Land Rights News*, no. 6, December 1976.

37 Silas Roberts, 'Evidence to the Ranger Inquiry', transcript pp. 9597–610.

38 Ibid.

39 L. Findlay reported in the *Age*, 18 September 1978, and tape recordings of the meeting played over ABC news. For a much more detailed account of these events see R. Groves, 'Ranger: The Events Behind the Signing of the Agreement', *Chain Reaction*, 4, nos. 2–3, 1979.

40 *National Times*, week ending 9 September, 1978.

41 T. Uren 'Speech by Tom Uren, MP to Activists' Conference', Sydney, 24 June 1978.

42 'U-export support "steady"', *Herald*, Melbourne, 28 November 1979, p. 25. This result was less dramatic than that obtained in a poll commissioned by the Japanese embassy in May 1979 which showed a change in opinion from 53 per cent 'for' and 34 per cent 'against' uranium mining in 1977, to 45 per cent 'for' and 44 per cent 'against' in May 1979. 'Uranium Poll', *Australian*, 20 October 1979.

43 'Anti-Uranium block', *Australian Financial Review*, 26 January 1979, p. 36.

44 L. Oakes, 'Dunstan Changes Uranium Tune', *Age*, Melbourne, 23 January 1979, p. 11; 'Dunstan rejects uranium switch', *Australian Financial Review*, 6 February 1979, p. 1.

45 T. Uren, 'No Change to Labor Uranium Policy', *Australian Financial Review*, 8 August 1979, p. 11.

46 'Amendment to the Executive Recommendations' moved at the

386 GLOBAL FISSION

ACTU Congress, 14 September 1979, by C. Dolan (ETU) and R. Taylor (ARU), and carried by 513 votes to 318.

47 I. Porter, 'QM uranium jump', *Age*, Melbourne, 31 October 1979, p. 25.

48 Australian Atomic Energy Commission, *Twenty-seventh Annual Report*, Australian Government Publishing Service, Canberra, June 1979.

49 'Unionists ban steel for Ranger', *Age*, Melbourne, 11 February 1980.

50 'Uranium: The policy of the ACTU explained', ACTU, Melbourne, Australia, 1980.

51 Australian Atomic Energy Commission, *Annual Report*, Australian Government Publishing Service, Canberra, June 1979, p. 18.

52 J. Hoare, 'U.K. wants to buy 1 300t of Australian uranium', *Australian Financial Review*, 17 October 1979, p. 3.

53 J. Hoare, 'Government waters uranium safeguards', *Australian Financial Review*, 26 June 1979, p. 1.

54 'Aust. U-ore in demand', *Herald*, Melbourne, 12 July 1978, p. 19.

55 'Loopholes gape in nuclear safeguards', *Sydney Morning Herald*, 28 October 1980, p. 2.

56 'Uranium import cut – US bid', *Australian Financial Review*, 24 September 1980, p. 31.

57 See, for example, Hayes, Falk and Barrett, *Redlight for Yellowcake*, p. 17.

58 T. Thomas, 'Canberra will see U plans', *Age*, Melbourne, 3 July 1978, p. 1.

59 T. Thomas, 'Study backs N-fuel plant', *Age*, Melbourne, 28 December 1978, p. 3.

60 J. E. Falk, 'Some Implications of Rapid Expansion to the Aluminium Industry', submission to the Senate Standing Committee on National Resources, 2 April 1981.

61 See, for example, H. Dick, *Aluminium Smelters and the Price of Electricity: Who Really Pays?*, Hunter Social Research Co-operative, Newcastle, 1981.

62 'Minister qualifies N-go ahead', *Age*, Melbourne, 30 December 1978, p. 4.

63 'Progress Report of South Australian Enrichment Committee', Interview with Premier Tonkin, *P.M.*, Australian Broadcasting Commission, 6 August 1980.

64 A. Lampe, 'Uranium enrichment team to offer Aust-France deal', *Australian Financial Review*, 8 February 1980, p. 16; and T. Ballantine, 'Govt may help Darwin build nuclear plant', *Sydney Morning Herald*, 18 April 1981, p. 2.

65 'U.S. study puts N-site in doubt', *Age*, Melbourne, 26 January 1980, p. 4.

66 Falk, 'Some Implications'.

NOTES 387

67 'U.S. study', *Age*, Melbourne, p. 4; State Energy Commission, Western Australia, *Annual Report 1979*, p. 10.

68 Senate Estimates Committee, 23 October 1978, p. 956; P. Sutton, *Victoria's Nuclear Countdown*, Community Energy Network, Melbourne, 1980.

69 J. Schulz, 'Nuclear power station lobby in Victoria', *Australian Financial Review*, 12 May 1978, p. 17.

70 'No N-power without vote', *Age*, Melbourne, 6 February 1979, p. 2.

71 For a detailed analysis of this evidence see P. Sutton, *Victoria's Nuclear Countdown*, Community Energy Network, Melbourne, 1980.

72 'Territory's Nuclear Quandry', *Australian*, 22 October 1980, p. 11.

73 '56% say "no" to N-power', *Herald*, Melbourne, 21 July 1979, p. 12.

74 'Anna Inherits Our Nuclear Earth: Port takes a stand', *Illawarra Mercury*, 8 August 1980, p. 1.

75 N. Wilson, 'Nuclear Fuel Treaty with France', *Sydney Morning Herald*, 8 January 1981, p. 1.

76 P. Kelly, 'N-bombs: The Australian safeguards are fragile', *Sydney Morning Herald*, 6 March 1981, p. 7.

77 K. Martin, 'Unions act on uranium exports', *Sydney Morning Herald*, 7 February 1981, p. 3.

78 P. Kelly, 'Plan for nuclear growth government advised', *Sydney Morning Herald*, 3 March 1981, p. 1.

79 'Forces unite against any nuclear step-up', *Sydney Morning Herald*, 4 March 1981, p. 9.

12   STRATEGY AND STRUCTURE

1 *Report of the No Nukes Strategy Conference*, Louisville, KY, 16–20 August 1978; and A. Gyorgy and Friends, *No Nukes: everyone's guide to nuclear power*, South End Press, Boston, 1979, pp. 381–458.

2 Australian Atomic Energy Commission, *Twenty-seventh Annual Report*, Australian Government Publishing Service, Canberra, June 1979, p. 15, table 3.

3 Calculated from Australian Atomic Energy Commission, *Report*, 1979, p. 22, table 5.

4 P. Cheeseright, 'Canadian uranium: a bonanza with reservations', *Australian Financial Review*, 4 June 1979, p. 20.

5 R. Dahonick, 'Port Hope: A Record of Success', Reprint from *The Sheaf*.

6 'Public participation in the energy debate: The people versus nuclear power', interview with Gordon Edwards, chairman of CCNR, in *Perception*, March–April 1979, p. 32.

7 Ibid., pp. 31–3.

388   GLOBAL FISSION

8 Ibid.

9 Much of the detail that follows is from R. D. Torrie, 'British Columbia Clamps a Seven-Year Moratorium on Uranium Mining and Exploration', *Not Man Apart*, March 1980; 'Uranium Inquiry for BC', *Spectrum*, published by the Canadian Scientific Pollution and Environmental Control Society, 22 February 1979, p. 1; 'Uranium ban "clearly political"', *Vancouver Sun*, 28 February 1980, p. 1.

10 'Greenland uranium goes to Denmark', *WISE*, no. 4, May–June 1980, p. 13.

11 See *Black Hills Report*, vol. 1, no. II, August 1979 for further details.

12 For example, *Chain Reaction*, FOE Australia; *La Gueule Ouverte*, France; *BBU Aktuell*, West Germany; and *Critical Mass*, USA; and the 'underground' French radio station 'Radio Verte'. An illegal Dutch station, 'Radioactivity', also transmitted at the Dodewaard blockade in November 1980. See 'Actie Dodewaard Zonder resultat', *Dutch–Australian Weekly*, 7 November 1980, p. 1.

13 *No Nuke News* may be obtained from Boston Clamshell Coalition, 1151 Massachusetts Avenue, Cambridge, Massachusetts, 02139, USA.

14 *WISE* may be obtained from Blasiusstraat 90, 1091 CW, Amsterdam, Netherlands.

15 Back copies of *International Nuclear News Service* may be obtained from Conservation Centre, 310 Angas Street, Adelaide, South Australia 5000. The service ceased publishing in June 1981.

16 'International action to stop waste ship', *WISE*, vol. 2, no. 3, March–April 1980, p. 12.

17 Press release, *ILWU Local 142*, 5 June 1979.

18 Press release, *United Public Workers Local 646 AFSCME, AFL–CIO*, 6 June 1979.

19 L. Gomes, 'Pacific Fisher ties up at P.H.', *Honolulu Star Bulletin*, 8 June 1979 and 'N-Ship refuels, leaves but future "fallout" remains', *Honolulu Advertiser*, 8 June 1979.

20 *Guardian*, 17 May 1979.

21 'Indian People vs. the Nuclear Industry', *Akwesasne Notes*, vol. II, no. 1, p. 5.

22 Gyorgy and Friends, *No Nukes*, p. 446.

23 E. W. Lammi, 'Uranium Miner Cancer Epidemic Charged to U.S.', *The Tennessean*, 20 June 1979; D. Liefgreen, 'Udall Files Damage Claims for 26 Disease-Stricken Miners', *Albuquerque Journal*, 29 July 1979.

24 Press release, National Indian Youth Council Inc., October 1978.

25 '"The Longest Walk": A Native Americans Trick for Justice', *No Nuclear News*, Special Issue, 30 May 1978, p. 1.

26 Ibid.

NOTES

27 'Native Americans to tour Europe', *WISE*, no. 6, October 1979, p. 12.

28 'Uranium mining: a threat to people of the third and fourth world', Copenhagen, 18–20 October 1979, reprinted in *WISE*, vol. 2, no. 1, November–December 1979, p. 12.

29 Calculated from Library of Congress, *Nuclear Proliferation Fact Book*, prepared for the Subcommittee on International Economic Policy and Trade of the Committee on International Relations, US House of Representatives and the Subcommittee on Energy, Nuclear Proliferation and Federal Services of the Committee on Government Affairs, US Senate, US GPO, Washington, 23 September 1977, p. 220.

30 For further details see W. C. Patterson, *Nuclear Power*, Penguin, Harmondsworth, 1976, p. 244.

31 Ranger Uranium Environmental Inquiry, *First Report*, Australian Government Publishing Service, Canberra, 1976, Principal Finding 3.

32 *Mobilisation for Survival Leaflet*, Philadelphia, September 1979.

33 For details of these particular demonstrations and further comment on the emerging alliance, see D. M. Alpern, M. Reese, and J. Walcott, 'Anti-Atom Alliance', *Newsweek*, 5 June 1978, p. 27.

34 'NATO attempts to force nuclear weapons against wave of public concern', *WISE*, vol. 2, no. 1, November–December 1979, p. 17.

35 'The neutron bomb affaire', *New Scientist*, 13 April 1978, p. 66.

36 'First nuclear strike by U.S. "possible"', *Sydney Morning Herald*, 19 August 1980, p. 4.

37 R. Gould (reporter), 'Nato-Nuclear Vote', Transcript of *Four Corners*, ABV2 Television Broadcast, Australia, Reference PN 81/575, 16 May 1981.

38 'Nuclear Debate in Dutch Forces', *Peace News New Zealand*, vol. 2, no. 1, March 1981, p. 2.

39 P. Toynbee, 'Unthinkable in all circumstances', *Guardian*, 6 July 1980, p. 19.

40 '50 000 in London Nuclear Protest', *New York Times*, 27 October 1980, p. 14.

41 'Labour's chaos complete', *Guardian*, 12 October 1980, p. 2.

42 No CANDU for Argentina Committee, *No Rights – No CANDU in Argentina*, Pamphlet, 1979.

43 Detail on the opposition to nuclear power in New Zealand, and on the history of the debate, is drawn largely from interviews with a number of New Zealand activists; *Consumer's Guide to Nuclear Power in New Zealand*, FOE New Zealand, June 1976; and T. P. McCarthy, (chairman), *Royal Commission on Nuclear Power Generation in New Zealand*, Government Printer, NZ, 1978.

44 The Planning Committee on Electric Power and Development first

390 GLOBAL FISSION

included a nuclear component in its power plan in its 1968 report to
the government; T. P. McCarthy (chairman), *Royal Commission on
Nuclear Power Generation in New Zealand*, Government Printer, NZ,
1978, p. 19.

45 *Consumer's Guide to Nuclear Power in New Zealand*, FOE New Zealand,
June 1976, p. 1.

46 See 'N.Z. Strike over nuclear ship!', *Sydney Morning Herald*, 28 August
1976 for details of *Truxton* strike. See also D. Bedggood, 'The Nuclear
Power Struggle in New Zealand', *Arena*, nos. 44–5, 1976, p. 20.

47 The Values Party had a detailed energy platform which spelled out an
energy policy based on sustainability, self-reliance, decentralization,
equity, conservation, and community control. See *The Values Party
Manifesto*, the Values Party, 1978, pp. 20–3. The party's policy also
included an explicit rejection of nuclear power. See *Beyond Tomorrow*,
Values Party Manifesto 1975, p. 7.

48 T. P. McCarthy, *Royal Commission on Nuclear Power Generation in New
Zealand*, Government Printer, NZ, 1978, p. 19. ·

49 Ibid., p. 20; and W. Green (private communication).

50 Ibid., pp. 20–1, gives considerable emphasis to the importance of
these reports in their deliberations.

51 Ibid., Conclusions, pp. 45–58.

52 S. Ichikawa, 'Struggles against Nuclear Power Plants in Japan', *International Congress Against Nuclear Power*, Gothenborg, 13–16 May 1976.

53 Australian Atomic Energy Commission, *Report*, p. 15, table 3.

54 M. Naoto and H. Kazuko, 'Japanese Nuclear Power Industry and Its
Problems', First Draft, Institute of Policy Studies, Pacific-Asia
Resources Centre, Tokyo, July 1979, table 12. Data is from the Japan
Nuclear Power Company.

55 Ibid.

56 D. D. Comey, 'Will Idle Capacity Kill Nuclear Power?', *Bulletin of the
Atomic Scientists*, November 1974, p. 23.

57 N. Matsuoka, 'For a Change of Social Structure', pamphlet, Jishu
Koza, Tokyo, 1977, pp. 5, 6.

58 Naoto and Kazuko, 'Japanese Nuclear Power', pp. 3–26.

59 'World List of Nuclear Power Plants', *Nuclear News*, August 1979,
pp. 69–87.

60 For more detail on the *Mutsu* see Gyorgy and Friends, *No Nukes*,
pp. 363–5.

61 Gyorgy and Friends, *No Nukes*, p. 362.

62 'Japanese port ban stays on disabled nuclear ship Mutsu', *Australian
Financial Review*, 14 May 1981, p. 21.

63 N. Matsuoka, 'For a Change'.

64 For further details of this flower and its properties see S. Ichikawa,
A. H. Sparrow, and K. H. Thompson, 'Morphologically Abnormal
Cells, Somatic Mutations and Loss of Reproductive Integrity In

NOTES

Irradiated Tradescantia Stamen Hairs', *Radiation Biology*, vol. 9, 1969, pp. 195–211; S. Ichikawa, 'A Radiobiological Study in the Stamen Hairs of Tradescantia blossfeldiana Mildbr', *Seiken Ziho*, 20, 1968, pp. 35–45; S. Ichikawa and A. H. Sparrow, 'Radiation-Induced Loss of Reproductive Integrity in the Stamen Hairs of a Polyploid Series of Tradescantia Species', *Radiation Botany*, vol. 7, 1967, pp. 429–41.

65 Matsuoka, 'For a Change', p. 9.

66 Ibid., p. 4.

67 Naoto and Kazuko, 'Japanese Nuclear Power'.

68 Detail of the aftermath of Three-Mile Island is drawn from Naoto and Kazuko, 'Japanese Nuclear Power', pp. 3–29 except where otherwise indicated.

69 The detail of the formation of Gensuikin and the attitudes of the two organizations is based on *A Summing Up on the Movement and Organisation of the Japan Congress Against Atomic and Hydrogen Bombs*, Gensuikin Press, Tokyo, 3 April 1970; interviews throughout Japan in 1977; and subsequent discussions and correspondence with Japanese activists.

70 Reported in Naoto and Kazuko, 'Japanese Nuclear Power', pp. 3–18.

71 F. Yamashita, International Secretary, Gensuikyo, (private communication), 25 January 1980.

72 Ibid.

73 M. Byrnes, 'Japan's secret power leak radiates to political scene', *Australian Financial Review*, 28 April 1981, p. 23.

74 M. Byrnes, 'Bungles and cover-ups set back Japan's nuclear program', *National Times*, 3–9 May 1981, p. 6. A leak in 1975 which exposed thirty-seven workers to radiation doses exceeding 1 rem was also later revealed: 'A-plant didn't disclose Leak', *Illawarra Mercury*, 16 May 1981, p. 6.

75 For more detail, see J. Albertini, N. Foster, et al., *The dark side of paradise: Hawaii in a nuclear world*, Catholic Action of Hawaii, September 1980; and references contained therein.

76 For further details see O. Wilkes, 'Nuclear Warfare in the Pacific and Nuclear Warships in New Zealand', *New Zealand Monthly Review*, July 1976; R. C. Aldridge, 'First Strike Via Australia', *Nuclear Countdown*, no. 2, Winter 1979, p. 1, and P. D. Jones, 'The Nuclear Pacific – an overview', *Nuclear Countdown*, no. 2, Winter 1979, p. 10.

77 Speech by member of Marshall Is. Delegation to Japan Congress Against A- and H- Bombs (Gensuikin) International Conference 1979, 'Marshalls: Every Island Affected by Tests?', reprinted in *Micronesia Support Committee Bulletin*, Spring 1980.

78 L. Pryor, 'Bikini islanders absorbing radiation', *Boston Globe*, 16 July 1977.

79 'Bikini Island – a radiation hotspot', *New Scientist*, 20 April 1978, p. 132.

80 Quoted from report of Mike Malone 'Bikini Relocation 1978' to 33rd

Atomic Bombing Anniversary Conference Against Atomic and Hydrogen Bombs, 3 August 1978, printed in *Gensuikin News*, special edn, 1978.

81 This detail is drawn primarily from 'French Polynesia: The Nuclear Tests', *Greenpeace*, Netherlands, 1980, and B. Danielsson and M. T. Danielsson, *Moruroa Mon Amour*, Penguin, Ringwood, Vic. 1977.

82 'Marshallese Occupy Missile Range: Legislator clubbed by security', *Micronesia Solidarity Committee Bulletin*, no. 4 (3), 1979.

83 'Northern Marianas Declared Nuclear Free', *Micronesia Support Committee Bulletin*, Spring 1980.

84 'Moruroa Update', *Epicentre News*, December 1979.

85 For more detail on this, see R. C. Aldridge, 'Palau Forward Base for Trident', in *Robert Aldridge on Trident*, Peace Office, Christchurch, 1976, pp. 13–15.

86 More detail on the superport proposal is given by D. R. Smith, 'Oil and Nukes Threaten Palau', *National Peoples News*, CIMRA, no. 2–3, March 1979.

87 'Palau Constitution – People Vote Yes: Legislature, U.S. Say No', *Micronesia Solidarity Committee Bulletin*, 4 (3), 1979; and interview with Senator Moses Uludong (Tia Beluad Movement, Republic of Belau), at the Pacific Trade Union Conference, Vanuatu, 3 June 1981.

88 Constitution of Belau, article XII, section 6, quoted in M. Uludong, 'Statement Before the Pacific Trade Union Conference', *Pacific Trade Union Conference*, 28–31 May 1981, Vanuatu.

89 'Palauans Win Nuclear Free Constitution', Pacific Concerns Resource Centre, *NFPC/1980 Action Bulletin*, no. 2, 24 July 1980.

90 For a description of this opposition see H. Wasserman, *Energy War: Reports from the Front*, Lawrence Hill, Connecticut, 1979, p. 160.

91 See also 'Nuclear Seen Largely Excluded From South East Asia Power Mix in Near Term', *Nucleonics Week*, 10 April 1980, p. 3.

92 Taiwan's target see 'A Bechtel Joint Venture in Taiwan Figures to be a Main Beneficiary', *Nucleonics Week*, 1 May 1980, p. 9; South Korea's target see 'Vendors Jockey for Position As Korea Prepares for New NSSS Bids', *Nucleonics Week*, 24 April 1980, p. 5.

93 See 'Briefs', *Nucleonics Week*, 10 May 1979, p. 8, (programme reduced from twenty-three to two reactors); 'Kraftwerk Union Is Throwing In The Towel In Iran', *Nucleonics Week*, 14 June 1979, p. 12, (programme effectively reduced to zero).

94 Ministry of Commerce and Industry, *Republic of Korea: your dynamic trading partner*, Republic of Korea, April 1978, p. 11.

95 Economic Planning Board, *Economic Management Plan for 1980*, Republic of Korea, February 1980, p. 33.

96 'A Leading U.S. A–E's Projections For Installed Nuclear Capacity in Asia', *Nucleonics Week*, 3 April 1980, p. 6.

97 This group is the Centre for Development Policy. A report is avail-

NOTES

able in their newspaper, 'U.S. Has O.K.'d Six Nuclear Reactors on Seismic and Volcanic Taiwan Sites', *CDP Monitor*, vol. 2:1, 1980, p. 4.

98 *New Philippines*, February 1976, pp. 25–6 cited in *Justice and Peace Notes*, November–December 1977, p. 3.

99 For a detailed summary of these, and the events surrounding the reactor see W. Bello, P. Hayes, and L. Zarsky, '500-Mile Island: The Philippine Nuclear Reactor Deal', *Pacific Research*, vol. X, no. 1, First Quarter 1979, and the extensive references contained therein.

1 Calculated in Bello, Hayes, and Zarsky, '500-Mile Island', *Pacific Research*, footnote 35, p. 23.

2 Letter from Congressman Clarence D. Long to the Honourable Cyrus Vance, Secretary of State, USA, 4 January 1978, p. 2.

3 D. Ford, Executive Director of the Union of Concerned Scientists, USA, Letter, 13 February 1978, cited in Bello, Hayes, and Zarsky, '500-Mile Island', *Pacific Research*, p. 2.

4 The author is indebted to a key anti-nuclear activist (who must remain anonymous) of the *Concerned Citizens of Bataan*, for a detailed interview in 1979 on the history of the movement in the Philippines.

5 'Military Harasses Anti-Nuke Protests', *TANOD*, Publication of the National Resource Centre on Political Prisoners in the Philippines, September 1978. This has been confirmed from sources which must remain anonymous for their own protection.

6 'Nuclear Opponent Shot Dead By Security Forces', *WISE* News Communique, reprinted in *No Nuke News*, Round 3, no. 7, mid April 1980, p. 18.

7 For two contemporary accounts see A. Doronila, 'Marcos rule is under seige', *Age* Melbourne, 21 September 1979, p. 11, and 'Powder Keg of the Pacific', *Time Magazine*, 24 September 1979, pp. 10–17.

8 'Philippine President Ferdinand Marcos Halted Construction', *Nucleonics Week*, 21 June 1979, p. 9.

9 'Bataan Site Construction to be resumed', *Bulletin Today*, 21 August 1979.

10 'U.S. State Department Approves Export of Reactor Vessel to Philippines', *WISE*, 11 October 1979.

11 'All Our Chips Are Riding on the Decision in the Philippines Case', *Nucleonics Week*, 14 August 1980.

12 'Nuclear Seen Largely', *Nucleonics Week*, p. 3.

13 'Japan Asked to Back Pacific N-Fuel Centre', *Mainichi Daily News*, 11 June 1978.

14 W. Chapman, 'U.S. Proposes Storing Spent Nuclear Fuel on Pacific Island', *The Washington Post*, 26 March 1979, p. A16.

15 'N.Z. Opposes N dump plan', *Age*, Melbourne, 21 August 1979, p. 6; for a more detailed overview see M. Indyk, 'Spent nuclear fuel dilemma looms for Australia', *Australian Financial Review*, 30 January 1980, p. 15.

16 M. Steketee, 'Nuclear Dump on island faces veto', *Sydney Morning Herald*, 24 March 1980, p. 4.

17 Wilkes, 'Nuclear Warfare', *New Zealand Monthly Review*, pp. 6–7.

18 R. Walsh and G. Munster, *Documents on Australian Defence and Foreign Policy, 1968–75*, Walsh and Munster, Australia, 1980, chapter III, pp. 117–29.

19 Foreign Affairs Section, Vanua Aku Pati, *Vanua Aku Pati Platform*, Vanuatu, October 1979, Clause II (iv) C, p. 25. Further confirmed in interview with Hilda Liny, Vanua Aku Pati Ministers Department, Vanuatu, on 25 September 1980.

20 See Gensuikyo, 'For a Nuclear Free Pacific: Our Determination and Actions', *Paper to the Nuclear-Free Pacific Forum*, Australia, 26–8 September 1980; details of this development in Gensuikyo's public stance were confirmed in an interview with Yoshikiyo Yoshida, Deputy Director-General of Gensuikyo, on 27 September 1980.

21 Pacific Trade Union Forum Preparatory Meeting 12–14 November 1980, Tanoa Hotel, Nadi, Fiji Islands, *Declaration and Documents*, Iterim Secretariat, C/– 174 Victoria Parade, East Melbourne, 3002, Australia, (Telex AA36789).

22 The author attended the Pacific Trade Union Conference as an observer on the invitation of the Australian Council of Trade Unions.

23 'Declaration, Adopted unanimously Port Vila, 31st May 1981', *Pacific Trade Union Conference*, 28–31 May 1981, Vanuatu.

24 'Proposals for Co-ordinated Activity', *Pacific Trade Union Conference*, 28–31 May 1981, Vanuatu.

25 J. Tinker, 'Uranium theft shatters nuclear safeguards', *New Scientist*, 5 May 1977, p. 251.

---

13 BEYOND NUCLEAR POWER

1 See D. Warnock and K. Bossong (eds), *Nuclear Power and Civil Liberties: Can We Have Both?*, Citizens Energy Project, Washington, 1978, pp. 84, 87, 94; and A. Jahnke, 'Clamshell Police Agent Uncovered', *Real Paper*, 5 February 1981.

2 A. B. Lovins, *Soft Energy Paths: Toward a Durable Peace*, Penguin, Harmondsworth, 1977.

3 See chapter 5. See also R. H. Murray and P. A. La Viollette, *Assessing the Solar Transition*, International Centre for Integrative Studies, US Office, 45 West 8th Street, New York, 10011, 1975, and reports referred to in Lovins, *Soft Energy Paths*.

4 This quote is taken from the 'prepublication draft' of the OTA's report, which is somewhat more forthright than the final edited version. See Office of Technology Assessment, Congress of the

NOTES

395

United States, *Application of Solar Technology to Today's Energy Needs*, Washington, June 1977, pp. I–10, I–11.

5 R. Reece, *The Sun Betrayed*, South End Press, Boston, 1979.

6 J. Grover, *The Struggle for Power*, E. J. Dwyer, 1980, in particular pp. 289–341.

7 C. Goldstein, 'The Opposition Movement and its Leaders', presented at the AIF–SVA *International Workshop, Nuclear Power and the Public*, Geneva, 26–9 September 1977.

8 See D. Massey, 'Survey: Regionalism: Some Current Issues', *Capital and Class*, 6, 1978, pp. 107–25 and references contained therein.

9 A. B. Lovins, 'Energy Strategy: The Road Not Taken', *Foreign Affairs*, vol. 55, no. 1, October 1976, pp. 65–96.

10 Lovins, *Soft Energy Paths*.

11 For a summary of much of this see activity see Lovins, *Soft Energy Paths*, pp. 219–22. See also V. Taylor, *The Easy Path Energy Plan*, Union of Concerned Scientists, Massachusetts, September 1979.

12 M. Flood and R. Grove-White, *Nuclear Prospects*, FOE, London, 1976; M. Flood, R. Grove-White, and K. Suter, *Uranium, the Law and You*, FOE Australia, 1977; Warnock and Bossong, *Nuclear Power*.

13 Declaration of the Salzburg Conference for a Non-Nuclear Future held 29 April–1 May 1977. Organized by the European Environmental Bureau, FOE International, Gensuikin, NRDC (USA), and Österreichische Naturschutzbund. Signed by delegates from Afghanistan, Australia, Austria, Belgium, Canada, Denmark, Finland, France, Fed. Rep. Germany, Israel, Italy, Japan, Malaysia, Netherlands, Sweden, Switzerland, UK and USA.

14 P. Kelly (chairperson of the Green Party in West Germany), 'Speech to the Harrisburg Rally', Trafalgar Square, FOE UK, March 1980.

15 For accounts of these see K. Coates (ed.), *The Right to Useful Work*, published for the Institute for Workers' Control by Spokesman, 1978; H. Beynon and H. Wainwright, *The Workers' Report*, Pluto Press, London, 1979; A. Roberts, *The Self-Managing Environment*, Alison and Busby, London, 1979, and references contained therein.

16 P. Kelly, *Letter on behalf of Die Grünen*, Bonn, 4 August 1980.

17 A. Huxley, *Brave New World*, Penguin, Harmondsworth. First published 1932; first published with a foreword 1950.

18 G. Orwell, *Nineteen Eighty-Four*, Penguin, Harmondsworth. First published 1949.

19 Huxley, *Brave New World*, preface.

INDEX

Aarau reactor, Switzerland 157
Abalone Alliance, USA 56, 132, 324
Aboriginal land rights 259, 268–71
Aboriginal Land Rights Act, Australia 259, 269
Aboriginals, Australian *see* Australia, Aboriginals and
Academy of Science, USSR 235
accident, nuclear 43, 52, 53, 232, 262, 292, 296–7, 308–9, 315; effects of 33; in France 197; in UK 196; in USSR 233–4, 235; in Yugoslavia 238–9; likelihood of 49–51; *see also* Three Mile Island reactor accident, USA
ACTU *see* Australian Council of Trade Unions
Adams, J. 132
Adelaide, Australia 262
AECL *see* Atomic Energy of Canada Limited
AFL–CIO (American Federation of Labour–Congress of Industrial Organizations), USA 247–8
Africa 318
Agesta, Sweden 142
Alamogordo Air Base, USA 14, 15
Albrecht, W. 223, 224
Aldermaston Atomic Weapons Establishment, UK 97, 195, 251
Alexandrov, A. 235, 236
Alfvén, H. 144
Almaraz nuclear reactor, Spain 209
Almelo uranium enrichment plant, Netherlands 211–12, 301
ALP *see* Australian Labor Party
Alsace, France 184
alternative energy 84, 154, 161–2, 252, 329
Alternative Production Congress, FRG 340
Altnabreac, UK 193
Amalgamated Metal Workers and Shipwrights Union (AMWSU), Australia 253, 277
American Physical Society 33
Andalusia, Spain 210
Anderson, J. 16
Angra-1 reactor, Brazil 217
Anthony, D.J. 281
Anti-Nuclear Campaign (ANC), UK 196
Antinuclear Group Representing York (ANGRY), USA 65

Approved Defence Projects Protection Act, Australia 133
Argentina 283, 292
Arhaus, FRG 224
Ariyoshi, G. 296
Armadillo Alliance, USA 56
Asco reactor, Spain 205, 209
ASEA-Atom 114–15, 167
Asse waste storage area, FRG 222
Association for Swiss Electrical Industries, Switzerland 164
Association of Atom Bomb Victims, Japan 307
Association of Industrial Power Producers, FRG 25
Astiz, I. 207
ATOM (Against Testing on Mururoa) Committee 318
Atom Law, FRG 216
Atomic Age 12
Atomic Energy Act, Australia 133; South Africa 133; Switzerland 161; USA 19, 116
Atomic Energy Agency, UK (UKAEA) 20, 132, 192–4, 234
Atomic Energy Commission, Argentina 122; Australia (AAEC) 113, 259, 281, 282; France 20, 181; Japan (JAEC) 21, 23; USA (USAEC) 19, 32, 33, 49, 66, 75, 84, 93–5, 115–17, 119, 123–4, 132
Atomic Energy of Canada Limited 26, 114–15
Atomic Industrial Forum, USA 82, 85, 90, 335
atomic pile 13
Atommash Project, USSR 231
Atoms for Peace 19, 231
Australia 10, 76, 120, 255, 256–84, 286, 294–5, 297, 303, 320, 337–8; Aboriginals and 174, 257, 259, 268–71, 277, 290, 298, 337; ACTU policy and 274; ALP policy and 258, 265, 267–8, 275–6; aluminium smelters and 281; citizen organizations in 257–8; civil liberties and 133, 134; CND and 98; demonstrations in 264–7, 272, 283; France and 262, 283, 302; Green Bans and 255, 340; international nuclear conflict and 257, 278–84; Land Rights Agreement 271;

INDEX

mining companies 273; national railway strike over uranium mining 261; nuclear-free Pacific, and 317, 318–20; nuclear legislation and 283; nuclear proliferation issue in 98, 262–4, 301; nuclear reactors and 259–60, 282–3; political parties in 245, 257–63, 265, 267–8, 271–6, 282; Ranger Uranium Environmental Inquiry and 76, 131, 133, 177, 259, 261–3, 299, 303; signature campaign in 264–5, 274; trade unions and 253, 254–5, 259–62, 260–2, 265–7, 272–7, 283; uranium enrichment and 280–1

Australian Council of Trade Unions (ACTU) 253, 260, 261, 265, 268, 272, 274–5, 282–3, 320; moratorium on uranium mining and 27

Australian Democrats 282

Australian Federated Union of Locomotive Enginemen 261

Australian Labor Party 57–9, 245, 261–2, 265, 267–8, 271–6, 282

Australian Railways Union (ARU) 261, 277, 283

Australian Workers' Union (AWU) 276

Austria 28, 142–6, 160, 202, 227, 320–1; information campaign in 152, 156; Nuclear Deterrent Act 167; nuclear programme in 142; political parties in 148, 156, 161–2; referendum 156, 161, 163–4, 165, 167–8; trade unions and 163, 246–7, 252; see also Zwentendorf reactor

Austrian Trade Union Confederation 247

Ayres, R., report of 131

Babcock and Wilcox 26, 52, 72–3, 79, 110–18

Baden Power Company 106

Baden-Württemberg, FRG, SPD branch, regional convention of 221; successes of Green Party in 224–5

Bailly Alliance, USA 56

Balfour, J. 281

Baner, J. 115

Bangor submarine base, USA 300

Bank of America 72, 78

Barcelona, Spain 200, 204–5

Barnaby, W. 167

Barnet, R. 135

Barnwell reprocessing plant, USA 65, 296

Barsebäck reactor, Sweden 157–8, 167, 170

Bartheld, K. 114, 217

Baruch Report 92

Basel, Switzerland 168, 201

Basque General Council, Spain 206

Basque region, France and 175; Spain 205, 336

Basque separatist movement (ETA) 206

Bataan reactor, Philippines 315–17

Battelle Institute, FRG 125

Battery Park rally, USA 87

Baxter, P. 259–60

Becker, I. 59

BEIR see Biological Effects of Ionizing Radiation Committee

Belau see Palau, Micronesia

Belgium 21, 127, 187, 279; demonstrations in 200–1; France and 175; reactor programme in 21

Belgrade, Yugoslavia 240

Beloyarsk reactor, USSR 235

Berlin, FRG 13, 224

Berne, Switzerland 144

Bertell, R. 49, 95

Berufsverbot law, FRG 129

Bethe, H. 246

Bikini Atoll, Micronesia 97, 310

Bilbao, Spain 205–7

Bilbis-C reactor, FRG 216

Billingen, Sweden 150, 292

Biological Effects of Ionizing Radiation Committee (BEIR), USA 67

Black Forest, FRG 291

Black Hills, USA 295, 297

Blayais reactor, France 184

Bogeda Head reactor site, USA 94

Bohn, E. 124

Bonn, FRG 223

Bordeaux, France 184, 296

Borken, FRG 224

Borssele reactor, Netherlands 203

Boston, USA 59, 93, 295

Bourgeois, Prof. 17

Boyes, R. 224

Bragg, L. 15

Bray, Philip A. 27

Brazil 26, 78, 217–19, 301; CND and 98; enrichment plant and 217; Netherlands and 212; nuclear programme in 217, 219

Breisach reactor site, FRG 104

British Columbia, Canada 292–4, 296

British Columbia Cattlemen, Canada 294

British Columbian Council of the Confederation of Canadian Unions 254

British Nuclear Fuels (Limited) (BNFL) 176

Brittany, France 175, 187, 336

Brokdorf reactor, FRG 225, 325–6; battles over 126–7, 128, 139, 215, 216

Brookhaven National Laboratories, USA 33

Bross, I. 48

Brower, D. 55

Brown, C. 81

Brown, J. 59

Brown's Ferry reactor accident, USA 43, 52

Buchanan reactor, USA 251

Bugey-1 reactor, France 184

Bugler, J. 177

Builders' Labourers Federation, Australia 340

Bull, T. 266

Bundestag, FRG, reactor safety subcommittee of 221

Bupp, I.C. 69

Buser, A. 44

BUU (*Bürgerinitiative Umweltschutz Unterelbe*), FRG 126, 127

Cactus Alliance, USA 56

Caecceres province, Spain 209

Calder Hall reactor, UK 20

Califano Jnr, J. 49

California, USA 53, 55, 59, 77, 86, 296, 298, 324; proposition 15 and 58, 247

GLOBAL FISSION

Calvert Cliffs reactor, USA 55
Cambridge Reports Inc., USA 86
Cambridge, UK 12, 52
Campaign Against Nuclear Energy (CANE), Australia 258
Campaign for Non-Nuclear Futures, NZ 302
Campaign for Nuclear Disarmament (CND) 97, 195, 196, 301
Campaign Opposing Nuclear Dumping (COND), Scotland 193
Canada 21, 74, 292–4, 337; native people and 296–7; nuclear programme in 21, 292; nuclear proliferation issue in 98, 301; Porter Inquiry and 177; reactor licensing times in 70; trade unions and 254; Whitsunday demonstrations and 200–1
Canadian Coalition for Nuclear Responsibility 293
Cap de la Hague, France 185; see also la Hague reprocessing plant, France
capacity factor 66
capital costs, generating stations 69
Carnsore Point reactor, Ireland 252
Carrick council, Scotland 193
Carroll, J. 252
Carson, R. 101
Carter, J. 23, 45, 81, 124, 317
Cass, M. 258
Catalan region, France 175
Catalonia, Spain 205, 209–10, 336
Catfish Alliance, USA 56
Catholics, nuclear opposition and 163, 316
Cavendish Laboratory, UK 12, 52
CCPAH (*Le Comité Contre la Pollution Atomique dans la Hague*), France 185
CDL see Centrala Driftsledningen, Sweden
CEGB see Central Electricity Generating Board, UK
Central Democratic Union (UCD), Spain 209
Central Electricity Generating Board (CEGB), UK 190–1
central Europe see Dreyeckland region
Central Intelligence Agency (CIA), USA 234
Centrala Driftsledningen (CDL), Sweden 142
centralization of power 285–9, 341; nuclear programmes and 171–99, 253, 328; opposition to nuclear power and 286–7, 335–41
centralized confrontation 183
centre 174–5, 196, 204
Centre for Development Policy (CDP), USA 317
Centre Party, Finland 147; Sweden 144, 148–51, 158–9, 161–3
CFDT (*la Confédération Française Démocratique du Travail*), France 186–7, 196–8, 251–4, 295
CGT (*la Confédération Générale du Travail*), France 187, 197–8, 250, 253, 296
chain reaction, nuclear 14, 31
Chalk River, Canada 21
Chamber of Manufacturers, USA 135
Chapple, F. 251
Charter 77, Czechoslovakia 238
Cherbourg, France 179, 295
Chicago Draft Convention for the Control of

Atomic Energy, USA 92
Chicago, USA 71
China 241–3, 244; cultural revolution and 101; France and 242; self-sufficiency in 101, 241
China Light Company, Hong Kong 243
China Syndrome 33
Chinon reactor, France 20
Chooz, France 187
Christian Democratic Party, FRG 223–4; Netherlands 211; Switzerland 147, 161
Christian League, Finland 146
Christmas Island 97
Chugoku Electric Power Company Union, Japan 306
churches, opposition to nuclear power and 55, 211, 238, 293–4, 299
Churchill, W. 13
Cicchetti, C. 82
CINPAC (Commander in Chief Pacific), USA 309
Citizens Committee for Environmental Concern, USA 94
Citizens for Energy (CITE), USA 86
Citizens for Energy Freedom (CEF), USA 86–7
citizens initiatives, FRG see Federal Republic of Germany (FRG), citizen action groups
Citizens Party, USA 339
civil liberties 129–34
Clamshell Alliance, USA 55–6, 132, 295, 324, 335
Clinch River, USA 14, 249
CND see Campaign for Nuclear Disarmament
Coalition for a Nuclear Free Australia 301
Cockburn, A. 234
COGEMA (*Compagnie Général de Matières Nucléaires*), France 158, 159, 186
Collie coal-field, Western Australia 281
Collins, B. 271
colonialism 124; see also Third World
Colorado, USA 58–9, 86, 296, 300
Columbia University, USA 13
Columbus Australis 266–7
Combustion Engineering 79, 110–18
COMECON (Council for Mutual Economic Aid) 237–9
Coming Into Force Law, Denmark 155
Commissariat à l'Énergie Atomique see Atomic Energy Commission, France
Commission sur la Production d'Électricité d'origine Nucléaire (PEON) 178
Committee for Energy Awareness, USA 85
Committee for Jobs and Energy, USA 86
Commoner, B. 60, 94, 102
Commonwealth Associates 23
Commonwealth Edison 70
Commonwealth Employees Act (1977), Australia 253
communist countries, nuclear programmes in 228–43
Communist Party 244; Australia 245, 258; China 101, 242; Finland 146; France 187, 197, 198, 246, 253, 275; Ireland 189, 245; Italy 245; Japan 307; Spain 204, 209, 245; Sweden 148, 163, 226, 245

INDEX

community, sense of, citizen action and
108–9
Campagnie Général de Matières Nucléaires, France
see COGEMA
Compagnie Républicaine de Sécurité (CRS) 179,
182–3, 186, 188
consensus decision-making 335
Conciliation and Arbitration Act, Australia
253
COND *see* Campaign Opposing Nuclear
Dumping, Scotland
Conference for a Nuclear Free Pacific 318–20
Congressional Budget Office, USA 78
Connecticut, USA 300
conservation *see* environmentalism
Conservative Party, Sweden 150–1, 162;
UK 176, 178
Consolidated Rexspar 293
consumer society 99
containment building 32; USSR and 232
control rods 31
Convention on the Physical Protection of
Nuclear Material 134
Cook Islands 318
Copenhagen, Denmark 157, 298
Cordiner, R. 116
Cornell University, USA 94
Cornwall, England 193
Corrur council, Scotland 193
Costello, V. 44
Council for the Status of Women, Ireland 189
Council for the US Small Business
Administration 136
Council of Salaried and Professional
Associations, Australia 260
counter culture 99
County Donegal, Ireland 189
Court, C. 281
courts, nuclear power and 106, 139, 216–19,
305
Coutre, Dr. 17
Covadonga 296
Creys-Malville, France 180–4
Creys-Malville reactor, France 181–4
CRILAN (*Le Comité Régional d'Information et de
Lutte Anti-nucléaire*), France 185
critical energy facility, USA 82
Critical Mass, USA 55, 137, 234, 291
Croatia, Yugoslavia 240, 336
Croatian Electricity Board, Yugoslavia 240
CRS *see Compagnie Républicaine de Sécurité*
Cruas reactor site, France 180
Cruise missile 214, 300
Cruz, J. 205
Culvert Cliffs, USA 94
Cumbria, UK 20
Curry, D. 43
Czechoslovakia 238–9, 245

Dalmellington, Scotland 193
Damaso, C. 296
Decatur, USA 43
Defense Department, USA 84
de Gaulle, C. 178
De Laguna, W. 93
Delporte, E. 17

Demona reactor, Israel 321
demonstrations, significance of 201–2;
pro-nuclear 86–7, 220
Denenberg, H.S. 62, 78
Denmark 28, 154, 155, 159, 291, 335, 338;
Barsebäck reactor and 157, 170; CND and
98; information campaign in 153–4;
referendum and 170
Denton, H. 40–1
Denver, USA 59, 86, 300
Department of Energy (DOE), USA 25, 67,
74, 88, 132, 136–7
Department of Energy, New Zealand 302
Derian, J.C 69
D'Estaing, G. 198
Diablo Canyon reactor, USA 53, 55, 58–9
Dickson, D. 229
Dietz, D. 6, 18
diffusion of concern 108, 297, 328
Document 22, of Charter 77 238–9
Dodewaard reactor, Netherlands 201, 203
Dolan, C. 275, 277
Dollezhal, N. 236
Dortmund stadium, FRG 220
Dounreay Nuclear Power Establishment,
UK 193
Dow Chemical plant, Spain 205
Drenthe, Netherlands 213
Dress, A. 129
Dreyeckland region 175, 104–9, 184, 336;
see also France, FRG, and Switzerland
Duff, P. 97
Dumond, R. 179
Dunstan, D. 274, 280
Duquesne Company 20
Dutch Catholic Trade Union Council
(NKV) 210
Dutch Council of Trade Unions (NVV) 210
Dutch Union of Policemen 213

East Anglia, UK 175
Eastern Europe 237–40
ECCS *see* Emergency Core Cooling System
Ecologists, France 179–80, 182, 339; *see also*
green lists, France; Green Party, FRG
economic determinism 62, 75, 84
economics of energy sources 84; of coal power
69, 70; of nuclear power 61–83, 80,
115–25, 142, 144, 152–3, 156, 160, 167,
247; *see also* nuclear industry; subsidies to
nuclear industry; of solar power 330–1
EDF *see Électricité de France*
Edinburgh, Scotland 47, 191
Edison Electric Institute, USA 85
Ehime prefecture, Japan 305
Ehrlich, P. 102
Einstein, A. 13
Eisenhower, D. 19, 231
Eldorado Nuclear Company 292
Electric Power Research Institute, USA 53
Electrical, Telecommunications and Plumbing
Union, UK 251
Electrical Utility Association, FRG 216
Electrical Workers' Union, Ireland 190
Électricité de France (EDF) 20, 178, 180, 184,
185, 188, 198, 253

Electricity Supply Board (ESB), Ireland 189
elites 121, 229
Elk River nuclear facility, USA 77
Elliott Lake, Canada 292, 297
ELSAM, Denmark 153
Emergency Core Cooling System (ECCS) 33
Employers Association, FRG 128
Endowment Committee on Atomic Energy 92
Energy and Water Appropriations Bill,
    USA 81
Energy Commission, Sweden 162
energy growth, projections and assumptions
    83–4
Energy Information Administration, USA
    23, 79
Energy Information Group, Denmark 154
Energy Mobilisation Board (EMB), USA 82
energy programme, requirements of 330–4
Energy Research and Development
    Administration (ERDA), USA 48, 76, 310
Engels, F. 138
England 175, 295; see also United Kingdom
    (UK)
Eniwetok atoll, Micronesia 310
enrichment see uranium enrichment
Environment Protection Council, Finland 146
Environment Protection Union, Finland 146
environmental regulation 102
environmentalism 95, 247, 101–3, 105
Equipos Nucleares factory, Spain 208
Esensham reactor, FRG 216
Eskuadi, Spain 205; see also Basque region
ETA see Basque separatist movement
Eurocommunism 245
Europe 120, 224–5, 298, 300, 309, 320–1;
    effective moratorium on reactor orders in
    114; growth of opposition in 144;
    Whitsunday demonstrations in 200–1
European Economic Community (EEC) 193
European Federation Against Nuclear
    Arms 97
Everingham, P. 282
Export-Import Bank (ExIm Bank), USA
    78, 316
Extremadura, Spain 205, 208–9
Exxon 333

Falldin, T. 144, 152, 159
Fangataufa atoll, French Polynesia 310
Farrell, T.F. 14
fast breeder reactor 127, 182, 215–19, 232,
    235, 249
FDP see Free Democratic Party, FRG
Federal Bureau of Investigation (FBI) 132
Federal Republic of Germany (FRG) 20–1,
    25–6, 28, 73, 95, 119, 124, 175, 203,
    214–17, 219, 224, 227, 275, 296, 321,
    325–6, 336, 339; Atom Law and 216, 219;
    Berufsverbot law and 129; citizen action
    groups in 215, 219; CND and 97, 98;
    courts and 139, 216–19; demonstrations in
    103–9, 125–7, 139, 179, 182–7, 200–1,
    215, 220, 222–4, 300; information
    campaign in 125, 128; K Groups and 215;
    nuclear industry in 110–17; nuclear
    programme in 20, 212, 216; nuclear
    weapons and 301; police actions in 126–8;

political parties in 124, 183–4, 215, 220–1,
    223–5, 246, 275; reactor closures in 225;
    reactor licensing times in 70; state and 127,
    221–5; 'terrorists' and 128–9; trade unions
    and 220; URENCO and 211; see also Wyhl
    reactor site; Brokdorf reactor; Gorleben
    reprocessing plant
federal structure, the state and 172
Federal Task Force on Energy, USA 84
Federation of Trade Unions, Switzerland 161
Fermi-1 reactor, USA 43, 53, 249
Fermi, E. 13
Fessenheim reactor, France 184
Fiji 318, 319, 320
Finland 146–7, 279; growth of opposition in
    145–6; nuclear programme in 145, 147–8,
    169; political parties 146–8
Finnish People's Democratic League,
    Finland 146
Fintown, Ireland 190
Flamanville reactor site, France 180, 185–6
Flanders, France 175, 185, 201
Flood, M. 131, 338
Flowers, B. 176
Flowers Commission, UK 131, 176, 303
Floyd, J. 40
FNV, Netherlands 254
FOE see Friends of the Earth
Folk Campaign Against Nuclear Power 164
Ford-Mitre group, USA 131
Forsmark-1 reactor, Sweden 159
Fouchard, J. 40
Framatome 110–15, 197, 242
France 20, 100, 127, 158, 172, 175, 178–89,
    197–9, 202, 215, 275, 279, 286, 295–6,
    337–8, 340; Australia and 283, 302;
    Belgium and 175; centralized structure in
    175, 178; centre, divisions within 187;
    CND and 98; demonstrations in 20, 180–4,
    197, 201–2, 187–8; Dreyeckland and
    103–5, 175; ecologists and 179–80, 339;
    information campaign in 181, 184;
    nationalism, role of 180; New Zealand and
    302; nuclear industry in 110–17; nuclear
    programme in 20, 178, 179, 180, 182–7,
    194–5, 200–1, 250; nuclear tests and 262,
    302, 310–2; periphery, of 175, 186–8,
    194–8; Spain and 175; trade unions
    and 182, 249–55, 295; see also Creys-
    Malville reactor; Plogoff reactor site;
    Flamanville reactor site
Franco, F. 203
Frankfurt, FRG 97
Fraser, M. 263, 272
Free Democratic Party (FDP), FRG 220–1,
    215, 275
free market, ideology of 61, 138, 329
Freedom of Information Act, USA 234
Freedom Party, Austria 148
Freiburg, FRG 104, 107
French Democratic Confederation of Labour
    see CFDT, France
French Polynesia 262, 310–12, 336
Friends of the Earth (FOE), Australia 258;
    France 179, 182; Ireland 189; Scotland
    192; Sweden 150; UK 74, 176–7, 338;
    USA 55

INDEX

Friesland, Netherlands 213
fuel rods 31, 42

Galloway, Scotland 193
Gas and Employees Union, Belgium 250
Gasselte, Netherlands 213
GDR *see* German Democratic Republic
General and Municipal Workers' Union,
    UK 196
General Atomic 27
General Confederation of Workers, France
    see CGT
General Electric 21, 26–7, 73, 79, 90, 95,
    110–18, 136–7, 333
General Public Utilities (GPU) 62, 65, 72
Geneva, Switzerland 168, 182
Genscher, H. 221
Gensuikin, Japan 307, 308, 319
Gensuikyo, Japan 97, 307, 308, 319
Geological Survey, US 55
geothermal energy, Iceland and 169
German Democratic Republic (GDR) 222,
    237–8
Gilinsky, V. 41
Giraud, A. 242
Glaser, P. 136
Glentrool Forest Park, Scotland 193
global economy 120–5, 135
Gofman, J. 94
Goldsboro, USA 44
Goldstein, C. 90, 335
Gorleben reprocessing plant, FRG 221–4,
    296
Gösgen reactor, Switzerland 143, 157
Gravelines reactor, France 185, 197
Green Bans 255
green lists, France 180, 183; FRG 220, 223
Green Party, FRG 183–4, 221, 224–5,
    339–40
Greenland 169, 170, 294–5, 298
Greenpeace 295
Grenoble, France 182
Grohnde reactor, FRG 127, 215–16
Groningen, Netherlands 213, 336
Grove-White, R. 131, 338
Grover, J. 335
Groves, L. 14
Guam, Marianas islands 310, 320
Gulf Oil 27
GWe (Gigawatt electrical), defined 20

Haegendorf, Switzerland 168
Haenschke, F. 221
Hahn, O. 13
Haifa, Israel 321
Hambraeus, B. 144
Hamburg, FRG 126, 220
Hamm reactor, FRG 216
Hanau, FRG 224
Hanford nuclear complex, USA 14, 47, 87
Harris, Scotland 194
Harrisburg, USA 29, 37, 59
Harvard Business School, USA 69
Harvest plan, UK 193
Hawaii, USA 58, 249, 295, 296, 309, 311,
    318–20

Hawke, R. 265, 267, 274, 275
Hayes, D. 130
Helsinki, Finland 145–6
Herbein, J. 37, 43, 44, 48
Hessen, FRG 216
Hikada, Japan 306
Hill, J. 234
Hiroshima, Japan 11, 15–18, 93, 96, 115,
    262, 299, 303, 307, 323
Hocevar, C. 95
Hogerton, J. 119
Hohoku reactor, Japan 306
Holding, C. 276
Hong Kong 243
House Committee on Government Operations,
    USA 26
Hutton, D.F. 26
Huxley, A. 341
hydro-electric power, in Norway 169; in
    Brazil 218; in China 241
hydrogen bomb 17, 97, 230–1

IAEA *see* International Atomic Energy Agency
Iberduero 205, 206, 207–8
Iceland 168
Ichikawa, S. 305
ICRP *see* International Committee on
    Radiological Protection
idea of progress 101, 123, 229–30
ideology of economic growth 107–8
Iguapé, Brazil 219
Ikata reactor, Japan 305
Illinois, USA 90
independence movements 205, 209–10, 241
India 21, 299
Indian Point reactor, USA 55, 68
Industrialists Association, Austria 164
INFCE *see* International Nuclear Fuel Cycle
    Evaluation
information campaign, in Austria 152; in
    Denmark 153–4; in France 181, 184; in
    FRG 125; in Ireland 189; in Netherlands
    211; in Sweden 149–55; in Switzerland 152
Ingå, Finland 145
Institute of Energy Economics, Japan 23
insurance, nuclear power and 63, 78
Interagency Task Force on Ionizing Radiation,
    USA 48
Inter-Church Peace Council (IKV),
    Netherlands 211
International Atomic Energy Agency (IAEA)
    25, 134, 263–4, 279
International Commission of Jurists 131
International Committee on Radiological
    Protection (ICRP) 67
International Consultative Group on Nuclear
    Energy 70, 73
International Court of Justice 262, 311
international crimes, nuclear power and 134
International Longshoreman's and Ware-
    housemen's Union (ILWU), USA 249
International Nuclear Fuel Cycle Evaluation
    (INFCE) 25
investment tax credits 77
Iowa, USA 58
Iran 26, 134, 279–80, 313
Ireland 175, 189–90, 200–1; Carnsore

Point reactor 189–90, 252; political parties in 189, 245; trade unions and 189–90, 254, 340; uranium mining and 190
Irish Congress of Trade Unions 189
Irish Republican Army (IRA) 131
Irish Sovereignty movement 189
Irish Transport and General Workers' Union (ITGWU) 189–90, 252, 254, 340
Isle of Man 175
Israel 234; uranium disappearance and 321
Italy 26, 201, 279; CND and 98; nuclear programme in 245
ITGWU *see* Irish Transport and General Workers' Union

Japan 11–12, 15, 17, 18, 20–1, 23, 255, 279, 281, 303–10, 319; Australia and 258, 281; citizens movement, history of 94–5; demonstrations in 200–1, 307; nuclear industry in 110–17; nuclear programme in 20–1, 95, 303–6 nuclear weapons and power, convergence of movements 303–9, 319–20; political parties in 307–8; safeguards agreements and 280; trade unions and 250, 254, 307; waste dumping in Pacific and 311, 312, 317
Japanese Council Against A and H Bombs *see* Gensuikyo
Jaslovske Bohunice reactor, Czechoslovakia 238–9
Jersey Central Power and Light Company, USA 22
Jervis Bay reactor, Australia, proposal for 260
Johnston, L.B. 99
Joint Committee on Atomic Energy, USA 19
Juda, T. 310
Jungk, R. *Foreword*, 130

Kagoshima prefecture, Japan 306
Kaiser Wilhelm Institute, FRG 13
Kaiseraugst reactor, Switzerland 143–4, 147, 152, 157, 168
Kaiserstuhl, FRG 103, 104
Kalkar fast-breeder reactor, FRG 127, 200, 215–17
Kansai Power Company, Japan 306
Kapiza, P. 235–6
Karlsruhe, FRG 224
Kasetsart University reactor, Thailand 315
Kashiwazaki reactor, Japan 304
KBS Report, Sweden 158
Kelly, P. 339
Kendall, H. 95
Kennedy, E. 46
Kennedy, J.F. 102, 312
Kennedy, R. 40
Kern County, USA 58
Kerr-McGee, USA 249–50
Kidder Peabody, USA 74
Kiribati 320
Kirkwall, Scotland 192
Kirsche, R. 23
Kitayama, F. 12
Kneale, G. 47
Kollontai, A. 228
Komanoff, C. 61, 70

Kopparnäs reactor, Finland 145–6
Koryakin, Y. 236
Kraftwerk Union (KWU), FRG 26, 114–15, 217, 219
Krause, H. 130
Kreisky, B. 152, 156, 160–3
Krsko reactor, Yugoslavia 239
Kuhn, S. 115
Kumihama reactor, Japan 306
Kurchatov Institute, USSR 232
Kwajolein atoll, Micronesia 311
KWU *see* Kraftwerk Union, FRG
Kyle council, Scotland 193
Kyoto, Japan 306–7
Kyushu Company, Japan 306

la Confédération Française Démocratique du Travail see CFDT, France
la Confédération Générale du Travail see CGT, France
la Hague reprocessing plant, France 185–7, 196–8, 249, 151
Labor Party, Australia *see* Australian Labor Party
Labour Committee for Safe Energy and Full Employment, USA 254
labour movement 226, 246, 247, 257, 258, 276, 299; *see also* left wing groups and parties; trade unions; and workers
Labour Party, Ireland 245; Netherlands 210–11, 213, 215, 226, 245; UK 97, 176–7, 196, 226, 251, 301
Lake Cayuga, USA 94
Land Rights Agreement, Australia 272
land rights, opposition to nuclear power and 55, 296–8, 309; *see also* Aboriginal land rights
Landestring Party, Switzerland 161
Larzac, France, and Plogoff 188
Latin America 318
Lawrence Berkeley Laboratory, USA 77
lead times, nuclear reactors *see* reactor licensing times
League of Communists, Yugoslavia 240
left wing groups and parties 163, 243–6, 275, 299; attitude to nuclear power of 226–7; China and 244; Eurocommunism and 245; European countries and 147, 212; K Groups 215; USSR and 244–5
legitimacy 10, 99, 106, 120, 125, 128, 130, 138–40, 226, 316, 324–7; centre and 174, 226; civil liberties and 134, 137, 191; inquiries and 149; nuclear technology and 230; periphery and 174, 226; the state and 138
Leibstadt reactor, Switzerland 143
Lemoniz reactor, Spain 205–8, 296
Lenin, V.I. 229–30, 244
Lester, R. 42
Leuschner, G. 238
Leventhal, P. 321
Lewis, C.S. 323
Lewis, H. 50
Liberal Democratic Party, Japan 308; Switzerland 161
Liberal Party, Australia 257–63, 268, 274,

INDEX

275; Denmark 170; Netherlands 211; Sweden 150–1, 159–60, 162
Liberal People's Party, Finland 147
licensing procedures *see* reactor licensing
Linz, Austria 144
Lip factory, France 340
Lisco, H. 92
LKAB uranium mining company, Sweden 150
LOCA *see* Loss of Cooling Accident
Local Assembly of Zadar, Yugoslavia 239–40
Loch Doon, Scotland 193–4
London, UK 12, 97, 177, 301
Long, C. 316
Longbeach, nuclear warship 302
Longest Walk, the, USA 298
Lonroth, M. 110–15
Los Angeles, USA 59
Loss of Cooling Accident (LOCA) 32–3, 40
Lothian, Scotland 191
Louisiana, USA 58
Louisville, USA 291
Lovejoy, S. 57
Lovins, A.B. 74, 329, 338
Lovisa reactor, Finland 145, 147–8
low level radiation *see* radiation, low level
Lower Cape Committee on Radioactive Waste Disposal, USA 93
Lower Saxony, FRG 220–4
Lucas Aerospace, UK 340
Lucas, N. 178
Lucens reactor, Switzerland 157
Luchow, FRG 222
Luxembourg 200–1
Lyon, France 179, 182, 184–5

Madoc, Canada 293
Madrid, Spain 204, 207, 209
Maine, USA 58, 87
Malmö, Sweden 157
Malone, M. 310
Malthus, T. 23
Malville reactor, France 181–4
Malwagu, D. 271
Manche, France 249
Mancuso, T. 47, 48
Manguia, Spain, council of 207
Manhattan Project 14, 17, 19, 75, 91, 116, 117, 119
Mannheim, FRG 107
Mans, France 185
Maoist parties 244
Marali No. 1, J. 271
Maralinga, Australia 262
Marcos, F. 316
Marcoule reactor, France 20
Marianas islands 311
Marin County, USA 59
Markolsheim lead factory site, France 104, 105
Marseille, France 20
Marshall Islands, Micronesia 97, 310–11, 313
Marviken reactor, Sweden 142
Marx, K. 244, 248
Mary Kathleen uranium mine, Australia 261
Maryland, USA 55, 324

Massachusetts Institute of Technology, USA 49, 95
Massachusetts, USA 49, 57, 59, 86, 93
Matthöfer, W. 126
MAUM *see* Movement Against Uranium Mining, Australia
Mazzochi, A. 249
McCormack, R. 27
McCoy, R. 74
McDonnell-Douglas 137
Mclure, J. 85
Mecklenburg, GDR 238
Medical Association, Canada 294
Medical Research Council, UK 47
Medvedev, Z. 233
Melbourne, Australia 264–6, 272, 283
meltdown of reactor core 32–3, 50–1
Mensing, B. 213
Meshoppen, USA 94
Metropolitan Edison 37–8, 43, 48, 62, 63–64, 65, 85
Michalon, V. 182
Michelson, C. 52
Michigan, USA 55, 58, 59, 86
Micronesia 310, 312, 313, 319
Middletown, USA 29, 37
Midland reactor, USA 55
Midway Island 317
MIGRI, Sweden 150
Milan, Italy 201
Miles, G.L. 281
Miliband, R. 135, 174
military industrial complex 67, 100
military, nuclear power and 19–20, 67, 75–6, 88, 115–16
Miljöcentrum, Sweden 150
Miljöforbundet, Sweden 150
Minimata, Japan 95
Ministry of Commerce and Industry, Austria 152
Minnesota, USA 58, 77
Miscellaneous Workers' Union, Australia 276
Missouri, USA 87
Mito city, Japan 307
Mitterand, F. 198
Miyagi, Japan 304
Mobilisation for Survival 299
Monmouth District Council, Wales 191
Montague, USA 57
Montana, USA 58, 87
Montpellier, France 179
moratoria on reactor construction, 27, 58, 114, 140, 168, 186, 197, 220–1; *see also* referendum
Morgan, K. 49
Morris reprocessing plant, USA 90
Mororoa atoll, French Polynesia 310, 312
Moscow, USSR 236, 244
Mothers Against Nuclear Power, Austria 163
Movement Against Uranium Mining (MAUM), Australia 258, 266
movements against nuclear power and nuclear weapons, convergence of 224–5, 298–309, 319–20
Muldoon, R. 317
Muller, R. 135

404   *GLOBAL FISSION*

Mullwarchar Valley, Scotland 193–4
Murray, T. 119
Mutsu, Japan 304–5
MWe (megawatt electrical), defined 20
MWt (megawatt thermal), defined 20

Nabarlek uranium mine, Australia 274, 276, 280
Nader, R. 55, 234, 291
Nagasaki, Japan 307
National Academy of Science, USA 67, 102
National Anti-Nuclear Power Struggle Committee, Japan 306
National Coalition Party, Finland 147–8
National Committee on Radiological Protection (NCRP), USA 67
National Coordination of citizens groups, Switzerland 164
National Country Party, Australia 275
National Economic Research Associates (NERA), USA 83
National Energy Advocacy Conference, USA 85
National Energy Supply Plan, Spain 203–4
National Indian Youth Council, USA 297
National Institute of Occupational Safety and Health, USA 47
national interest, nuclear power and 120
National Opinion Research Centre, USA 19
National Party, Scotland 193
national prestige, role of 119
National Union of Mine Workers, UK 196, 254
nationalism, in France 180
Nationalist Party, Scotland 192, 196; Wales 196
Native Americans Against Uranium Mining, USA 298
native peoples, Australian Aboriginals 174, 257, 259, 268–71, 290; North American 174; periphery and 174, 337; uranium mining and 296–8
NATO 214, 300
Natural Resources Defence Council (NRDC), USA 55, 74, 317
Nazareno, E. 316
NCRP *see* National Committee on Radiological Protection
Neckar-2 reactor, FRG 216
Neporozhniy, P. 235
Netherlands 28, 127, 203, 210–14, 226, 296, 336; Brazil and 212, 217, 301; churches and 211–12; CND and 98; demonstrations in 200–1, 211–12; nuclear programme in 201, 203; nuclear weapons and 98, 211, 214, 300–1; regionalism and 210, 212–13; trade unions and 210, 254
networks 57, 287, 324
Neupotz-1 reactor, FRG 216
neutron bomb 211, 300
New Caledonia 320
New England, USA 61
New Hampshire, USA 55–6, 58, 86, 324
New Jersey, USA 59, 77
new left, the 244

New Mexico, USA 58–9, 296, 297
New Peoples Army, Philippines 316
New South Wales, Australia 260, 281
New York, USA 48, 55, 58–9, 61, 68, 71, 77, 83, 86, 87, 94, 248, 251, 300
New Zealand (NZ) 301–3, 320; French nuclear weapons testing and 311; nuclear-free Pacific and 317, 318–20; nuclear programme and 302; trade unions and 302
New Zealand Federation of Labour 320
Nicolin, C. 167
Niels Bohr Institute, Denmark 154
Nixon, R. 23, 99, 121
NKV, Netherlands 210; *see also* Dutch Catholic Trade Union Council
NLC *see* Northern Lands Council (NLC), Australia
No Nuke News 295
Nogent-sur-Seine reactor site, France 180
non violent action 55–7, 193
Non-Violent Action Kaiseraugst, Switzerland 143
Nord reactor, GDR 238
Normandy, France 175, 336
North America 321
North American Indians *see* USA, native peoples and
North Marianas Commonwealth Legislature, Marianas islands 311
Northern Ireland 175
Northern Lands Council (NLC), Australia 269–71
Northern Territory, Australia 261, 267, 270–1, 277, 281–2
Northgate, uranium mining and 190
Northrhine Westphalia, FRG 224
Northumberland, England 193–4
Norway 169; CND and 98; energy needs of 169; trade unions and 254
Noto reactor, Japan 306
Novoronezh reactor, USSR 232, 236
NRC *see* Nuclear Regulatory Commission
NRX reactor, Canada 21
nuclear accident *see* accident, nuclear
Nuclear Deterrent Act, Austria 167
Nuclear Energy Agency, of OECD 85
Nuclear Energy Women (NEW), USA 86
nuclear enterprise, defined 172
Nuclear-Free Pacific Forum, Australia 319
nuclear-free zones 282, 309–14, 318–20
nuclear industry 22, 110–18, 294; centralized form of 285–7; decline in uranium demand and 279–80; history of 19–28, 115–25; participation of state in 110–25; political power and 119; possibility of collapse 73–5, 112–15, 327; response to Three Mile Island 81–8; strategy of 85–8
Nuclear Installations (Licensing and Insurance) Act, UK 78
Nuclear Non-Proliferation Treaty 308
Nuclear Objectors for a Pure Environment (NOPE), USA 57
Nuclear Power Corporation 110–15
Nuclear Power Inspectorate, Sweden 159
Nuclear Regulatory Commission (NRC),

INDEX

USA 36, 38–9, 44, 52, 59, 64–5, 68, 71, 76, 78, 80–2, 88, 95, 132
Nuclear Ship Development Agency, Japan 304
Nuclear Siting and Licensing Bill, USA 81
nuclear state 130–4, 181–4
nuclear submarines 47, 116, 298, 300, 309, 312
nuclear technology, capital intensiveness of 124; centralized form of 174; political relations inherent in 123; vulnerability of 130, 137
nuclear waste *see* waste, nuclear
nuclear weapons 86, 88, 131, 260, 261, 262–4, 280, 298, 300–1, 310; FRG and 212, 224–5; Netherlands and 211, 214; nuclear power and, convergence of movements 224–5, 298–309, 319–20; Pacific and 309–14; proliferation of 321; Argentina and 283; India and 21, 299; safeguards 98, 263–4, 279–80, 290, 301, 321; US policy and 300
NVV, Netherlands 210

Oak Ridge National Laboratory, USA 48, 93, 234
Oak Ridge, USA 249
Occitanie region, France 175
occupations 56, 103–9, 125–6, 139, 143–4, 181–6, 191, 208, 222–3
OECD *see* Organization for Economic Co-operation and Development
Offenburg Agreement, FRG 107
Office of Economic Analysis, USA 79
Office of Technological Assessment (OTA), USA 84, 330–1
Offices for the Protection of the Constitution, FRG 129
Ohio, USA 65
Ohira, M. 308
oil corporations 79, 136
oil price rises 103, 178, 203
Oklahoma, USA 59, 249
O'Leary, J. 27
Olkiluoto reactor, Finland 145, 147
O'Mally, D. 190
Onagawa reactor, Japan 95, 304, 306
Ontario, Canada 292, 293, 297, 303
OOA (*Organisationen til Oplysning om Atomkraft*), Denmark 7, 153–5, 170, 298, 335
OPEC *see* Organization of Petroleum Exporting Countries
opinion polls 145, 150, 155, 169, 184, 214, 219, 248, 265, 273, 282, 300–1
Oppenheimer, R. 15
opposition to nuclear power 290, 291–322; basis of 173; centralization and 335–41; history of 90–109, 286; infiltration of 324–5; structure of 57, 286–91; *see also* entries for individual countries
Oregon, USA 58, 87
Organization for Continuous Energy, Denmark 154
Organization for Economic Co-operation and Development (OECD) 25, 85, 145

Organization for Information on Energy, Denmark *see* OOA
Organization of Petroleum Exporting Countries (OPEC) 103, 178, 203
Orkney Islands 175, 192, 336
Orwell, G. 341
Osaka, Japan 306–7
Ostroweski, R. 248
Outer Hebrides, Scotland 194
Oyster Creek reactor, USA 22, 77

Pacific Concerns Resource Centre, Hawaii 319
Pacific Fisher waste transport ship 187, 295
Pacific islands, regionalism and 311
Pacific Ocean 10, 256, 303, 308–18, 321; co-operation within 314–18; waste dumping and 311, 317
Pacific People's Action Front 318
Pacific Swan waste transport ship 295
Pacific Trade Union Conference, Vanuatu 320
Pacific Trade Union Forum 320
pacifism, opposition to nuclear power and 55
Paddlewheel Alliance, USA 56
Palau (Belau), Micronesia 312–14, 319, 320, 336
Paley Commission, USA 84
Palme, O. 160
Palmyra Island 317
Papua New Guinea 319, 320, 338
Paris, France 179, 187
Parker, Mr Justice 177
participatory democracy 332–5
peace movement 290, 299–308; *see also* Campaign for Nuclear Disarmament (CND); and nuclear weapons
Peace Squadron, NZ 302
peaceful atom 303, 307
Pearl Harbour, USA 296
Peking, China 242
Pennsylvania Emergency Management Agency 38
Pennsylvania, USA 28–9, 59, 62, 64, 66, 78, 94
Penya, D. 206
Peoples Party, Austria 156
People's Treaty for a Nuclear Free Pacific Zone 318–19
periphery, defined 173–5; export of pollution and 174–5, 337; federal structure and 174–5; legitimacy and 174; native peoples and 174, 337; of France 175, 186–8; of Spain 204–10; of UK 175, 189–94, 195–9
Perlmutter, H. 135
Pershing Two missile 300
Perth, Australia 267
Pesonen, D. 94
Pesoyants, A. 232
Philadelphia, USA 44
Philippines 78, 279, 310, 315–17, 319; Bataan reactor and 315–17
Planning Committee on Electric Power Development, NZ 302

Plogoff reactor, France 187–8, 198, 201–2
plutonium 124, 182, 239, 321; civil liberties
    and 130–2, 327
Plymouth, USA 59
police, nuclear opposition and 106, 207, 213
political strikes 255
Polk, J. 135
Pollard, R. 95
pollution 174–5, 337
Pompidou, G. 179
Ponape, Micronesia 319
Port Hope, Canada 292
Port Kembla, Australia 255, 266, 282
Porter Commission, Canada 177, 293, 303
Portland, Australia, nuclear reactor and 282
Portskewett reactor, Wales 191
Portsmouth submarine yard, USA 47
Portugal 79, 205
Potomac Alliance, USA 56
Prevention of Terrorism Act, UK 131
Price-Anderson Act, USA 78, 249
primary cooling system 31, 34
progress see idea of progress
Project Independence 23
Proletkult 230
Protestant church of Mecklenburg, GDR 238
Provisional Irish Republican Army 189
PSU see Unified Socialist Party, France
Public Accounts Committee, House of
    Commons, Canada 292
Public Citizens' Health Research Group,
    USA 68
public ownership 333–4
public power, threat of 117
Public Utilities Commission, Pennsylvania,
    USA 64

Quakers 55, 56
quality of life 329
Queensland, Australia 261, 267, 281
Qun, C. 242

Radford, E. 67
radiation, effects of 47–8, 310; exposure of
    workers 48, 67–8, 250, 297; low level
    46–8; model of 47; permitted levels of
    67, 233
Radiation-protection Law, Austria 144
Radiation Safety Institute, GDR 238
Radical Party, Italy 245; Switzerland 147
Rajasthan desert, India 21
Rancho Seco reactor, USA 59
Ranger Uranium Environmental Inquiry,
    Australia 76, 131, 133, 177, 259, 261–3,
    299, 303; First Report of 261–3, 264;
    Second Report of 263, 269
Ranger uranium mine, Australia 270–1, 274,
    277, 279–80
Rasmussen, N. 49
Rasmussen Reactor Safety Study 49–50
Rauma, Finland 145
reactor 13, 29–33, 195; capital costs 70; codes
    and standards 70; construction lead times
    69–70; core 29–31; development, pace of
    22, 91; licensing of 157, 216–17; licensing
    times 69–71, 81–2, 88, 113, 216; market

110; mass production of 72, 231–2; orders
    21–5, 73, 88, 110–15; safety, costs 71;
    ideology of 51; sales 23–5; standardization
    of 72, 231; types 31, 182, 195; see also
    under individual countries
Reactor Safety Commission, FRG 225
Reagan, R. 88
Red Army Faction, FRG 128
Red Lillies Lagoon, Australia, meeting at 271
Reece, R. 333
referendum, Brazil and 218–19; Denmark
    and 170; in Austria 141–72; in Czecho-
    slovakia, 239; in Sweden 141–72; in
    Switzerland 141–72, 161; in USA 58, 87,
    247; industry expenditure on 163–5;
    Ireland and 190; object of 141, 166–7;
    Palau and 313–14
Regional Committee for Information and
    Anti-Nuclear Struggle, France see CRILAN
regional identity see periphery
regionalism 175, 179, 187, 192, 195, 204–10,
    212–13, 215, 240–1, 311, 336–7; see also
    periphery
rem (röntgen equivalent man) 39
reprocessing plants 65, 77, 90, 176–7, 185–7,
    196, 221–4, 251, 295–6, 308
Rettig, J. 184
revolutionary transition 244
Reykjavik, Iceland 169
Rhinghals-3 reactor, Sweden 158–9
Rickhover, H. 20, 115
Rio de Janiero, Brazil 218
Rio-Tinto Zinc 190
Robert Morton (D.G.) Ltd 26
Roberts, A. 116, 228, 247
Roberts, S. 270
Rockefeller Foundation, USA 73, 113
Rockwell International 87, 137
Rocky Flats nuclear weapons plant, USA 86,
    88, 300
Rodgers, P. 48
Roissman, A. 74
Ronay, G. 240
Roosevelt, F.D. 13, 15
Roosevelt, T. 101
Rosenbaum Report, USA 132
Ross, D. 59
Roswell Clark Memorial Institute, USA 48
Roxby Downs uranium deposit, Australia 265
Royal Commission of Inquiry, NZ 302
Royal Institute of International Affairs, UK
    113
Royal Investigative Commission, Sweden 148
Rum Jungle uranium mine, Australia 261,
    263
Russia see Union of Soviet Socialist Republics
    (USSR)
Rutherford, E. 12
Rüthi reactor, Switzerland 156, 168
Ryle, M. 52

Sabato, J. 122, 123
Saclay nuclear plant, France 251
Salzburg Conference for a Non-Nuclear
    Future, Austria 320–1, 338
San Francisco, USA 59, 94

INDEX

Saskatchewan, Canada 279, 292
Scargill, A. 196
Schleswig-Holstein, FRG 139–40
Schleyer, H. 128
Schmidt, H. 221, 223, 224, 225
science, struggle for access to 96; in China 101; in USSR 230
scientists, nuclear power and 16–19, 91–6, 119, 163, 218, 235–7, 243, 291, 329
Scotland 47, 175–6, 190–4, 336, 340; demonstrations in 191, 193, 200–1; Torness reactor and 191–2; trade unions and 254; *see also* Nationalist Party, Scotland
Scottish Campaign to Resist the Atomic Menace 191, 193
Scottish Trade Union Congress 193
Scourrie, Scotland 193
SCRAM *see* Scottish Campaign to Resist the Atomic Menace
SCRAM-South West (SCRAM-SW), Scotland 193
Scranton, W. 38, 44
Seabrook reactor, USA 55–6, 325
Seamen's Union, Australia 283
secondary cooling system 31, 34
Secretariat for Future Studies, Sweden 162
Secretary of State for Scotland 190, 192, 194; for Wales 190, 191
Sedov, V.K. 232
Sendai reactor, Japan 306
Severn Estuary, UK 176, 191
Seville, Spain 209
Shad Alliance, USA 248
Shimane-2 reactor, Japan 306
Shin Forest council, Scotland 193
Shiprock, USA 297
Sibbo group, Finland 145
Sieghart, P. 131, 134
Sierra Club, USA 55
SIFO, Sweden 145, 150
signature campaigns 155, 157, 179, 193, 211, 264–5, 274, 295, 300, 302–3, 306
Silkwood, K. 249–50
Simut Party, Greenland 295
Sinn Fein, Ireland 189
Skövde, Sweden 150
smiling sun symbol 7
Smith, A. 75
social democracy 244
social democratic parties 245
Social Democratic Party, Austria 226; Denmark 170; Finland 146; FRG (SPD) 124, 215, 220–1, 225, 246, 275; Sweden 142, 148–9, 160, 163–4, 226, 246; Switzerland 147, 161
socialism 228–30, 244
Socialist Party, Australia 245; Austria 148, 156, 161–2, 164; France 182, 187, 197, 198, 226, 245, 253; Japan 307; Spain 204, 209
Socialists against Zwentendorf 163
Society for Scientific Progress, Brazil 218
SOHYO, Japan 250, 254, 307, 320
solar power 84, 330–1; centralization and 136, 331; transnational corporations and 136

Solomon Islands 320
Somner, T. 220
Sorenson, B. 154
South Africa 133
South Australia 262, 267, 274, 280–2
South Carolina, USA 65
South Coast Trades and Labour Council, Australia 260
South Dakota, USA 58, 87, 295, 296, 297
South Korea 199, 279, 315
South of Scotland Electricity Board (SSEB) 191–2
South West African People's Organization of Namibia 298
Soviet Baltic Republics 239
Spain 203–10, 214, 226, 296, 336; demonstrations in 200, 204, 206–9; France and 175; nuclear programme in 203–5, 204–5; reactors in 205, 208–9, 296; regionalism and 204–10
SPD *see* Social Democratic Party (SPD), FRG
Spiderwort strategy 305
SSEB *see* South of Scotland Electricity Board
St Maurice l'Exil reactor site, France 180
St Pantaleon reactor proposal, Austria, 144
Stalin, J. 244
state, the 100, 120, 138, 181–3, 194, 202, 226, 247, 301, 325; centralization of power in 135–6, 138, 171, 253, 324; coercive apparatus of 214; confrontation with 215; courts and 219; federal structure and 172; fracture of by nuclear issue 221–5; government Inquiries and 263; interests of 117–20; legitimacy and 128, 138–40; limits to power of 137–40; nuclear power and 120; regional government and 174–5; relationship to nuclear industry 110–25, 137–40; relationship to opposition to nuclear power 125–34; role of 109; structure of 137–40
State Committee on Atomic Energy, USSR 232
State Department, USA 317, 319
State Electricity Commission (Australia), in New South Wales 260; in Victoria 282; in Western Australia 281
State Power Board, Sweden 158
steam generator 32, 34
Sternglass, E.J. 95
Stewart, A. 47
Stewart, M. 136
Stipulation Law, Sweden 158, 159
Stockholm, Sweden 142
Stop the Neutron Bomb, Netherlands 211
Strasbourg, France 179, 336
Strathclyde, Scotland 340
Strauss, L. 117
strikes over nuclear issues 186, 197, 253, 255, 261, 266–7, 302, 306
Stromness, Scotland 192
Subic Bay, Philippines 310
subsidies to nuclear industry 75–80, 125, 115–18
Sunsat project 136
Super Phoenix reactor *see* Creys-Malville reactor, France

Suva, Fiji 318
Svintsev, Y. 232
Swanston Dock, Australia 266–7
Sweden 21, 28, 74, 142–3, 144–5, 159–60, 161–4, 279, 286, 338; alternative energy and 161–2; CND and 98; demonstrations in 157–8, 167, 170; Denmark and 170; Falldin government and 152, 157, 159, 162; information campaign in 149–55; KBS Report and 158; Liberal Party government and 159–60; nuclear industry in 110–17; nuclear programme in 21, 142, 157–9, 167, 170; political parties 144, 148–51, 158–9, 161–4, 226, 245–6; Stipulation Law and 158–9; study circles in 149, 151; waste disposal issue in 145; *see also* Billingen, Sweden
Switzerland 28, 142, 143–4, 147–8, 152, 160, 163–4, 168, 182, 279, 286, 321, 336; demonstrations in 143–4, 200; Dreyeckland and 103–6; nuclear programme in 142, 143, 156–7, 168; political parties in 142, 147; signature campaign in 157; *see also* Kaiseraugst reactor
Sydney, Australia 266, 272, 319
Szilard, L. 13, 91

Tahiti 310–11, 320
Taiwan 199, 315
Tamplin, A. 94
Taylor, V. 83
Teachers Against Nuclear Power, Austria 163
technical neutrality 17, 92, 229
technocracy, USSR and 229
technological dependence 122, 123, 338–9
technological optimism 16, 17–19, 91, 117, 120
technology 228–30; opposition to nuclear power and 251–2, 326
Tennessee, USA 94, 249
Tennessee Valley Authority, USA 52, 53, 117
terms of trade, nuclear power and 120
Terreno, G. 207
Thailand 315
Tham, C. 159
Third World 102, 121; nuclear lobby in 123; technological dependence and 122, 218; trade relations and 98, 120
Thornburgh, R. 37, 41, 64
Three Mile Island reactor accident, USA 9, 29–42, 52, 53, 63–4, 71, 232, 240; cost to Nuclear Regulatory Commission 76; costs of clean up 64; effect on nuclear economics 63–80; inquiries into 45, 54; nuclear industry response to 81–8; political fallout 43–60, 159–60, 179–80, 187, 200–1, 214, 223, 240, 242, 245, 308, 316; radiation effects and 39–41, 46–9; reactor orders and 73, 88; waste disposal and 64–5
Tihange reactor, Belgium 250
Tokai Mura nuclear complex, Japan 21, 308
Tokyo, Japan 306–7
Tolman group, USA 115
Tornashean council, Scotland 193
Torness reactor, Scotland 191

Toulouse, France 179
trade, nuclear power and 120
Trade Union Confederation, Sweden 164
trade union organizations 161, 164, 186, 189–90, 193, 196–8, 210, 213, 247–55, 260–2, 265, 272, 277, 282–3, 295–6, 302, 306–7, 320, 340
Trade Unionists Against Nuclear Power, Austria 252
trade unions 220, 227, 246–55, 286, 291, 293; anti-nuclear movement and 247–55; leadership and 275–6; *see also* under individual countries
transnational corporations 121, 253, 336; centralization of power in 135–6; solar power and 136
Transport and General Workers' Union, Scotland 254
Traube, Dr 130
Treaty of Paris 211
Treaty of Tlatelolco 318
Tricastin reactor, France 197
Trident submarine 298, 300, 309, 312
Trombay reactor, India 21
Trotsky, L. 244
Truman, H. 16, 84
Truxton, nuclear warship 302
Tsuruga reactor, Japan 305, 308–9
Tudela, Spain, council of 207
Tummerman, L. 234
Tvind, Denmark 154

Ullsten, O. 160
Unified Socialist Party, France 226, 245
unilateral disarmament 97, 301
Union of Concerned Scientists (UCS), USA 52, 55, 83
Union of Construction, Allied Trades and Technicians, UK 252
Union of Soviet Socialist Republics (USSR) 110, 180, 228–37, 244, 299, 300, 307, 309, 334; debate over nuclear power in 235–7; Finland and 147; nuclear accidents in 233–4, 235; nuclear programme in 20, 231–2, 235–6; nuclear safety in 232–3, 236; *Proletkult* and 230; technical bureaucracy in 230–1; waste disposal and 233–4
Union of Yugoslav Electrical Enterprises 239
United Auto Workers, USA 249
United Church, Canada 294
United Kingdom (UK) 13, 20, 172–8, 184, 189–96, 198, 202, 279, 296, 336, 338; advanced gas cooled reactors and 195; Aldermaston marches in 195; Australia and 261, 262; central developments in 175–8, 194–9; citizens movement, history of 97; civil liberties and 131–2; CND and 196, 301; demonstrations in 177, 301, 200–1; nuclear industry in 110–17; nuclear programme in 175–7, 194–5; nuclear weapons and 301; periphery of 175, 189–99; regionalism in 175; political parties in 97, 176–8, 196, 226, 251, 301; trade unions and 251, 252, 254; Windscale Inquiry and 176
United Mine Workers, USA 249

INDEX

United Nations (UN) 19, 92, 310, 318–19
United Public Workers Union, USA 249, 296
United States of America (USA) 227, 231–2,
242, 249–50, 279, 291, 296–9, 325, 327,
334, 337–8; Carter policy on plutonium
recycling 124; citizens movement, history
of 54–8, 291; civil liberties and 132–3;
CND and 98; demonstrations in 56, 58–9,
86–7, 94, 200–1, 300; economics of nuclear
power in 62; energy growth projections
83–4; Hiroshima bombing and 13–19;
military and 19–20, 75–6, 84, 88; native
people and 296–7; nuclear industry in 78,
85–8, 110–20; nuclear programme in
19–27; nuclear waste and 58; nuclear
weapons and 211, 298, 300, 310–11;
Pacific Ocean and 309–10, 313–14,
317–18, 319; Philippines and 315–17;
reactor licensing times in 69–70; reactor
orders in 22–7, 73, 88, 114; reactors in 22,
43, 52–3, 55–6, 59, 68, 77, 94, 249, 251,
325; referenda on nuclear power 58, 87,
247; terms of trade and 120–4; trade
unions and 247–8, 251, 252, 254, 296;
uranium mining and 280; women and
nuclear industry in 86; see also Three Mile
Island reactor accident
University of Birmingham, UK 47
University of California, USA 50
University of Pennsylvania, USA 135
Uppsala, Sweden 159
Ural mountains, USSR 233–4
uranium enrichment 123–4, 180, 211–12,
217, 280–1
uranium mining 255–84; cancers and 48,
297; demand and 279–80; in Australia
256–84, 291; in Canada 279, 291–2; in
France 188, 291; in FRG 291; in Green-
land 170, 295; in Ireland 190; in Italy
291; in Orkney Islands 192, 292; in Scot-
land 192, 291; in Spain 291; in Sweden
150, 162, 291–2; in USA 291, 295
Uranium Moratorium, Australia 264, 301
Uranium-235 13, 14
Uranium-238 14, 182
URENCO 211, 280
Urey, H. 91
US Oil, Chemical, and Atomic Workers
Union (OCAW) 249
Utrecht, Netherlands 214

Valdecaballeros reactor, Spain 208
Val-de-Saône enrichment plant site, France
180
Valencia, Spain 204
Values Party, NZ 302
Vanuatu 319–20, 336
Varberg, Sweden 158
Venice, Italy 201
Vereinigung Industrieller Kraftwirtschaft see
Association of Industrial Power Producers
Vermont, USA 55, 58–9, 86
Vermont Yankee reactor, USA 55
Vickers Ltd 340
Victoria, Australia 258, 267, 281, 282
Vienna, Austria 134, 142

Vietnam War 60, 98, 99, 100, 255, 272
Villanueva de la Serena, Spain 208
Viner, I. 270
violence, question of 56, 191, 325–6
Vir Island, Yugoslavia 239
Virginia Electric and Power Company,
USA 72
vitrification of waste 193
Voice of Energy (VOE), USA 86

Waitham, R. 16
Wake Island 317
Wald, G. 95
Wales, UK 175–6, 190–1, 194, 336
Walker, W. 110–15
Warnock, D. 132
Warsaw Pact 235
WASH-740 33
Washington, USA 47, 59, 81, 85, 87, 94,
296, 298, 300
waste, nuclear 58, 65, 93, 145, 158, 193, 196,
221–4, 233–4, 296, 308; disposal costs and
76–7; Pacific and 311; reprocessing
and 65, 77, 90, 176–7, 185–7, 196, 221–4,
251, 295–6, 308; salt domes and 213–14,
222; test drilling and 168, 192–4; transport
and 58, 295
Waterside Workers' Federation, Australia
265–7
Weaver, J. 54
Weber, M. 99
Weish, P. 152, 167
Wellington, NZ 302
West Valley reprocessing plant, USA 77, 250
Western Australia 281
Westinghouse 22, 26, 79, 110–18, 136–7,
147, 239, 292, 317
Wexford Nuclear Safety Association, Ireland
189
Wharton School, USA 135
Whitlam, G. 267
Whitsunday demonstrations 200–1, 203, 210,
213, 216, 217, 307, 321
Wicker, T. 76
Wilster, FRG 127
wind energy, in Denmark 154
Windscale Inquiry, UK 176
Windscale nuclear complex, UK 176–7, 196,
251, 295
Wisconsin Public Service Commission,
USA 82
Wisconsin, USA 58, 82
women, and Metropolitan Edison 63; Austria
and 163; nuclear industry and 86; opposition
to nuclear power and 55, 189; role of 102
women's movement 102, 258
workers 331; and Seabrook reactor, USA 247;
environmental movement and 247; France,
farmers and 186; self-management and,
Yugoslavia 239; see also trade unions
World Information Service on Energy (WISE)
7, 295
Wyhl reactor site, FRG 104, 106–8, 139,
143, 151, 216, 224, 290, 326; occupation
of 125–6; regionalism and 215
Wyoming, USA 296

Young Liberals, UK 196
Yugoslavia 98, 239–40, 336
Yunupingu, G. 271

Zadar reactor, Yugoslavia 239–41, 336

Zangger, C. 152
zirconium fuel cladding 36, 42, 52
Zwentendorf reactor, Austria 142, 144,
   152–3, 160, 167, 247